# Natural Radioactivity in Water Supplies

# About the Book and Author

There is little disagreement that the potential effects of water contamination on human health and the environment should not be ignored, even though the exact nature of those effects is not yet fully understood. That permanently incapacitating and even lethal substances (asbestos, for example) have, in ignorance, been introduced into the environment may become apparent only decades after their introduction. A new principle in water quality regulation is emerging in response to awareness of these dangers: An individual or organization can be held accountable for hazards to human health or for degradation of the environment created by the introduction of a substance, even if the individual or organization is not the source of that substance, even if no regulation of the substance currently exists, and even if the substance is not known to be hazardous or to degrade the environment at the time its release occurs.

This book outlines the scientific aspects of the control of natural radioactivity in water supplies, as well as the labyrinthine uncertainties in water quality regulation concerning natural radiocontamination of water. The author provides an introduction to the theory of natural radioactivity, addresses risk assessment, describes sources and effects of natural radiocontamination of water, surveys federal water law concerning natural radiocontamination, and presents an account of how one city dealt with the perplexities that mark this rapidly evolving area of water quality regulation.

Jack K. Horner is a senior scientist at Science Applications International Corporation in Colorado Springs, Colorado.

# Natural Radioactivity in Water Supplies

Jack K. Horner

Routledge
Taylor & Francis Group
NEW YORK AND LONDON

First published 1985 by Westview Press, Inc.

Published 2021 by Routledge
605 Third Avenue, New York, NY 10017
2 Park Square, Milton Park, Abingdon, Oxon OX14 4RN

*Routledge is an imprint of the Taylor & Francis Group, an informa business*



**Library of Congress Catalog Card Number: 85-51917**

ISBN13: 978-0-3670-0534-4 (hbk)
ISBN13: 978-0-3671-5521-6 (pbk)

DOI: 10.4324/9780429035203

dedicated to my parents,
who gave me the chance to learn

and

to the memory of Alberto Coffa,
who taught me

# Contents

# Figures

# Foreword

In the last fifteen years, the United States has come to realize that radiocontamination of surface water and groundwater imposes highly unusual and difficult water quality management problems. Because they often persist for thousands of years and are practically impossible to remove from water supplies, radiocontaminants create risks which are more complex than risks from contaminants traditionally regulated under federal and state programs.

This book, the sixth in Westview's Water Policy and Management series, provides a unique opportunity for the water professional or student to learn the theory and nature of radioactive substances, their occurrence in the environment, and their biological effects. The author's fascinating historical presentation of the atomic theory and his lucid, non-technical description of radioactivity and its effects on man permit even the layperson to understand these phenomena.

The author forcefully argues that our understanding of the biological consequences of radiocontamination of water is still in its infancy. He shows how our ignorance of those consequences, transmitted through our water quality laws, generates serious legal and ethical problems for public water providers. This legal environment is described from an historical and philosophical perspective. Finally, he provides an instructive case history describing how one city, in the face of this scientific and legal uncertainty, dealt with the prospect of radiocontamination of its water supplies.

*Charles W. Howe*
General Editor

---

*Charles W. Howe is professor of natural resources and environmental economics at the University of Colorado.*

# Acknowledgments

No book is the work of a single mind; few, the work of a single hand. I owe much to many for the present volume.

Throughout this effort, my wife, Clancey Maloney, patiently tolerated my inattentions, repeatedly provided moral support, and spent several long weekends and evenings to see the manuscript to completion. She also did all of the artwork in Chapters 2 and 5 and the Appendix, adapted the artwork for Chapter 7, and guided the camera-ready typescript through the throes of production.

Buzz Ferguson and Tony Pawlicki of Science Applications International Corporation read the draft of Chapters 1 and 2 and made excellent suggestions for improving the readability of those materials.

Jim Mariner of Fountain Valley School thoroughly critiqued Chapter 3, saving me from several infelicities.

Frear Simons reviewed Chapter 6 for legal sense and made a number of superb stylistic suggestions.

Kate Bemis created the index; if the book is useful as a reference, it is largely due to her efforts.

Tad Foster of the city attorney's office, Colorado Springs, Bob Schukle of the Colorado Department of Health, and Charles Weir of the Fountain Valley Authority supplied me with most of the primary documentation for Chapter 7.

And not least, Professor Charles Howe, the editor, gave the project proportionate, thoughtful, and clear guidance from its inception; without his encouragement, I probably would not even have attempted it.

*J.K.H.*

# Acknowledgments

[illegible]

# 1 Foundations of the Atomic Theory

## INTRODUCTION

The theory of natural radioactivity, which forms a crucial part of contemporary management of natural radioactivity in water supplies, springs from nineteenth-century (Daltonian) atomism. This chapter recounts very briefly the development of the Daltonian view and can be skimmed by anyone who has familiarity, at least at the level of an introductory college chemistry course, with it.

## EARLY ATOMISM

Every reflective person must face a formidable paradox in the material world: on the one hand lies prodigious change; on the other, profound permanence. Waves roll ceaselessly toward the shore, though no wave is quite like its neighbors; chalk remains chalk no matter how finely pulverized, but no piece of it is exactly the same as any other. Any theory of matter which hopes to claim our respect must attempt to resolve this paradox.

Modern science has no privileged possession of this problem: the early Greek philosophers (600-100 B.C.) devoted great effort attempting to solve it. Among them was Democritus[1] (about 400 B.C.), who argued that if the permanence of matter is real, then there must be something in the composition of matter which ensures that permanence. Democritus supposed that the ensurer of this permanence was itself a material entity or entities. Whatever entities ensure the permanence we observe must themselves be eternal and unchangeable, he argued, because something more permanent cannot come from something less permanent.

Furthermore, these entities must be unlike any we directly observe, Democritus urged. A piece of salt, for example, can be divided in two, the resulting pieces again divided in two, and so on. Any visible pieces, again divided in two, and so on. Any visible pieces, therefore, are subject to change and hence cannot be the fundamental unchangeable entities. At the least, then, such entities must be so small as to be invisible. These fundamental, unchangeable, invisible entities Democritus called atoms, from the Greek atomos, meaning "indivisible."

Differences in the types of matter we observe, he further speculated, are caused by differences in the form, position, and arrangement of the atoms which compose that matter. Thus, for example, one might conjecture that iron is a strong, hard solid because the atoms of iron are hard and strong and have spines which lock them together; water flows freely because its atoms are smooth and slippery, allowing them to roll easily over one another; salt has a sharp taste because its atoms are sharp and prick the tongue.

In many respects, the atomic theory of Democritus is remarkably mature. It openly embraces the paradox of material change and permanence. It explains at least one kind of material change: the decomposition of sensible matter into smaller parts. It accepts the fact that except possibly in superfine divisions, matter retains the very properties we find in ordinary sized lumps of it (except shape and size). And, in a rudimentary fashion, the theory provides an accounting for differences among various kinds of matter.

Democritus' view of matter did not go without challenge, however. Other Greeks, notably Parmenides,[2] argued that change, including apparent change in matter, is illusory. That we see water freeze and thaw, that the sky is now one color and later some other, that material objects appear to move at one time and rest at another are tricks which our senses play on us, he argued.

The view of Parmenides in some sense "solves" the paradox of material permanence and change. It does this by denying that any change in matter which we observe is significant. Such a stand found a wide audience in early Greece. But whatever its strengths, Parmenides' theory pays a formidable price. It denies, in particular, that ordinary observation of the material world is an acceptable tool for adjudicating disputes between different theories of matter. And without this tool, we have little, if any, way of coming to agreement about what matter is. Whether Parmenides' view seemed convincing to Democritus' contemporaries, we may never know. In any case, as the classical era of Greece drew to a close, the atomic theory passed into obscurity, largely ignored for nearly twenty centuries thereafter.

A pervasive revival of interest in the works of the early Greeks occurred in central and southern Europe between the fifteenth and seventeenth centuries. With this revival, the theories of Democritus and the early atomists were once again entertained by people of a philosophic and scientific bent. An articulate formulation of seventeenth-century atomism, for example, can be found in the writings of the English philosopher John Locke.[3] In 1690, Locke took up the question of how the properties of atoms could account for the properties of matter which the senses can detect. This is a fundamental problem for any atomic theory: since atoms are invisible, and, in general, so small that they cannot be individually detected by any of the senses, whatever properties atoms have may not be the properties of matter we directly observe. An adequate atomic theory of matter therefore has to explain how we get from the properties individual atoms actually have to the properties of matter we actually perceive. Locke attempted to solve this problem by postulating the existence of two types of qualities. The first type he called primary qualities of atoms, which he identified as shape, size, motion, and situation. These qualities, he claimed, act on our sense organs, our sense organs transmit information to our minds, and this information produces in our minds the secondary qualities "of matter" such as color, odor, taste, and so on, which we actually observe. Properly speaking, secondary qualities are not properties of matter itself, but just the results of matter acting on our sensory and mental apparatus.

For all its sophistication, the atomism of Locke still suffers from one of the same limitations as the atomism of Democritus. In particular, Locke's theory sharply distinguishes the properties of matter-by-itself from matter-as-perceived. The connection between the two is at best postulated, not convincingly explained by the theory. If atomism is to make any advance past this stage, therefore, some less arbitrary connection between the properties of individual atoms and the properties of matter-as-perceived has to be drawn. (In fairness, we should say that this problem is still a problem today, though less so: we now understand more clearly some of the components of the perceptual chain at the atomic level. Others await further experimental or philosophical investigation.)

Some headway was in fact made along these lines by Locke's contemporaries. Pierre Gassendi (1592-1655),[4] a French philosopher, proposed that gases are composed of tiny rigid bodies (atoms) which move at random in a void and collide with one another. Robert Hooke (1635-1703)[5] elaborated this idea, suggesting that the elasticity of gases was the result of collisions of the atoms of those gases with the walls of their containers. In this formulation, at least one observable ("secondary") property (elasticity) of matter is a natural, immediate consequence of some "primary" properties (the impact of atoms on the walls of their containers).

The atomic theory was further articulated by Robert Boyle (1627-1691),[6] who proposed that the continued identity of certain substances (e.g., pure metals) derives directly from (what he considered to be) the fact that the atoms of those substances are all of the same kind. In support of this view, Boyle, like many of the atomists before him, sharply distinguished elements from compounds. Elements, he asserted, are substances which are made exclusively of a single kind of atom; compounds are substances which are made of combinations (of some sort) of two or more kinds of atoms.

Boyle's distinction between elements and compounds in terms of the identity of their atomic constituents lends credibility (explanatory power) to the atomic theory only if we believe that there are distinct kinds of substances which are preserved through chemical processes. But not all of Boyle's contemporaries were convinced of this conjecture. For at the least, there was no general recognition among them that chemical processes involve conservation of anything, particularly weight. As a result, by modern standards relatively little effort was made by seventeenth and early eighteenth-century chemists to control experimental conditions. And the consequence of this inattention was that experimental results often appeared irreproducible, except in a very loose qualitative sense.

## MODERN ATOMISM

At first, it was probably the failure of alternatives to the atomic hypothesis to predict useful results, more than the intrinsic clarity and force of atomism, which contributed to the decline of non-atomistic theories. For whatever reason, by the late eighteenth century atomism had found a wide audience. The work of Antoine Lavoisier (1743-1794),[7] perhaps more than any other, contributed heavily to its acceptance. In a series of carefully controlled experiments, he showed that the oxidation of sulfur, phosphorous, lead, and tin involves weight conservation. He further demonstrated that exactly the same weight of mercury is regenerated throughout repeated oxidation/reduction cycles. These results strongly suggest that pure substances (that is, elements) do not react in arbitrary proportions with other pure substances, but in fact combine in quite definite ratios of their weights. This hypothesis is called the Law of Definite Proportions.

The atomic theory was not long in providing an explanation of the Law of Definite Proportions. Indeed, by 1807, John Dalton,[8] an English mathematics tutor, had published an derivation of the Law from the atomic theory. Let us suppose, he argued, that all atoms of an element have a characteristic (fixed) weight. If compounds are just fixed combinations of atoms of elements, then pure substances (collections of atoms of the same kind) will combine in fixed, definite ratios of their weights to produce compounds. But this is precisely the Law of Definite Proportions.

There is not a little irony in Dalton's abstraction of the Law of Definite Proportions from Lavoisier's data, for the very elements which Lavoisier investigated contain some of the worst violators of the Law. We now know, for example, that the behavior of the transition metal oxides and sulfides only roughly approximates that predicted by the Law of Definite Proportions; the composition of cuprous sulfide, in particular, can range from 1.7 to 2 atoms of copper to one atom of sulfur.

It has been conjectured that the atomic theory of the late eighteenth century would have been crushed if Lavoisier's measurements had been more accurate and if Dalton had paid more attention to minor deviations in the behavior of actual substances from the Law of Definite Proportions. The question is no doubt ripe for speculation, but, in some sense, it is fortunate that Dalton ignored the details. There was an escape from the problem of deviations from the Law, but not a ready one in the eighteenth and early nineteenth century. We now know, in particular, that some elements do not always combine in simple integer ratios, that is, they do not, strictly speaking always form <u>molecular</u> compounds. If one admits that there are non-molecular structures of matter (say, non-molecular crystals), the atomic hypothesis can be saved. What these structures are, however, was elucidated only long after Dalton died.

Undaunted, and perhaps unaware, of the limited applicability of the Law of Definite Proportions, Dalton pressed the theory further. A consequence of the theory, he argued, is that if two elements form more than one compound, then the different weights of one which combine with the same weight of the other stand in the ratio of small whole numbers. The weights of nitrogen which combine with 16 grams of oxygen in $N_2O$, NO, and $NO_2$, for example, are respectively 28, 14, and 7 grams, which stand in the ratio of 4:2:1. This prediction, known as the <u>Law of Multiple Proportions</u>, was established experimentally shortly after Dalton published it. This confirmation of the Law lent further support to the atomic theory. And, perhaps for the first time, the predictive utility of the theory was thereby demonstrated.

In spite of this success, Dalton's atomism was not without problems. It rested, in particular, on the then-unconfirmed assumption that atoms of a particular element have a characteristic weight. Dalton and the early nineteenth-century atomists had to show that this belief was reasonable, or leave themselves open to the charge of arbitrariness.

From the outset, this proved to be a rather nasty task, because the only data available to the early nineteenth century atomists at best gave only the composition by elemental weight for compounds (that is, the total weight of each element as it occurs in the compound), not actual atomic weights. Consider just how inadequate this kind of data is for the task of determining atomic weights. Suppose we wished to determine the the relative weights of hydrogen and oxygen from experimental data which shows that there are 8 grams of oxygen to each gram of hydrogen in water. If we assume that the empirical formula for water is HO, we will conclude, using the Law of Definite Proportions that

$$\frac{w_H}{w_O} = \frac{1\ g}{8\ g}$$

or

$$w_O = 8\ w_H$$

where

$w_O$ is the atomic weight of oxygen
$w_H$ is the atomic weight of hydrogen.

But if we assume that the empirical formula for water is $H_2O$, then we will correspondingly conclude that

$$\frac{2w_H}{w_O} = \frac{1\ g}{8\ g}$$

or,

$$w_O = 16\ w_H.$$

That is, what we conclude about the relative atomic weights of hydrogen and oxygen drastically depends on what we assume the empirical formula of some compound of those elements to be. Composition-by-elemental-weight data alone and the Law of Definite Proportions, therefore, cannot yield the atomic weights of the elements in a compound. In the mathematician's jargon, the solution to the problem, given the constraints, is "underdetermined:" we have three variables (composition-by-elemental weight for a compound, the ratio of numbers of atoms of the respective elements in a molecule of that compound, and the atomic weights of the elements which comprise that compound), but only two independent equations (the Law of Definite Proportions and the composition-by-elemental-weight data). Dalton's approach to this problem was to conjecture molecular formulas, then use these conjectures as working hypotheses to determine relative atomic weights. To help formulate these conjectures, he invoked Newton's principle of simplicity: if no evidence to the contrary is at hand, assume the simplest possible hypothesis. For example, since the only compound of hydrogen and oxygen known to Dalton and his contemporaries was water, Dalton, having no other evidence, accordingly assumed that the ratio of the number of hydrogen atoms to the number of oxygen atoms in a molecule of water was 1:1. That is, he assigned water the formula HO. This assumption, coupled with the experimental fact that water contains 8 grams of oxygen for each gram of hydrogen, obviously led Dalton to assign oxygen a weight of 8 on a scale on which hydrogen has a weight of 1 (see calculation in the example above). It was a false step, but the best which Dalton (or anyone else at the time) could have been expected to muster.

Dalton's approach to assigning atomic weights is obviously risky, because is requires that we assume, apart from experiment, the empirical formulas for compounds. If atomism is to stand on its own merits, some additional, independent corroboration of these assignments is needed. Just such evidence is to be found in the chemical behavior of certain gases. In 1808, for example, Joseph Gay-Lussac (1778-1850)[9] reported the results of studies on various gas and gas/liquid reactions, including reactions of ammonia with carbonic acid. Those look like:

1 volume ammonia + carbonic acid => ammonium hydrocarbonate

2 volumes ammonia + carbonic acid => ammonium carbonate

The volumes of ammonia consumed in these two reactions are in the ratio 1:2. On analogy with the Law of Multiple Proportions, Gay-Lussac proposed that volumes of gases which form different compounds with the same substance stand in the ratios of small whole numbers. Just to have a name for this hypothesis, let us call it the Law of Proportional Volumes.

Gay-Lussac further noted that in a given reaction, volumes of reacting gases tend to combine in the ratio of small whole numbers. Applying this generalization, which we might by analogy call the Law of Definite

Volumes, to data obtained by others on the production of ammonia, he concluded that ammonia is composed of three volumes of hydrogen to one volume of nitrogen.

To Gay-Lussac, the volume combination laws were confirmation of the atomic hypothesis. Properly speaking, of course, the Law of Proportional Volumes and the Law of Definite Volumes are only analogues of the Law of Multiple Proportions, and the Law of Definite Proportions, respectively. The latter, furthermore, are at best consequences, not a proof, of atomism.

Ironically, Dalton rejected Gay-Lussac's proposals. Dalton saw, in particular, that Gay-Lussac's observations implied that the numbers of particles contained in equal volumes of reacting gases were either equal or integral multiples of one another. To this, which we will call the "equal-volumes, equal numbers" hypothesis, Dalton had two serious objections.

First, Dalton knew that the mass density of the water vapor which results from the reaction of hydrogen with oxygen is less than the original density of oxygen. He was disposed to think of the creation of water as the addition of one atom of hydrogen to one atom of oxygen. As Dalton saw the problem, therefore, a given number of water particles would have to weigh more than the same number of oxygen particles (which he thought consisted of single atoms (i.e., were "monatomic")). Accordingly, there was only one way in his eyes that a volume of water could be less dense than a volume of oxygen: fewer molecules of water than oxygen had to be contained in a volume (at a given temperature and pressure). And this flatly contradicted the "equal-volumes, equal-numbers" hypothesis.

Second, Dalton argued, we know that from equal volumes of nitrogen and oxygen, two volumes of nitric oxide will be produced. If equal volumes of different gases contained equal numbers of particles, and if, he urged, each particle of an elemental gas were an indivisible, single atom, we would be forced to write

nitrogen + oxygen => nitric oxide

1 volume + 1 volume => 2 volumes

n atoms + n atoms => 2n molecules.

The first two lines represent experimental facts. The last, Dalton thought, is an impossibility: the reaction of *n* indivisible atoms can never produce more than *n* new particles.

Of course, this reasoning is based on the wholly arbitrary prejudice that the particles of pure gases are individual atoms. Dalton either was not aware of the insidious role this assumption played in his reasoning, or chose not to consider it. For whatever reason, he rejected the notion the "equal volumes, equal numbers" hypothesis and the data on which it was based.

Atomism would have reached an impasse if these objections could not have been answered. In 1811, however, Amadeo Avogadro (1776-1856)[10] advanced precisely the suggestion that Dalton had ignored or suppressed: molecules of gaseous elements might consist of more than one atom. Avogadro noted that once this hypothesis is accepted, the paradoxical (at least through Dalton's eyes) nature of the combining volumes data can be resolved. With the correct molecular formulas in hand for water ($H_2O$) and oxygen ($O_2$), for example, it is easy to explain why a volume of water vapor at a given temperature and pressure is less dense than the same volume of molecular oxygen at the same temperature and pressure, even though both volumes contain the same number of "particles" (molecules). In particular, the relative molecular weight of oxygen is about 32, whereas the molecular weight of water is only about 18 on the same scale. Thus, if fixed volumes of gaseous water and oxygen contain equal numbers of particles, the (mass) density of the volume of oxygen should be greater than that of the water.

The problem of determining atomic weights languished at this stage for over 40 years; curiously, Avogadro's proposals remained largely unexplored by his peers. Between 1811 and 1858, the problem of determining the relative atomic weight scale became more and more vexing. Many solutions were proposed, only to fall prey to seemingly lethal objections or limitations. Some chemists even began to believe that a solution to the problem of determining the characteristic weights of atoms would be impossible to obtain.

These difficulties were ingeniously overcome by Stanislao Cannizzaro in 1858.[11] Cannizzaro first assumed, following Avogadro and Dalton, that a molecule must contain a whole number of atoms of each of its constituent elements. Cannizzaro further assumed, again following Avogadro, that at fixed temperature and pressure, equal volumes of gases contain equal numbers of molecules. From this it follows that the relative weights of the molecules of these gases volumes stand in the ratio of the weights of their volumes. Cannizzaro then defined the relative (molecular) weight of hydrogen gas to be 2. This convention, conjoined with the "equal volumes, equal numbers" hypothesis, immediately fixes the relative weights of all other gaseous molecules. For example, the weight of a given volume of nitric oxide is about 15 times that of hydrogen at a fixed temperature and pressure. Consequently, the relative weight of nitric oxide is 30 on a scale on which the weight of hydrogen is 2. The relative weight of a compound determined by Cannizzaro's method, which we will call the molecular weight of that compound, is very close to the molecular weight we assign to that compound today.[12]

Let us define a gram mole of a compound to be that quantity of the compound whose mass is a molecular weight number of grams of that compound; similarly we will define a gram atom of an element to be that quantity of the element whose mass is an atomic weight number of grams of that element. Suppose we analyze a gram mole of a gas to determine the weight of a given element that quantity of gas contains, Cannizzaro continued. If we examine a series of (gaseous) compounds of the element in this way, we will in fact find that there is some weight, which when multiplied by various integers, will give precisely the weights of the elements which analysis shows are contained in a gram mole of the various compounds. That weight is very probably an atomic weight number of grams of that element, Cannizzaro argued.

Although it is a fundamental advance in placing the atomic theory on quantitative footing, Cannizzaro's method is limited to determining the atomic weights of elements which occur in gaseous compounds. This is a serious limitation, because relatively few elements occur in gases under everyday conditions on this planet. To help overcome this limitation, Cannizzaro revived a method which had been introduced several decades earlier but which had failed to gain the appreciation it deserved. In 1819, Dulong and Petit[13] attempted to measure the specific heats of fixed numbers of atoms of various metals. Such an attempt presumes that we have a way of measuring fixed numbers of atoms. If one had an accurate table of atomic weights, of course, one could simply measure the specific heat of gram-atomic weights of various metals, because gram-atomic weights of different elements contain the same number of atoms, by the definition of a gram atom and (the extension to the case of atoms of) the "equal volumes, equal numbers" hypothesis. The number of atoms in a gram atom (or equivalently, the number of molecules in a gram mole) is called Avogadro's number.

This procedure, unfortunately, presupposes an accurate table of atomic weights. Dulong and Petit had a table of atomic weights, but some entries in the table appeared to be in profound disagreement with expectations. In particular, several values in the table were near integer multiples of what Dulong and Petit thought they should be, assuming that the quantity of heat required to raise a fixed number of atoms of a metal a unit temperature is a constant. And there was good reason to believe that there is such a constant: for the majority of the entries in the table, Dulong and Petit noted that this "gram-atomic" specific heat was (in modern parlance) about 6 cal/gram-atom/degree Celsius. In 1819, however, no one had a convincing explanation of the integer-multiple variations from the law which some elements exhibited. Hence the claim that there was such a thing as a universal gram-atomic specific heat was regarded at best problematic.

Cannizzaro conjectured that the apparent discrepancies in Dulong and Petit's table might have arisen because of a confusion between atomic and molecular weights. Some elements may occur as polyatomic molecules, he suggested. If they do, and if we were to erroneously assume (as Dalton had) that all elements occur as monatomic molecules, we would get just the kind of integer multiple discrepancies that the table of Dulong and Petit contains. Cannizzaro realized that Dulong and Petit's universal gram-atomic specific heat hypothesis, if taken as a given, could solve the problems in the atomic weight table. Suppose, for example, that analysis has shown that in a compound of copper and chlorine there are 0.3286 grams of chlorine and 0.5888 grams of copper. Suppose we know (say, from Cannizzaro's method) that the atomic weight of chlorine is 35.46. This data tells us that one gram-atomic weight of chlorine will combine with ((0.5888/0.3286) x 35.46) = 63.54 grams of copper. If the formula for copper chloride is CuCl, then the atomic weight of copper is 63.54. But if the formula is $Cu_2Cl$ or $CuCl_2$, the atomic weight of copper is 31.77 or 127.08, respectively. This embarassment of riches is, of course, just another example of the problem which plagues Dalton's method (see the discussion about simultaneously determining the empirical formula of water and the atomic weights of hydrogen and water, above). Now Cannizzaro knew that the specific heat of copper is 0.093 cal/gram/degree Celsius. Using the constant of Dulong and Petit, he could calculate the atomic weight of copper to be approximately

$$\frac{6 \text{ cal/gram-atom/degree}}{0.093 \text{ cal/gram/degree}} = 64 \text{ gram/gram-atom.}$$

This result very strongly suggests that the appropriate atomic weight for copper is 63.54. Thus Dulong and Petit's universal gram-atomic specific heat hypothesis provides a check of atomic weights which is independent of our assumptions about the empirical formulas of compounds in which given elements occur, thereby solving the problem which plagues Dalton's method.

This approach proved wildly successful. By 1869, Mendeleev[14] published a table of atomic weights established largely by Cannizzaro's method. Although the table contained some gaps and miscalculations, it contained credible entries for nearly all known elements, thereby convincingly demonstrating the power of the atomic hypothesis to characterize atomic weights.

Mendeleev's table had at least one further virtue. The table attempted to organize elements of "similar" chemical properties into the same column. For example, the alkali metals (e.g., potassium and sodium) were in one column together; the noble gases (e.g., neon and argon), in another column together. The table was constructed so that atomic weights increased across each row and down each column. This organization showed that many important properties recur periodically as a function of "atomic number" (where the "atomic number" of an element is just its ordinal position in Mendeleev's table). Thus, when the mere position of a heretofore undiscovered element is identified in the table, it is possible to predict some of the element's (relative) chemical properties.

None of atomism's competitors could remotely approach this performance. What the theory could not explain, however, was how atoms can combine to form reasonably stable compounds. Answering that question, we now know, required refining a fundamental assumption of atomism: that atoms are indivisible and hence have no interior structure. But that story is intimately wed to the theory of natural radioactivity, which the next chapter claims for its own.

## NOTES

1. We know very little about Democritus, because none of his work survives. We would know even less about him save for the fact that his views were zealously championed by the Roman poet Lucretius. See, for example, Lucretius, On the Nature of Things, translated by H. J. Munro, in W. J. Oates, ed., The Stoic and Epicurean Philosophers (Modern Library, 1940).

2. See, for example, Plato, Parmenides, translated by F. M. Cornford, in E. Hamilton and H. Cairns, ed., Plato: The Collected Dialogues (Princeton, 1961).

3. See, for example, J. Locke, An Essay Concerning Human Understanding, in E. A. Burtt, ed., The English Philosophers from Bacon to Mill (Modern Library, 1939).

4.-14. See A. R. Hall, The Scientific Revolution: 1500-1800 (Beacon Press, Boston, 1962).

# 2 An Introduction to Natural Radioactivity

## INTRODUCTION

By virtue of its predictive power, late nineteenth-century atomism buried its competitors. Nevertheless, a crucial and difficult task remained: to quantitatively explain how atoms can combine to form relatively stable compounds. That explanation required, it turned out, recognizing that atoms are not indivisible lumps, but are themselves composed of smaller, particle-like entities. This chapter surveys how that recognition matured into the understanding of natural radioactivity we have today.

## ELECTRICITY IN MATTER

The early Greeks knew that a piece of amber, when rubbed with the hair of an animal, would attract small bits of dry leaves or pith. This fact notwithstanding, a theoretical grasp of the electrical nature of matter did not begin to evolve until the beginning of the nineteenth century. We would not be far wrong, in fact, to say that much of our present knowledge of the electrical nature of matter begins with the work of Michael Faraday, born in 1791 in a small village near London. His formal education was extremely meager but he possessed great native talent for organizing and conducting experimental work. In 1812, Faraday made a particularly bold and presumptuous request for employment at the Royal Institution. The Director of the Institution, Sir Humphrey Davy, hired the largely self-taught scholar as an apparatus and lecture assistant. Sir Humphrey's gamble paid handsomely: between 1816 and 1819 alone Faraday published 37 papers on the nature of electricity, an average of one per month![1]

To Faraday we may attribute the discovery of the generator and motor principles, electromagnetic induction, and a variety of conceptual models for predicting the behavior of electrical and magnetic fields. He demonstrated that electricity from a friction machine would deflect a galvanometer and cause electrolysis in the same way that electricity from chemical sources (batteries) would. To describe electrolysis, Faraday developed the terminology still used by electrochemists today: "anode," "cathode," "ion," "anion," "cation." (An ion is "just" an atom which carries a net (positive or negative) charge. A cation is a positively charged ion; an anion is a negatively charged atom. Similarly, an anode is a positively charged electrode (of an electrolytic cell), whereas a cathode is a negatively charged electrode of an electrolytic cell. In a cathode ray tube (described later), the (negative) electrode which is the source of the cathode rays is called the cathode; the (positive) electrode at which cathode rays are collected is called the anode. Thus, an anion is attracted to an anode; a cation, to a cathode.

The phenomenon of electrolysis suggested to Faraday that the nature of material bonding is both atomic and electrical. He even went so far as to propose a deep connection between certain weights, which he called equivalent weights, of chemical species, and the quantity of electricity associated with those weights:[2]

> Equivalent weights of bodies are simply those quantities which contain equal quantities of electricity...it being the electricity which determines the combining force (between atoms of a molecule). Or, if we adopt the atomic theory or phraseology, then the atoms of bodies which are equivalent to each other in their ordinary chemical action, have equal quantities of electricity naturally associated with them.

Or, to put these ideas in a modern idiom,

1. The weight of a given material deposited at an electrode by a given amount of electricity is always the same.

2. The weights of various materials deposited, evolved, or dissolved at an electrode by a fixed amount of electricity are proportional to the equivalent weights of those substances.

We can, in a fashion much like the way relative atomic weights are assigned to the elements, construct a table of relative equivalent weights of the sort Faraday had in mind. In particular, let us by convention say that the equivalent weight of elemental hydrogen is just the weight of one gram-atom of hydrogen. Then by this convention and Faraday's definition of equivalent weight, the mass of any element X which is consumed or liberated by the same amount of electricity (charge) as that required to produce or consume (e.g., by electrolysis) 1 gram-atom of elemental hydrogen will be called X's <u>equivalent weight</u>. The quantity of electricity (charge) which will produce or consume a gram-atom of elemental hydrogen is called a <u>Faraday</u> (F); the currently accepted value of this quantity is about 96,500 coulombs.

Suppose we were to determine by experiment the quantity of various elements liberated or consumed by a Faraday. We would soon discover that a gram atom of some elements is liberated by that charge. 22.98 gm of sodium, for example, is liberated from sodium chloride by a Faraday. For other elements, twice that charge is required to liberate a gram atom. Two Faradays, for example, are required to liberate a gram-atom of calcium (which has an atomic weight of about 40). For yet other species, three Faradays are required to liberate a gram-atom of that material. Now suppose that X is an (electrically neutral) atom of an element. An <u>ion</u> of X is a chemical species formed by adding a definite positive or negative charge to X (for the moment, we won't bother with just what the phrase "adding a charge to a neutral atom" really means). Some neutral chemical species can be converted to several different ions, distinguished by the magnitudes of the charges they carry. We will define a <u>gram-ion</u> of an ionic species to be an equivalent weight number of grams of that species. Let us call ions which require a Faraday to generate a gram-atom of the corresponding neutral species <u>univalent</u> or <u>monovalent</u>. Ions which require two Faradays will be called <u>divalent</u>; those for which three Faradays are required, <u>trivalent</u>, and so on.

Let the number of ions in a gram-ion of hydrogen ions be $N_0$. This number is known as <u>Avogadro's number</u>, and it also represents the number of atoms in a gram-atom of hydrogen or any other material. If we adopt the view that the charge taken up or released when a gram-ion of hydrogen ions is consumed or produced is uniformly distributed among the hydrogen ions, it follows that there is a unit charge acquired or released by each hydrogen ion (or, equivalently, by each ion in a gram-ion of a univalent species). On this view, that unit charge is

$$e = F/N_0 \tag{2-1}$$

In 1891, G. J. Stoney proposed the name "electron" (from the Greek <u>elektron</u>, meaning "amber") for this natural unit.[3] Determining e from Equation 2-1 requires knowing $N_0$, unfortunately, and the best estimate of $N_0$ (Stoney's own) available in 1891 was about 16 times larger than the currently accepted value. Accuracy notwithstanding, the very notion that there is a definite, atomic, fundamental unit of charge was a giant leap for atomism in explaining the electrical nature of matter. The idea became a cornerstone, in fact, of the theory of natural radioactivity.

## THE DISCOVERY OF RADIOACTIVITY[4]

In the late fall of 1895, Wilhelm Konrad Roentgen,[5] then professor of physics at the University of Wuertzburg, was investigating electrical discharges through rarefied gases. The apparatus he used in these experiments, called a <u>gas discharge tube</u>, is shown schematically in Figure 2-1. It consists of a sealed glass tube containing a gas at low pressure. Electrodes are embedded in either end of the tube and extend into its interior. The exterior leads of these electrodes are connected to a high-voltage source such as an induction coil. When a high-voltage pulse is applied to the leads of the tube, the gas in the tube will, under certain conditions (pressure, temperature, and applied voltage), flouresce. The flourescence of the gas is accompanied by a current passing through the gas between the electrodes (hence the name "gas discharge tube").

To make the flourescence easier to see, Roentgen had shrouded the tube with a piece of thin black cardboard an placed the apparatus in a dark room. Quite by accident, he discovered that when the tube was activated, a nearby paper screeen which had been washed

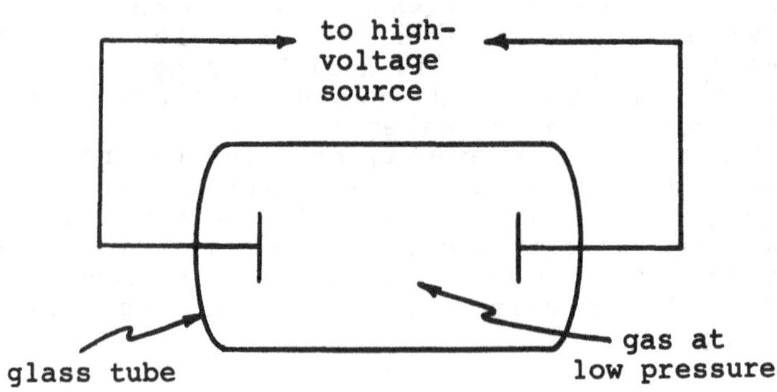


Figure 2-1. Schematic of Roentgen's gas discharge tube.

with barium-platino-cyanide flouresced brightly. The phenomenon was detectable when the screen was up to two meters from the tube.

A few experiments convinced Roentgen that the flouresence phenomenon was induced by radiation coming from a specific region of the discharge tube. In this region, he hypothesized, emanations from the tube's cathode ("cathode rays") were striking the tube's walls, thereby generating the invisible radiation which caused the barium-platino-cyanide screen to flouresce. Roentgen at once proceeded to investigate the properties of this mysterious radiation, which he called "x-rays" ("x" = "unknown"). Among the properties he discovered were

1. All substances are more or less transparent to x- rays.

2. Photographic emulsions are sensitive to x-rays.

3. X-rays cannot be reflected or refracted by conventional optics.

4. Unlike cathode rays, x-rays cannot be deflected by a magnetic field.

5. Many substances flouresce under "illumination" by x-rays.

6. X-rays are generated when cathode rays strike any solid body, although heavier elements seem to be more efficient producers of the rays than lighter ones.

7. X-rays can discharge charged bodies.

By any standards, Roentgen was a careful and thorough investigator. His work was to form the foundation for several new techniques for investigating the structure of matter. Moreover, as we shall shortly see, his efforts motivated the discovery of, and helped to provide a model for, at least one feature of natural radioactivity.

Roentgen's unexpected discoveries suggest that x-rays are associated with the phenomenon of flourescence or luminescence. This hint inspired an aggressive search for materials which could be induced to generate x-rays by exposure to visible light. In February 1896, Antoine Henri Becquerel[6] reported to the Académie des Sciences of Paris an odd discovery which seemed to show just this connection. He had placed a "thin crust" of crystals of a double sulfate of uranium and potassium on a piece of black paper, which in turn enclosed a photographic plate that had not been exposed to light.

Thinking that x-rays might be produced when these crystals were induced to flouresce by exposure to light, Becquerel exposed the entire package to sunlight for several hours. The silhouette of the crystal appeared on the plate. Furthermore, Becquerel reported, when a perforated metal plate was interposed between the salt crust and the paper, the outline of the metal appeared on the plate.

The experiments, we now know, were somewhat misconceived, because the images would have appeared without exposure to the sun. By a fortuity of the weather, Becquerel soon discovered this fact. Because sunlight was intermittent at that time of year in Paris, he stored the paper-wrapped photographic plate and the uranium-potassium sulfate crust in a desk drawer together without first exposing the package to the sun. On developing the plate, Becquerel still found strong images of the crust. Clearly this phenomenon was not caused by visible-light induction of flourescence. Becquerel accordingly proposed that the exposure of the plate was due to radiations actives[7] from the salt crust, an obvious forerunner of the modern term "radioactivity."

Within two months, Becquerel reported to the Académie that uranium and all compounds of that element he had studied using the photographic technique exhibited radioactivity.[8] He also noted that metallic uranium appeared more active than any of its compounds when distributed over the same surface area on the photographic plate. He further determined that the paper-wrapped photographic plates were less exposed when thin films of aluminum, copper, and other metals were interposed between the uranium salt and the plate.

Inspired by Becquerel's discovery, Marie Curie (born Marie Sklodowska), a brilliant young Polish chemist, began a systematic search for radioactive elements other than uranium. At the least, she realized, some technique more discriminating than Becquerel's photographic emulsion procedure for measuring radioactivity would be highly desirable for this purpose. The heart of her method was an ingeniously simple detector connected to an sensitive electrometer. The detector was a parallel-plate condenser, on one plate of which the substance under test is placed. If that substance is radioactive, it will cause ionization of the air between the plates. This ionization is detectable as a change in the current which flows across the condenser. Thus the current through the condenser provides a measure of the activity of the substance under test.

Of the substances she analyzed, Mme. Curie found that thorium was the only element which had approximately the same radioactivity as uranium. Some substances, notably the minerals pitchblende, chalcolite, autinite, and carnotite, were more radioactive than metallic uranium; pitchblende in particular showed from 0.75 to nearly 4 times the radioactivity of uranium. These results strongly suggested the existence of an as-yet-undiscovered radioactive element.[9] Using strictly chemical isolation techniques, Mme. Curie and her husband Pierre were able to identify two new radioactive elements among these minerals, one chemically like bismuth, and the other akin to barium. The former Mme. Curie named polonium and the latter (which she discovered in collaboration with G. Bémont) radium. The chemical purification of radium, given the techniques available to the Curies, proved difficult; that of polonium, impossible. Eventually the Curies were able to isolate about 0.1 gram of spectroscopically pure radium. Using this sample, Mme. Curie determined the atomic weight radium to be 225, remarkably close the currently accepted value.[10]

## THE CHARACTERIZATION OF RADIOACTIVE EMANATIONS

By 1900 a variety of radioactive elements had been discovered, largely by the Curies and Ernest Rutherford's group at the Cavendish laboratory in Cambridge. Rutherford and his colleagues were able to show that there are at least two types of emanations from radioactive substances, distinguished at least by their penetrating power. The first of these, called alpha rays, could easily be stopped by a sheet of paper or a few centimeters of air. The second type, called beta rays, could pass through several millimeters of aluminum. In 1900 a third and much more penetrating type of emanation, gamma rays, was discovered by P. Villard.[11]

The task of understanding the mechanism of radioactivity remained. Were these strange radiations like particles, like waves, or unlike anything heretofore explored?

### The characterization of beta rays

Beta rays proved the easiest of these strange emanations to comprehend. This was due in no small part to the fact that a substantial understanding of beta radiation was already at hand, although no one at the time fully appreciated it. It was the the work of J. J. Thompson, perhaps more than any other, which provided the basis for identifying beta rays with cathode rays. To identify two entities or processes, of course, we need to show that they have the same properties. Thus, identifying beta radiation with cathode rays requires having a reasonably precise quantitative characterization of cathode rays.

Thompson conjectured that cathode rays were streams of particles. Perrin,[12] the Curies,[13] and Becquerel[14] had, he knew, demonstrated that cathode rays carry a charge. If cathode rays could be thought of as charged particle streams, Thompson reasoned, there should at least be an average charge-to-mass (e/m) ratio assignable to the particles in this stream. Furthermore, as particles, cathode rays should have some velocity. To determine these properties, Thompson constructed a modified cathode ray (gas discharge) tube like that schematically shown in Figure 2-2. In this apparatus, a potential difference of a few thousand volts is maintained between anode A and cathode C. The tube contains air at greatly reduced pressure. One end of the tube, S, is coated with a cathode-ray-sensitive phosphor. Any positive ions present in the tube (say, from residual air) are accelerated toward the cathode C by the electric field in the anode-cathode region. When

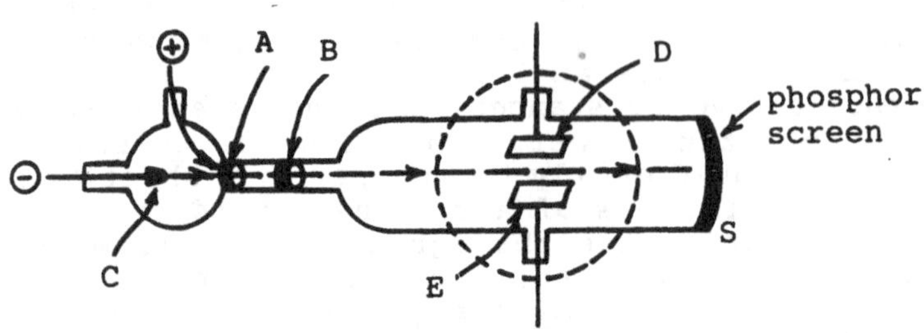


Figure 2-2. Schematic of Thompson's apparatus for determining the charge-to-mass ratio of cathode rays.

these ions strike the cathode, they liberate cathode rays in the direction of anode A. A few of these rays pass through a small hole in A, and are able to continue through a hole in a plug at B. Some of the rays which pass through B arrive at the phosphor screen S, creating a luminous spot on the screen.

Cathode rays, Thompson knew (from Becquerel and others), can be deflected by electric and magnetic fields. Exploiting these properties, Thompson fitted his apparatus with a pair of parallel plates, D and E, across which an electric potential was applied, thus inducing an electric field normal to the plates; in this same region of the tube he established a magnetic field normal to the electric field by means of a pair of Helmholtz (electromagnetic) coils. By careful adjustment of the magnetic and electric fields, the deflection of the cathode beam (as visualized by the spot on the screen) can be made zero. It can be shown that under these conditions

$$v = \frac{E}{B} \tag{2-2}$$

and

$$\frac{e}{m} = \frac{E}{RB^2} \tag{2-3}$$

where

v is the velocity of the rays/particles

E is the (magnitude of the) electric intensity between E and F

R is the radius of curvature of the cathode beam in the absence of an electric field between E and F

B is the (magnitude of the) magnetic induction

e is the charge on the cathode ray/particle

m is the mass of the cathode ray/particle.

Thompson found e/m for cathode rays to be about $1.75 \times 10^{11}$ coul/kg, and their velocity to be about 0.8c, where c is the speed of light. The very fact that e/m can be measured, since it is a property of a particle, very strongly hints that cathode rays are particle-like streams. While realizing that Thompson's experiment is not ironclad proof that cathode rays are particle streams, let us for the purpose of further discussion assume that Thompson's work established the existence of a negatively charged particle-like entity, the electron. Cathode rays, on this view, are electron streams.[15]

In 1899 F. Giesel,[16] S. Meyer,[17] E. von Schweidler,[18] and Becquerel[19] succeeded in characterizing the behavior of beta rays in a magnetic field, among other things showing that beta rays are deflected in the same sense as cathode rays.[20] This result strongly suggests that beta rays are electromagnetic in nature, and are presumably negatively charged like cathode rays.

A pictorial of the apparatus used by Becquerel is shown in Figure 2-3. A sample of radioactive material is placed at the bottom of a narrow well drilled in a small lead container. The lead container is placed on a photographic plate. A magnetic field parallel to the photographic plate is established by means of a pair of Helmholtz coils. Beta rays emerging from the container in the plane of the Figure are deflected by the magnetic field and strike the plate at various distances from the source, depending on their momenta. It can be shown that the distance x from the source at which the rays

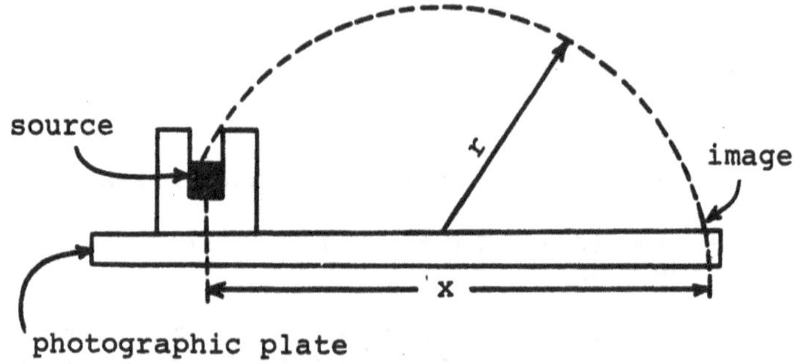


Figure 2-3. Pictorial of Becquerel's apparatus for measuring the magnetic deflection of beta rays.

strike the plate is given by

$$x = 2r = 2\frac{mv}{eB} \qquad (2\text{-}4)$$

where

r is the radius of curvature (the particle's trajectory is circular)

m is the particle's mass

v is the particle's velocity

B is the magnitude of the magnetic induction

e is the charge on the particle.

If we solve Equation 2-4 for Br (called the magnetic rigidity), we get

$$Br = \frac{mv}{e} \tag{2-5}$$

Becquerel found a wide variation in Br for beta rays from the same source; since e, B, and m in Equation 2-5 are presumably fixed (a slightly incorrect assumption, since beta rays travelling near the speed of light will have a relativistic mass gain), this result implies a wide variation in the velocities, and hence, a wide variation in the the kinetic energies of beta rays from the same radioactive source. Becquerel was able to demonstrate with this apparatus, in fact, that the kinetic energies of beta rays from the same source range continuously from 0 to some maximum $E_{max}$ which is characteristic of the source.

Becquerel's magnetic deflection experiments strongly suggested that beta rays have electromagnetic properties much like cathode rays. Using an apparatus similar to the one which Thompson used to determine the velocity of electrons, W. Kaufmann showed that beta rays having velocities of about 2.4 - 2.8 x $10^{10}$ cm $sec^{-1}$ have a charge to mass ratio of about 0.63 - 1.3 x $10^{7}$ emu per gram (1 coul = 0.1 emu = 2.998 x $10^{9}$ esu).[21] These values agree well with the values for the same properties of cathode rays which J. J. Thompson had obtained earlier. Thus, by 1900 the charge-to-mass ratio, the velocities, and interactions of beta rays with electric and magnetic fields were reasonably well understood, without doubt showing that beta rays are very similar, if not identical to, cathode rays. Since Thompson's work strongly hints that cathode rays can be thought of as electron streams, it accordingly seems natural to identify beta particles with electrons.

## The characterization of alpha particles

The characterization of alpha particles proved more elusive. From the outset, Rutherford was disposed to think of alpha particles as some fragment of an atom. There was at least one strong hint that he was right. He and a colleague, Frederick Soddy, noticed that helium was invariably associated with radioactive minerals. To demonstrate that the connection was not accidental, they collected the gas generated by a radium bromide solution in a gas discharge tube. After the collection had run some four days, the characteristic spectral lines of helium were detected in the discharge tube.[22]

Although this experiment shows that helium is intimately associated with certain kinds of radioactivity, it does not show precisely what the connection between the two is. Clearly enough helium is not identical with alpha particles, because alpha particles can ionize air, whereas helium cannot.

Showing what the connection between helium and alpha radiation was required getting a fairly precise characterization of alpha radiation itself. Toward this end Rutherford first demonstrated that alpha radiation can be deflected by strong magnetic fields in the sense opposite beta rays. He then adapted the experimental setup that Thompson had used to determine the velocity and charge-to-mass ratio for cathode rays to measure the same properties for alpha particles. Rutherford's measurements showed that the velocity of alpha particles from Radium C' (a preparation designated such by early experimenters and originally distinguished from radium by its preparational history) is about $2.5 \times 10^9$ cm $sec^{-1}$ and that e/m for this same preparation is about $6 \times 10^3$ emu $gm^{-1}$. Unlike the energies of beta rays, Rutherford noted, the energies of alpha particles from a single source are, within the limits of observation, the same (we now know that this is not quite correct: occasionally a very high-energy alpha particle is ejected from the atomic nucleus. But to first approximation, Rutherford's measurements are excellent.)

With the velocity and and charge-to-mass ratio of alpha particles in hand, in 1908 Rutherford enlisted the support of Hans Geiger, one of many very bright young physicists attracted to the Cavendish laboratory under Rutherford's administration,[25] to determine the mass of the alpha particle. In these experiments, Rutherford and Geiger first directly counted the number of alpha particles emitted by radium C' (RaC') in equilibrium with 1 gram of radium, obtaining a count of about $3.4 \times 10^{10}$ alpha particles per second per gram of radium. In a second experiment, Rutherford and Geiger determined by direct measurement the total charge collected on a plate suspended above the same RaC'/radium preparation; beta rays and secondary electrons were excluded by placing the apparatus in a strong transverse magnetic field. The result of the charge-determination experiment showed that the alpha particles in the preparation carried about 31.6 esu per gram of radium per second. Dividing the result of the charge-determination experiment by the result of the counting experiment, they obtained a value for e/m for alpha particles from the RaC'/radium mixture of about $9.3 \times 10^{-10}$ esu per particle.

The relative mass of the alpha particle could then be obtained by comparing the charge-to-mass ratio which Rutherford and Soddy had obtained for alpha particles with the charge-to-mass ratio for hydrogen ions (which had been determined earlier by Faraday), assuming that the charge on the alpha particle was some integer multiple of the unit charge e. In particular, suppose that the charge on the alpha particle is ne, where n is some positive integer. Then in a fashion similar to the way relative atomic weights can be determined, we would have

$$\frac{m_H}{m_a} = \frac{(ne/m_a)}{n(e/m_H)} \tag{2-6}$$

where

$m_H$ is the mass of the hydrogen ion

$m_a$ is the mass of the alpha particle

e is the magnitude of the fundamental unit charge

n is the number of unit charges carried by the alpha particle.

But such a method presumes that we know the integer n, that is, know how many fundamental unit charges are carried by the alpha particle. This, in turn, requires having a reasonably good idea of the magnitude of e. An estimate of e, albeit a crude one, was available from the work of J. J. Thompson,[26] H. A. Wilson,[27] and Robert Millikan.[28] These first attempts to determine e exploited the fact that it is possible to create a cloud of water droplets, place a measurable total charge on this cloud, and, at least roughly, determine the number of droplets in the cloud. Then, assuming that each droplet has a single univalent ion as its nucleus, the unit charge e is just the total charge on the cloud divided by the number of droplets in the cloud.

There are obviously risky assumptions in this method, not the least of which are the assumptions that all droplets are of the same size and carry the same charge. Nevertheless, if carefully executed, the technique can give somewhat better than order-of-magnitude results. Millikan eventually refined this technique to provide a much more precise and accurate estimate of e. The apparatus he used[29] is shown schematically in Figure 2-4. In this experiment, tiny droplets of atomized oil or some other low-volatility liquid are allowed to enter the region between two charged plates. The electric field between the plates is adjusted so that the electrical force on the drop just balances the net weight (gross weight minus the buoyant force of the air on the drop). It can easily be shown that under these conditions, the charge on a droplet is given by

$$q = \frac{w - B}{E}$$

where

E is the (magnitude of the) electric intensity

w is the gross weight of the drop

B is the buoyant force on the drop

q is the charge on the drop.

Millikan's experiment produced a remarkable result. Every oil drop measured was found to carry an integer multiple of a fixed unit charge e. That is, each drop was determined to carry a charge of 2e, 4e, 3e, 150e, and so on, but never 0.3e, 0.72e, 1.1e, and so on. Millikan accordingly inferred that e is the fundamental unit charge, indeed, the charge of the electron itself.

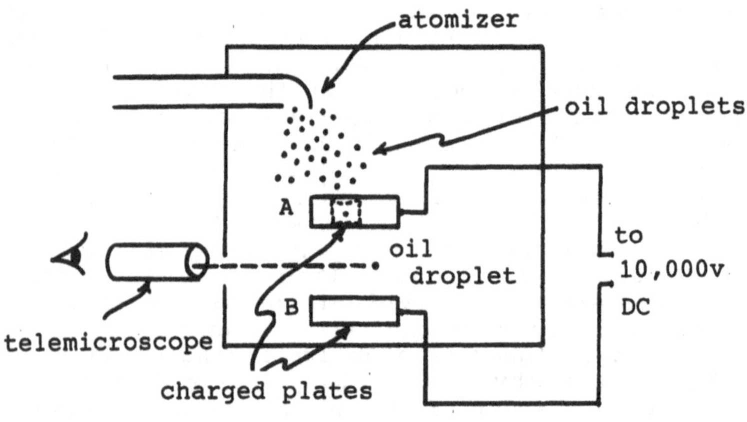


Figure 2-4. Schematic of Millikan's oil-drop apparatus.

This value is about[30]

$$e = 1.61 \times 10^{-19} \text{ coulomb.}$$

Since the oil-drop experiments were first performed, a number of other techniques have been developed to measure e. None of them have ever demonstrated that e does not exist or that it has a significantly different value from that which Millikan obtained.

Notice that with e/m and e for the electron from Thompson's and Millikan's experiments, we can get a good handle on the mass $m_e$ of the electron. In particular

$$m_e = (e/m_e)^{-1} \times e$$

or,

$$m_e = (1/1.75 \times 10^{11} \text{ coul kg}^{-1}) \times (1.61 \times 10^{-19} \text{ coul})$$

or,

$$m_e = 9.1 \times 10^{-31} \text{ kg.}$$

Knowing e, furthermore, allows us to compute the number $N_0$ of electrons whose collective charge is a Faraday, or equivalently, the number of monovalent ions which must be generated to neutralize or deposit an equivalent weight of a substance. By the definition of gram-atom, furthermore, this number is also the number of atoms in a gram-atomic weight of an element. In particular, from Equation 2-1 we have

$$F = N_0 \times e$$

or,

$$F/e = N_0$$

or,

$$N_0 = (9.65 \times 10^4 \text{ coul})/(1.61 \times 10^{-19} \text{ coul})$$

or,

$$N_0 = 6.02 \times 10^{23}.$$

Rutherford and Geiger noted that if the charge on the alpha particle were an integer multiple of e (a bold assumption), then its charge must be 2e or 3e. Now the best value of the electronic charge available at the time was $3 - 4 \times 10^{-10}$ esu.[31] For various inconclusive reasons, Rutherford and Geiger argued this value of e was too low and that a better value was $4.4 \times 10^{-10}$ esu. Given this value, they accordingly concluded that the number of (positive) unit charges on the alpha particle was 2. This means that n = 2 in Equation 2-6. Thus substituting $6 \times 10^3$ emu $gm^{-1}$ for $2e/m_a$ and $0.96 \times 10^4$ emu $gm^{-1}$ for $e/m_H$ and 2 for n in Equation 2-6, they concluded that the weight of alpha particles relative to hydrogen ions is about 3.84. At the time, the mass of the helium atom relative to hydrogen ions was accepted as 3.96. Rutherford and Geiger concluded that, within the limits of experimental error, an alpha particle "after it has lost its positive charge, <u>is</u> a helium atom."[32] Rutherford's faith in the deep similarity of helium and alpha radiation was borne out.

Knowing the velocity and mass of alpha particles allows us to compute the kinetic energy of the particles. If the mass of an alpha particle is about 6.6 x $10^{-24}$ gm, then the kinetic energy of the particle is[33-35]

$$\begin{aligned} K.E. &= (1/2)\ mv^2 \\ &= (1/2)(6.6 \times 10^{-24}\ gm) \times \\ &\qquad (2.5 \times 10^9\ cm\ sec^{-1})^2 \\ &= 2 \times 10^{-5}\ erg. \end{aligned}$$

In contrast, the amount of energy liberated by individual molecules in <u>chemical</u> reactions is on the order of $10^{-12}$ erg, at most. Thus, the emission of a single alpha particle involves an energy change several million times that involved in a typical <u>chemical</u> reaction. If there had been any doubt that natural radioactivity was different from ordinary chemical phenomena, the characterization of kinetic energy of the alpha particle laid that doubt to rest.

### <u>The characterization of gamma rays</u>

The nature of gamma rays more proved elusive than that of alpha particles. No known magnetic or electric field could deviate them. It was not until about 1910 that the light-like character of these rays was established. Experiments by M. von Laue,[36] W. Friedrich,[37] P. Knipping[38], and work by W. H. Bragg and W. L. Bragg[39] established, in particular, that gamma rays could be diffracted like light. And in 1914 Rutherford and E.N. da C. Andrade[40] showed that gamma rays were quite like X-rays which Roentgen had discovered twenty years earlier.

## THE THEORY OF RADIOACTIVE TRANSFORMATIONS[41]

In the late fall of 1899, R. B. Owens demonstrated that thorium-containing preparations emit a radioactive gas which has chemical properties distinct from thorium, thus showing that radioactivity can involve actual changes to the chemical identity of the emitters.[42] In this demonstration, Owens placed a radioactive solid

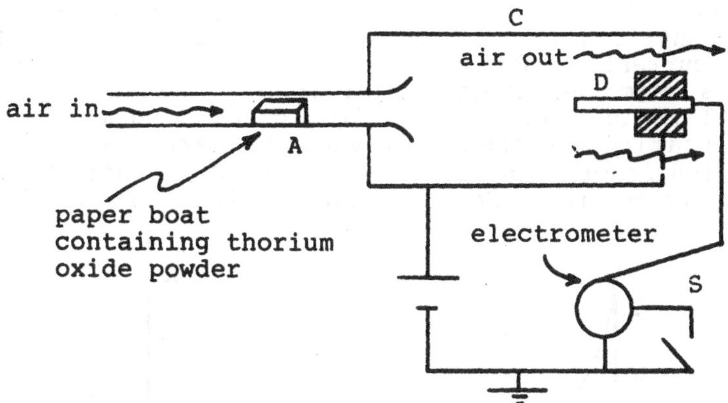


Figure 2-5. Rutherford's apparatus to determine the decay of radioactivity from the gas emitted from thorium.

on the lower plate of a parallel-plate capacitor. The capacitor was placed in a metal box through which a stream of air could pass. The ionization current which flowed between the plates (when the potential across the plates was raised sufficiently) was measured with an electrometer both with, and without, air flowing over the radioactive solid. When air was allowed to pass over the solid, Owen noted, the ionization current dropped to about two-thirds of the current observed when the air was not flowing.

Rutherford was quick to recognize the significance of the apparent change in the activity caused by the passage of air in Owens' experiment. In February 1900 he published a paper which described experiments in which the activity transferred from thorium preparations could be concentrated onto wires which had a negative potential relative to their surroundings.[43] In that publication, Rutherford reported what was to become the foundation of the quantitative law of radioactive change. The experiment consisted of two parts. The first part used an apparatus much like Owens'; the setup is shown schematically in Figure 2-5. In this setup a stream of air passes over a paper boat which contains a "thick" layer of thorium oxide. The air then passes into a large cylindrical insulated vessel C and flows out through a small hole in the end of C. An insulated electrode D is attached to an electrometer. After the air has been allowed to flow for a while, it is stopped and the ionization current between C and D is measured for about 10 minutes. The results look like curve A of Figure 2-6.

Rutherford then placed the thorium oxide between two concentric insulated metal cylinders along which a copious current of air was passed to prevent any gaseous products from the oxide powder from accumulating (see Figure 2-7). On stopping the flow of air and measuring the ionization between the two cylinders, curve B of Figure 2-6 is obtained. Curves A and B are normalized so that the maximum of B is equal to the initial value

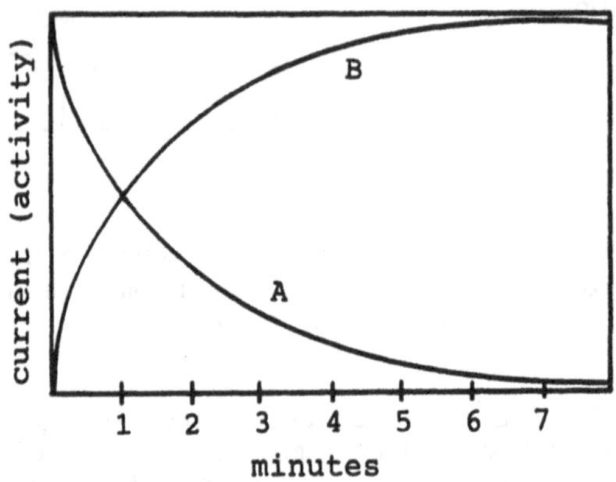


Figure 2-6. Decay (A) and growth (B) curves for thoron.

of A. Rutherford found that the sum of the quantities on curves A and B is a constant at any time t. This result strongly suggests that at equilibrium the gaseous emanation from thorium will be produced as fast as it disappears.

From these results and chemical analyses of the emanation from thorium, Rutherford confidently concluded that the thorium emanation ("thoron") is a chemically inert but radioactive gas akin to members of the argon family. He further concluded that the law of decay of activity is simply that a "substance alters at a rate proportional to the amount remaining."

Rutherford's distinction between thorium (oxide) as a source and thoron as an offspring of that source immediately suggests some natural terminology. Let us call thorium the "parent" (element) and thoron the "daughter" (element). In this jargon, when a parent decays by alpha emission, the resulting products are a daughter species and an alpha particle (plus some energy.)

Rutherford saw that if emission of an alpha particle were just emission of a helium ion of charge

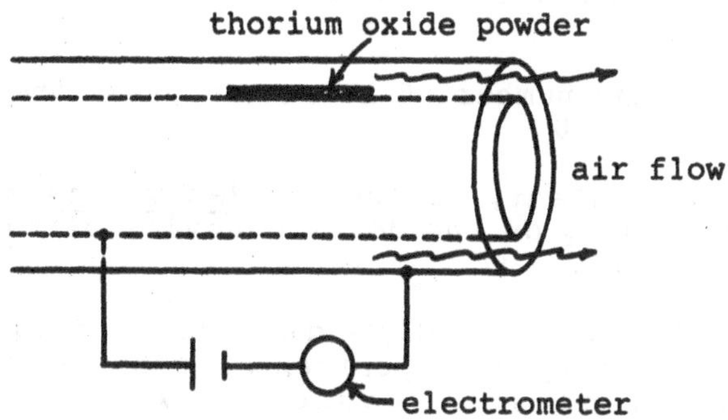


Figure 2-7. Schematic of thoron growth apparatus.

+2e, then one could predict the atomic weight of the daughter which results when an alpha particle is ejected. That weight would be the weight of the parent element minus the mass of an alpha particle. Using this model, Rutherford (together with Soddy) predicted that the gaseous emanation from radium (radon gas) must be 222 and that radium A (the product obtained when radon-222 emits an alpha particle) must be 218.[44]

Notice what Rutherford's and Soddy's work implies. The emission of alpha particle yields (in addition to the alpha particle) a new, possibly radioactive, species, which has a mass that is very nearly equal to the mass of the parent species minus the mass of a helium atom (or, more precisely, minus the mass of an alpha particle). Since this offspring behaves chemically like the element which appears two positions earlier in the periodic table, we can describe the offspring which results from the emission of an alpha particle as an element with atomic number two less than the atomic number of its parent. The emission of a beta particle, in contrast, yields an offspring which has very nearly the same weight as the parent but behaves chemically like the element immediately to the right of the parent in the periodic chart.

By 1903 Rutherford and Soddy were able to derive the quantitative form of the law of radioactive change. This law is[45]

$$N(t) = N(0)\exp(-kt) \qquad (2\text{-}7)$$

where

N(t) is the number of atoms (or mass) remaining at time t

N(0) is the number of atoms (or mass) present at at time t = 0

k is a constant, called the decay constant, which depends only on the decay behavior of the element of interest.

Rutherford's and Soddy's derivation of Equation 2-7 was based strictly on experimental considerations. An alternate approach (based on work by E. von Schweidler[46]) which is more commonly used today assumes that the law of radioactive decay is a statistical law in which the rate of change (- dN(t)/dt) of the number (or mass) of atoms at time t is proportional to the number N(t) of radioactive atoms present, that is

$$\frac{-\,dN(t)}{dt} = kN(t) \qquad (2\text{-}8)$$

Solving Equation 2-7 (using calculus) for N(t) yields

$$N(t) = N(0)\exp(-kt)$$

where, as before, N(0) is the number of atoms present at time t = 0. This is, of course, just Equation 2-7.

Equation 2-7 allows us to define the notion of the "half-life" for an element which undergoes spontaneous radioactive decay. The half-life of a species is just the time it takes half of a given quantity of that material to decay. Suppose, in particular, that there are N(0) atoms of an element at time t = 0. Then the time $t_{1/2}$ at which N(t) = (1/2)N(0), the half-life, is by substitution into Equation 2-7

$$(1/2)N(0) = N(0)\exp(-kt_{1/2})$$

or, solving for $t_{1/2}$

$$t_{1/2} = \frac{\ln 2}{k}$$

or

$$t_{1/2} = \frac{0.693}{k} \qquad (2\text{-}9)$$

The half-lives of the spontaneously radioactively decaying species (the <u>radionuclides</u>) range from about $10^{-9}$ sec to $10^{9}$ years.

Let's look at an example of the concept of half-life (which incidentally motivates the definition of a natural unit of radioactivity). Suppose we wish to calculate the number of atoms disintegrating per second in one gram of pure radium. Let's suppose that the atomic weight of radium is 226, so that one gram of radium contains

$$\frac{1 \text{ gram}}{226 \text{ gram (gram-atomic-weight)}^{-1}} \times \frac{6.023 \times 10^{23} \text{ atoms}}{\text{gram-atomic weight}}$$

$$= 2.67 \times 10^{21} \text{ atoms.}$$

The half-life of radium-226 is 1622 years. The decay constant k is, from Equation 2-9

$$k = \frac{0.693}{t_{1/2}}$$

or

$$k = \frac{0.693}{(1622 \text{ yr})(3.156 \times 10^{7} \text{ sec yr}^{-1})}$$

or

$$k = 1.35 \times 10^{-11} \text{ sec}^{-1} .$$

The disintegration rate of one gram of pure radium-226, from Equation 2-8 is

$$-dN(t)/dt = kN(t)$$
$$= (1.35 \times 10^{-11}\ sec^{-1}) \times (2.67 \times 10^{21}\ atoms)$$
$$= 3.61 \times 10^{10}\ sec^{-1}.$$

In earlier experiments this quantity was thought to be nearer $3.7 \times 10^{10}\ sec^{-1}$. The latter value has in fact become the basis of the present-day definition of the <u>Curie</u>, which is $3.7000 \times 10^{10}$ disintegrations per second.[47]

The law of radioactive transformation, as expressed by Equation 2-7, considers the decay of only a single radionuclide. Typically, however, we are interested in the concurrent decay of a series of species. Let's look, in particular, at some important features of situations in which a parent nuclide A decays to a daughter B, which in turn decays to a stable element C.[48] (A more general and technical treatment of this phenomenon is given in Appendix A.)

First, it can be show that the quantity Q(t) of radioactive atoms of B at time t is given by

$$Q(t) = \frac{k_A}{k_B - k_A} P_0(\exp(-k_A t) - \exp(-k_B t)) \qquad (2\text{-}10)$$

where $k_A$ and $k_B$ are the decay constants for A and B, respectively, and $P_0$ is the quantity of A present initially. It is clear from Equation 2-10 that the quantity Q of daughter B present at time t = 0 and at time t equal infinity is zero. These features are just what we would expect from a situation in which B begins to be produced at t = 0 and ultimately decays to some other species. Given this fact, calculus shows that there is some time $t_m$ between zero and infinity at which Q reaches a maximum value. At $t_m$

$$t_m = \frac{\ln(k_B/k_A)}{k_B - k_A} \qquad (2\text{-}11)$$

If A is very much longer-lived that B, then $k_B >> k_A$. In this case, it can be shown that if B is separated (say, chemically) from A

$$\frac{dN_B}{dt} = \frac{dN_A}{dt}(1 - \exp(-k_B t)) \qquad (2\text{-}12)$$

For t equal to or greater than just a few half-lives of B, Equation 2-12 becomes

$$\frac{dN_B}{dt} = \frac{dN_A}{dt} \qquad (2\text{-}13)$$

Thus if A is very much longer-lived than B and t is large relative to the half-life of B, B decays with the half-life of A, which we would intuitively expect. By the time a few half-lives of B have passed, furthermore, the activities of A and B have become equal (Equation 2-13).

The equilibrium expressed in Equation 2-13 is called <u>secular equilibrium</u>. For example, radium-226 ($t_{1/2}$ = 1622 yr) decays to radon-222 ($t_{1/2}$ = 3.82 da) and strontium-90 ($t_{1/2}$ = 28 yr) to yttrium-90 ($t_{1/2}$ = 64 hr). A graph of the growth of radon-222 which has been freshly separated from radium-226 is shown in Figure 2-8.

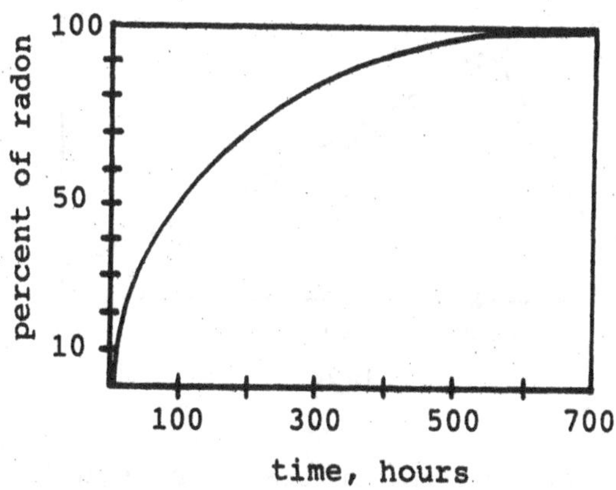


Figure 2-8. Growth curve for radon-222 which has been freshly separated from radium-226.

If $k_B > k_A$, but not greatly so, we can show that

$$Q(t) = \frac{k_A}{k_B - k_A} P_0 \exp(-k_A t) \qquad (2\text{-}14)$$

so that for large values of t

$$\frac{dN_B}{dt} = \frac{k_B}{k_B - k_A} \frac{dN_A}{dt} \qquad (2\text{-}15)$$

Thus, for large values of t, the daughter B decays with the half-life of its parent, but its activity is greater than that of the parent by a factor of

$$\frac{k_B}{k_B - k_A}$$

This type of equilibrium is called <u>transient equilibrium</u>. An example of transient equilibrium is shown in Figure 2-9.

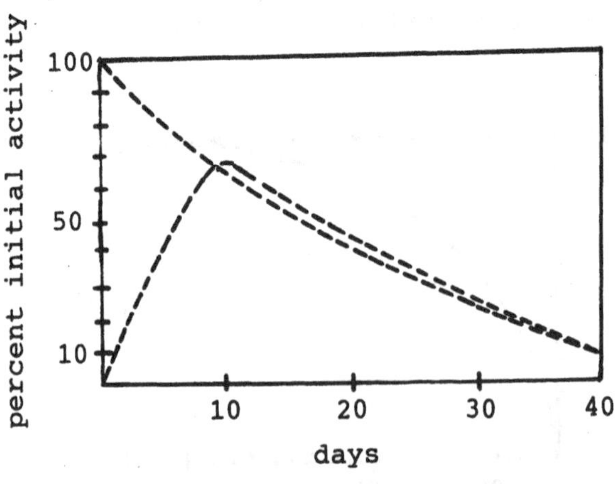


Figure 2-9. An example of transient equilibrium. Growth and decay of lanthanum-140 ($t_{1/2}$ = 40 hr) from freshly separated barium-140 ($t_{1/2}$ = 12.8 da).

If $k_A > k_B$ and t is sufficiently large, Equation 2-10 becomes

$$Q = \frac{k_A}{k_A - k_B} P_0 \exp(-k_B t) \qquad (2\text{-}16)$$

The number of atoms Q of B, that is to say, decays with a half-life characteristic of the daughter, while the number of atoms P of A decreases more rapidly with a shorter half-life. An example of this kind of activity is shown in Figure 2-10.

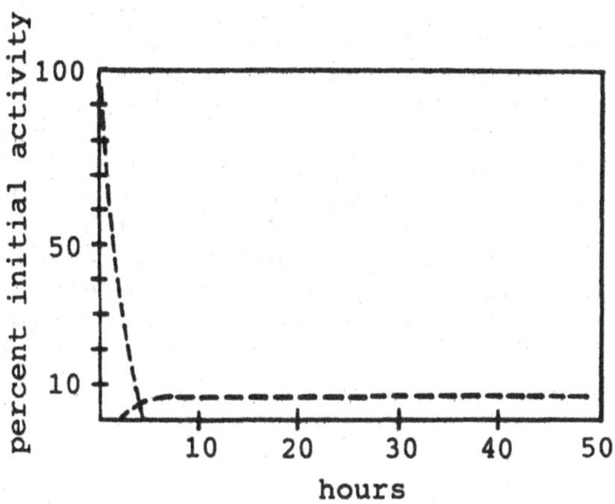


Figure 2-10. Growth and decay curves for iodine-131 ($t_{1/2}$ = 8.1 da) from freshly separated tellurium-131 ($t_{1/2}$ = 25 min).

The law of radioactive transformation (Equation 2-7) and the nature of the chemical and mass transformations which occur during alpha or beta particle ejection allow us to organize radioactive species into natural groupings, called "families" or "series." These groupings are families in the sense that each member of such a group lies along the chain of radioactive decompositions of a given species or parent. There are just four such families or series of strictly terrestrial origin known, and they are distinguished by their first members:

1. The uranium-238 series (sometimes called the uranium-radium series).

2. The thorium-232 series (sometimes called the thorium series).

3. The actinium series (sometimes called the uranium-235 series).

4. The potassium-40 series.

Suppose, for a given family, we place these species in a chart which is arranged with the species with the highest atomic number at the top and lowest atomic number at the bottom, and for any given atomic number contains the species in descending order of atomic mass. The result is a chart like Figure 2-11. A segment of this graph involving a decrease in atomic number and a decrease in mass represents an alpha particle emission. A segment involving an increase in atomic number and a decrease in mass represents a beta particle emission.

By 1910 it had become quite clear that there are a variety of small groups of elemental species which have distinct atomic weights but which are nevertheless chemically identical. Species which differ in atomic weight but are chemically identical are said to be isotopes of the element to which they are chemically identical. For example, there are three isotopes of radium, having atomic weights of about 223, 224, and 226, respectively. Isotopes of an element need not be radioactive.

## A NEW MODEL OF THE ATOM EMERGES

The chief aim of atomism is to explain the known properties of matter in terms of a coherent, unified model based exclusively on the properties of atoms. The picture of atomic structure which had evolved by about 1910 was this:

1. Atoms have an internal structure of some sort which presumably consists of particle-like entities.

2. With some subatomic components a negative charge can be associated, and with some, a positive charge.

Since even at the atomic level matter is electrically neutral, it seems reasonable to believe that the (opposite) charges in individual atoms are present in equal numbers. What the distribution of the charges is, is of course another matter. One might propose, for example, a model in which both charges are more or less uniformly distributed throughout the atomic volume,

| Nuclide | Half-life | Principal radiation types<br>alpha | <br>beta | <br>gamma |
|---|---|---|---|---|
| U-238 | $4.5x10^9$ y | x | | |
| Th-234 | 24.1 d | | x | |
| $^{m}$Pa-234 | 1.17 min | | x | |
| Pa-234 | 6.75 h | | x<br>(from $^{m}$Pa-234) | x |
| U-234 | $2.47x10^5$ y | x<br>(from $^{m}$Pa-234) | | x |
| Th-230 | $8.0x10^4$ y | x | | x |
| Ra-226 | 1600 y | x | | x |
| Rn-222 | 3.8 d | x | | x |
| Po-216 | 3.05 min | x | x | |
| Pb-214 | 26.8 min | | x<br>(from Po-216) | x |
| At-218 | 2 sec | x | x<br>(from Po-216) | |
| Bi-214 | 19.7 min | x | x | x |
| Po-214 | 164 microsec | x<br>(from Bi-214) | | x |
| Tl-210 | 1.3 min | | x<br>(from Bi-214) | x |
| Pb-210 | 21 y | x | x | x |
| Bi-210 | 5.01 d | x | x | |
| Po-210 | 138.4 d | x<br>(from Bi-210) | | x |
| Tl-206 | 4.19 min | | x<br>(from Bi-210) | |
| Pb-206 | Stable | | | |

Figure 2-11. The uranium-radium (U-238) series. The times for each transformation are half-lives.

something like raisins in pudding. But in 1911, Rutherford[49] performed an experiment which clearly showed that

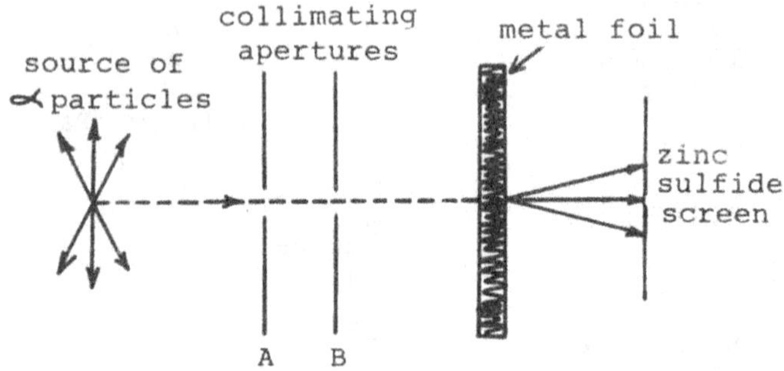


Figure 2-12. Rutherford's alpha-scattering apparatus.

this model was untenable. In this experiment, a stream of alpha particles is passed through a set of collimating slits (which geometrically define a reference ray) and allowed to strike a thin foil target (see Figure 2-12). The angular distribution of the trajectories (relative to the reference ray) of the alpha particles which are scattered by the foil is measured by intercepting the particles with a zinc sulfide screen. When an alpha particle strikes the screen, it causes the zinc sulfide to emit a tiny flash of light. Most of the alpha particles pass through the foil undeflected. But a few, Rutherford found to his amazement, are scattered at large angles, up to $180^{o}$.

Now the kinetic energy of alpha particles, Rutherford knew, is extremely large. Consequently, anything which could significantly deflect an alpha particle must be the seat of a large, presumably electrical, force. If this force is electrical, moreover, it must arise from the interaction of the positively charged alpha particle with some other positively charged particle (s), Rutherford reasoned. Furthermore, the latter would have to be massive compared to an electron, because an electron would be brushed aside by an alpha particle. And lastly, the fact that only a few alpha particles are scattered strongly suggests that the atomic volume is largely empty space. Taken together, then, these results suggest at atom consisting largely of space, containing a small, relatively massive, posi-

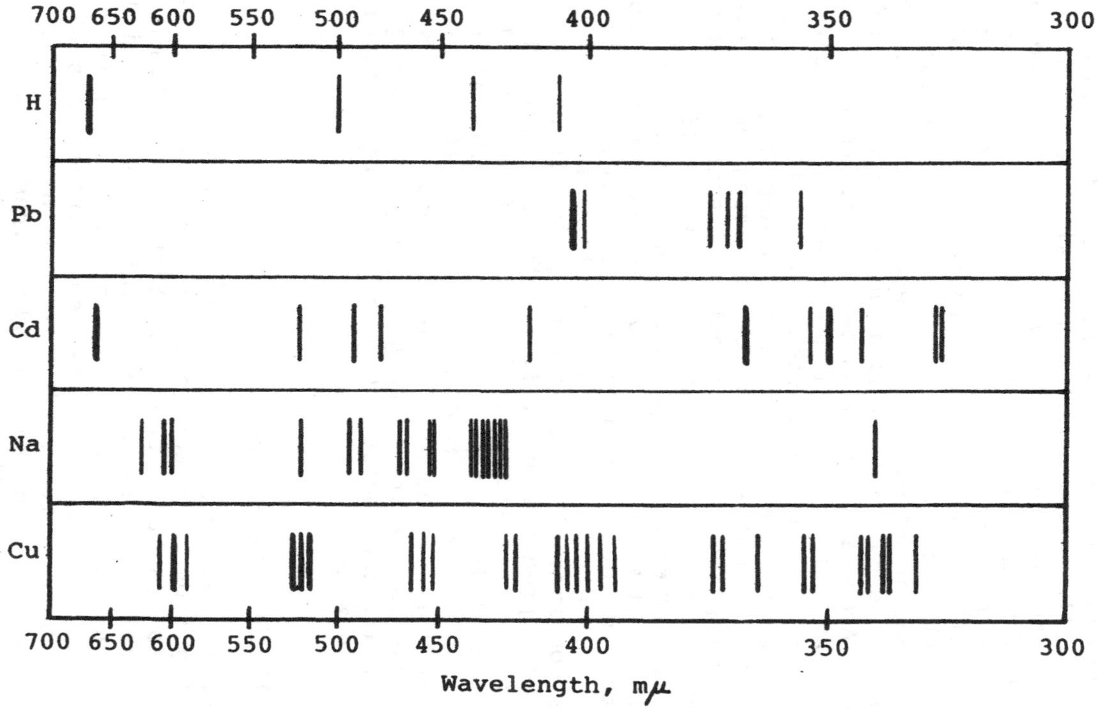


Figure 2-13. Line spectra of various elements.

tively charged nucleus, surrounded by a shell(s) of negatively charged electrons.

Rutherford's alpha-scattering experiment demonstrates that the positive and negative charges are well segregated in the atomic volume. But the experiment raises the rather nasty question of how this segregation is even possible. If, for example, the electrons are initially at rest, the electrostatic attraction of the nucleus will, according to classical physics, soon drag them inward and the atom as such will cease to exist. To meet this problem, we could propose that electrons orbit the nucleus much in the same fashion as the planets orbit the sun. A closed (say, circular or elliptical) orbit of the sort envisioned requires, by definition, an acceleration, because the direction of motion of the particle in such an orbit must continually change. According to classical electromagnetic theory (a la Maxwell) accelerating a charged particle will cause the particle to continuously emit energy. Thus, according to the classical electromagnetic account of the atom, all matter should be radiating energy even if none is supplied to it from external sources. Now such energy can come only from the atom itself. But this means that eventually the energy which keeps the electron in orbit will be spent and the electron will spiral into the nucleus. On its way to the nucleus, furthermore, the atom will emit radiation whose frequency varies continuously over some frequency interval.

These predictions, unfortunately, collide directly with known facts about the behavior of matter. Indeed, there are at least three very general, intimately related problems which the classical (Maxwellian electrodynamic) view of the atom faced by about 1910. We will call these the _discrete spectrum problem_, the _blackbody problem_, and the _photoelectric problem_, respectively. To these we now turn.

## The discrete spectrum problem

If we examine with a spectrograph the light emanating from an electrically excited gas (say, light from a gas discharge tube), or light from a flame into which a volatile salt has been introduced, we will not find a spectrum consisting of the continuous spectrum of wavelengths which composes, say, ordinary sunlight. Instead, the spectrum will consist of a set of discrete, reasonably well isolated parallel lines. (Each line is an image of the spectrograph slit, deviated through an angle which is characteristic of the wavelength of that spectral component.) A spectrum of this sort is called a _line spectrum_. The line spectra of several elements are shown in Figure 2-13. Line spectra are characteristic of the elements which give rise to them.

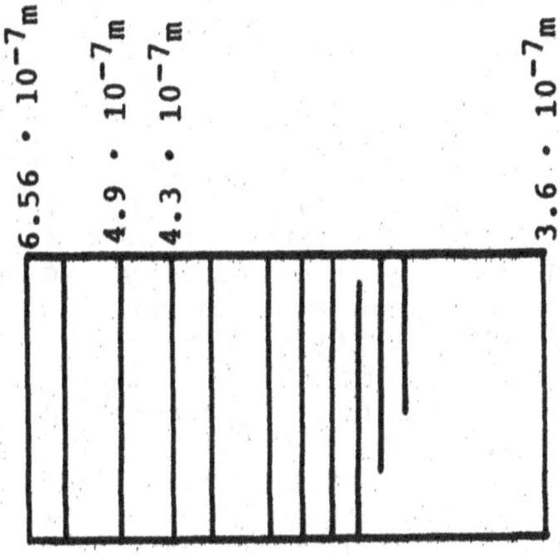


Figure 2-14. Balmer series for atomic hydrogen.

In 1885, Johann Jakob Balmer (1825-1898) found a very simple formula[50] which gives the frequencies of a group of lines emitted by atomic hydrogen. Balmer's formula is

$$1/L = R((1/2^2) - (1/n^2)) \qquad (2\text{-}17)$$

where

L is the wavelength of a spectral line

$n = 3, 4, 5, \ldots$

R is a constant, called the Rydberg constant.

The spectral lines corresponding to Equation 2-17 are called the Balmer series of (atomic) hydrogen. A spectrogram of this series is illustrated in Figure 2-14.

Other spectral series for hydrogen have since been found and named after their discoverers. They are, respectively, the

Lyman series:[51] $$1/L = R(1/1^2 - 1/n^2), \quad n = 2, 3, \ldots \qquad (2\text{-}18)$$

Paschen series:[52] $$1/L = R(1/3^2 - 1/n^2), \quad n = 4, 5, \ldots \qquad (2\text{-}19)$$

$$\text{Brackett series:}^{53} \quad 1/L = R(1/4^2 - 1/n^2), \quad n = 5, 6, \ldots \qquad (2\text{-}20)$$

One feature of the Balmer, Lyman, Brackett, and Paschen series leaps to our attention: it very strongly appears that only fixed, definite characteristic frequencies are allowed to be emitted by electrically excited elements. Any theory of radiation by matter must explicitly account for this peculiarity; in particular an adequate theory of radiation by matter should have Equations 2-17 through 2-20 as consequences. And any such theory, we should note, will collide directly with the prediction of classical electromagnetic theory that electrically excited atoms ought to be able to emit radiation at *any* frequency within a continuous interval.

## *The blackbody problem*

A second serious difficulty faced by the classical (Maxwellian electrodynamic) view of the atom concerns the energy distribution of radiation from matter when it is in energy equilibrium with its environment. It is an experimental fact that the emission energy spectrum of a body in energy equilibrium with its surroundings at a given temperature is a characteristic function of the temperature of the body and the material of which that body is made. Classical theory was unable to account for this distribution; in fact, the theory predicted that at very short wavelengths the rate at which energy would be emitted from an idealized radiator, called a *blackbody*, would be infinite. This problem was (and should have been) a direct threat the credibility of the classical view of the atom, because it violates a fundamental tenet of classical physics: the conservation of energy. Either something was deeply wrong with classical theory, many physicists concluded, or the general view of the atom was fundamentally mistaken.

The solution to this problem, we should not be surprised to learn, involved a rejection of some of the most widely held views about the nature of matter. In December 1900, Max Planck, a highly respected but iconoclastic German physicist, published a theory of energy emission from matter which took just this tack.[54] In that paper, Planck assumed that the energy emitters (he did not claim that they were atoms) in a blackbody were absolutely incapable of emitting energy except in discrete packets, or energy *quanta*. The energy of a quantum, Planck proposed, is proportional to the frequency of the radiation emitted. Intuitively, this is somewhat like saying that humans can jump only 2 feet, but not 1 or 1.5 feet. The idea is fundamentally perverse. Indeed, upon hearing Planck's thesis, one of Planck's faculty colleagues remarked that it was a good thing that Planck had tenure, because such speculation would not be forgiven a younger (presumably less demented) man.

Planck's hypothesis, however rash, had one very great virtue: from it one could derive the correct blackbody spectral distribution. One can argue with this kind of success, but not very forcefully.

## The photoelectric effect

The third serious difficulty faced by the classical view of the atom came from rather different quarters than the first two. In 1887 Heinrich Hertz performed a series of experiments aimed at understanding the behavior of electromagnetic waves.[55] He generated these waves by forcing electricity to jump a spark gap between two electrodes. A second circuit, also containing a spark gap, was placed near the first. Electromagnetic energy generated in the first circuit, he discovered, could cause a spark at the gap in the second circuit, even though no visible physical connection between the two circuits existed. Thus, the gap in the second circuit serves as a detector of energy generated by the first. (Of course, we take this mechanism (sans sparks) for granted today: it is the basis of all radio and television communication.) To visualize the sparks at the detector gap more easily, Hertz enclosed it in a black box. To his surprise, he discovered that when the detector gap was enclosed in the box, it had to be made <u>shorter</u> than when it was left out of the box and exposed to the <u>light</u> of spark gap in the first circuit! Now Hertz happened to know that sparks emit ultraviolet light. He accordingly turned his efforts to showing that it was the ultraviolet light itself, not the sparks as such, which were responsible for the odd effect he was seeing at the detector gap. His hunch proved correct: by illuminating the detector gap with an ultraviolet source, detection was enhanced (that is, the detector gap could be made wider and still exhibit a spark).

Hertz's discovery immediately attracted the attention of other investigators. Hallwachs,[56] for example, found that a freshly polished, negatively charged zinc plate connected to an electroscope and illuminated by ultraviolet light would lose its charge. There was no effect if the plate was positively charged. He further observed that an electrically neutral plate would acquire a slight positive charge if illuminated.

The work of Hertz, Hallwachs, and others suggests that the illumination of polished metal surfaces facilitates the liberation of electrons from those surfaces. It appears that these electrons can, under the influence of an electric field, pass from cathode to anode, thus creating a photoelectric current. This phenomenon is called the photoelectric effect.

In 1900, P. Lenard[57] showed that the photoelectric effect has a further very curious property. The experimental apparatus he used looks like that shown in Figure 2-15. This apparatus consists of an evacuated glass tube containing an aluminum cathode C which can be illuminated by ultraviolet light passing through a quartz window S. The cathode can be charged to any potential. A screen A, connected to the earth, serves as an anode. $P_1$ and $P_2$ are metal electrodes connected to electrometers. When C is illuminated and charged to a negative potential of several volts, photoelectrons are accelerated toward A. A few electrons pass through

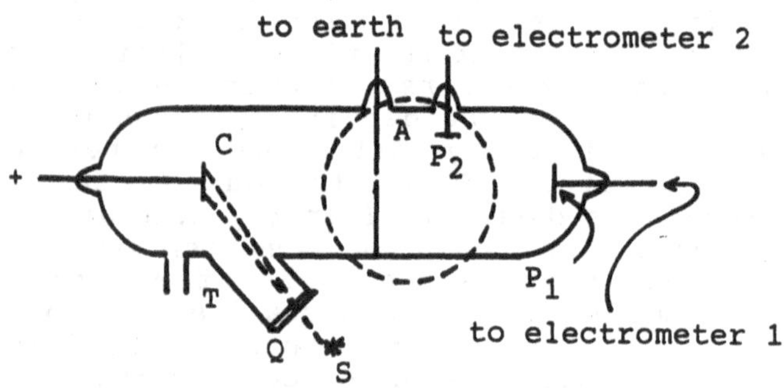


Figure 2-15. Lenard's apparatus for determining e/m for electrons.

a hole in the center of A and proceed at uniform velocity to $P_1$. If these electrons are subjected to a magnetic field produced by a pair of Helmholtz coils (shown by the dotted lines in Figure 2-15) so that the field is directed toward the reader in the region between A and $P_1$, the electrons are deflected upward in a circular path and can be made to strike electrode $P_2$.

Lenard first investigated the relation between the current reaching the anode and the potential V applied to C. A was assumed to always be at zero potential. No current was observed when V was several (about 10) volts positive. But, strangely enough, when V was dropped to about 2 volts positive, a small current was observed. This indicated that the photoelectrons were not simply freed from the cathode, but some of them were ejected with velocity sufficient to enable them to overcome the retarding potential of 2 volts. The current increased when V was reduced to zero, and increased still more rapidly as V was made more negative, up to a point. The current then reached a limiting or "saturation" value when V was about 15 to 20 volts negative.

Let us denote the positive potential which is required to prevent the escape of electrons by $V_0$. Suppose a negative potential V, large compared with $V_0$, is applied to cathode C. Then the photoelectron on reaching the anode will have a kinetic energy given approximately by $eV = (1/2)mv^2$. Lenard's experiment shows that the kinetic energy of the photoelectrons as they leave the cathode does not exceed a definite maximum, $eV_0$, where e is the unit charge.

Now a very simple linear relation exists between the maximum energy of photoelectrons and the frequency f of light which causes their emission. If we plot $eV_0$ against f, a straight line results which has some intercept $f_0$ on the f-axis. The quantity $f_0$ is characteristic of the emitting electrode, but the slope of the curve is constant for all electrodes. The equation of this curve is given by

$$(1/2)mv^2 = h(f - f_0) = hf - w_0 \qquad (2\text{-}21)$$

where h (called <u>Planck's constant</u>) is the slope of the curve and $w_0 = hf_0$.

Equation 2-21 says something very odd: photoelectrons are not emitted with just any energy in an interval, but take on only very <u>distinct, discrete values</u>. This result sounds very much like Planck's hypothesis, and indeed, the deep connection between Planck's proposal and the photoelectric effect was derived by Einstein.[58] It is totally at odds with the predictions of classical theory of atomic structure.

In the discrete spectrum problem, in the phenomenon of blackbody radiation, and in the photoelectric effect, early twentieth-century atomism was thus handed a formidable challenge; if the theory were to survive, some rather drastic departure from the classical view of the atom was clearly demanded. In 1913 a young Danish physicist named Niels Bohr[59] proposed a new model of the hydrogen atom aimed at addressing these problems. At the heart of Bohr's proposal were four radical postulates:

1. The electron in the hydrogen atom has certain definite states of motion allowed it; each of these states has a definite, fixed energy.

2. When an atom is in one of these states it does not radiate. When changing from a high-energy state to a lower-energy state, the atom emits a fixed quantity (a quantum) of radiation which is

equal to the energy difference between the two states.

3. In any of the states the electron moves in a circular orbit about the nucleus.

4. The states of allowed electronic motion are those in which the angular momentum of the electron is an integer multiple of $h/2\pi$, where h is Planck's constant.

We will call these four postulates the Bohr model of the hydrogen atom.

The Bohr model has four great virtues. First, it it correctly predicts the line spectra of hydrogen. Second, the model provides an almost geometric picture of the atom as a reasonably stable "planetary" system. Third, the model allows us to distinguish the potential energies of electrons as a simple function of their distance from the nucleus. Fourth, it explicitly incorporates Planck's quantum hypothesis, thereby taking a strong stand (against) on the issue of whether classical electromagnetic theory can provide an adequate description of (bound) electron behavior.

In spite of these features, the Bohr model does not do a particularly good job of explaining the electronic properties of the heavier elements. Out of fairness, we may say that Postulate 1 and 2 of the Bohr model have survived the test of experiment. Postulate 3 has been abandoned, while Postulate 4 is partially correct--the angular momentum of an electron is fixed, but not with the value Bohr assigned.

Further refinement of the Bohr model by Bohr himself,[60] Heisenberg,[61] Schroedinger,[62] Dirac,[63] and others produced by 1926 a model of electronic behavior which promises to provide, although at great computational cost, a quite general picture of the electronic structure of atoms and molecules. In particular, it shows how and to what extent molecular bonds can and do form, and to what extent they are stable. This theory is collectively known as the quantum theory. The quantum theory preserves Postulates 1 and 2 of the Bohr model of hydrogen. But in contrast to the postulates of the early Bohr model, the theory describes at best only the probability that an electron with a given energy state will be found in a given region in space. For an electron with a given energy state, this probability (distribution) is called an orbital. Although the notion of an orbital is not as intuitively forceful as the idea of an actual electron orbit, it turns out that this is the most parsimonious interpretation consistent with a now large body of experimental work.

To put this story in other terms, the quantum theory says that the energy associated with the transition of an electron from one orbital to another is emitted or absorbed only in definite quanta. The magnitudes of these quanta are given by

$$E = hf \qquad (2\text{-}22)$$

where

h is a constant (= $6.56 \times 10^{-27}$ erg sec)

and

f is the frequency of the radiation.

If we assume that energy is emitted (or absorbed) only when electrons jump discontinuously from a higher to a lower (from a lower to a higher) potential energy orbital, then the energy emitted (absorbed) as a result of a single electron moving from an energy level $E_1$ to an energy level $E_2$ will be a photon (a light-energy quantum) with magnitude

$$hf = E_1 - E_2 \qquad (2\text{-}23)$$

In order account for the discrete lines of the optical spectrum, Bohr postulated that only certain discrete energy levels are permitted for electrons bound to an atom. It follows, then, that electrons can occupy only distinct orbitals or shells, about the nucleus. Historically, these shells were labeled, in increasing order of distance from the nucleus, K, L, M, N, and so on. Let's suppose that a hydrogen atom has an electron in shell K with potential energy $W_K$. Suppose this electron is moved to another shell farther away from the nucleus by the addition of energy (by some means) to the atom. In the L shell, for example, the electron has potential energy $W_L$. Now suppose that the electron drops from the L shell back to its original (K) orbit. A quantum with energy

$$E = hf = W_K - W_L \qquad (2\text{-}24)$$

will be emitted by the atom as a result of this transition.

The alert reader will probably have noted that the account of natural radioactivity given to this point appears to describe only the emission of an alpha or beta particle from an atom. Yet clearly, the work of Rutherford and P. Villard demonstrates the existence of a third type of radiation from naturally occurring radioactive substances, namely, gamma radiation. How does this fit into the picture we have presented so far?

We now know that beta or alpha emission in the

naturally occurring radioactive series is often accompanied by some gamma radiation. Gamma radiation arises when the nucleus of an atom is left in an excited (energetically unstable) state following beta or alpha emission. The nucleus can bleed some of the "excess" energy in this state by emitting it in the form of gamma radiation.

### Other decay mechanisms

Since the discovery of natural radioactivity, several other mechanisms of radioactive decay have been discovered, although they typically occur in non-terrestrial (say, in stars) or artificial (man-made) conditions, and are certainly more exotic than any of the mechanisms described so far.

First, radionuclides can decay by the emission of a positron--a particle the magnitude of whose charge and mass are the same as those of an electron, but which carries a positive charge. When positron emission occurs, the daughter element is one place down in atomic number, or in the periodic table.

Second, a nucleus can capture an electron, then can emit a gamma ray. This transition also results in a loss of one in atomic number. The excitation energy resulting from a nuclear capture of an electron can also interact with other electrons in the same atom, giving them sufficient energy to escape the atom.

Third, the nuclear excitation energy caused by an alpha or beta particle emission, or by a positron ejection, can be transferred to electrons which remain bound to the atom even after the transfer. This process is called internal conversion.

## PASSAGE OF ALPHA, BETA, AND GAMMA RAYS THROUGH MATTER

Among the principal concerns over radioactivity in water supplies lies the question how radiation of this sort interacts with biological tissues, particularly our own. The answer to this question is just beginning to be understood in enough detail to provide a basis for public policy. And that answer, whatever its fine details, will be built on an understanding of how radioactive emanations behave while passing through physically simpler systems. In this section we survey the present understanding of the passage of natural radioactive emanations through air, water, and through solids of relatively simple structure (e.g., elemental metals) in particular. To do this, we must first get a somewhat more precise handle on the energies involved in the interaction of radioactivity with matter.

A natural unit which is widely used today to describe the energy associated with radioactivity is the electron volt (eV). An electron volt is the energy acquired by an electron when it is accelerated through a potential difference of one volt. One electron volt is about 1.602 x $10^{-12}$ erg. (The unit is natural in the sense that the electron energy transitions which give rise to visible light are roughly 1-10 eV.) Transitions which generate x-rays range from 50 eV to $10^5$ eV. The energies of alpha, beta, and gamma rays are on the order of $10^3$ eV ($10^3$ eV is abbreviated keV) to $10^6$ eV ($10^6$ eV is abbreviated MeV).

## Passage of alpha particles through matter

The electrometric experiments of the Curies, Owens, and Rutherford's group clearly show that the passage of radiation from radioactive substances through air causes the air to become conducting, that is, ionizes the air. To ionize atoms in air, work must be done (energy expended), because ionization involves separating charges against an electrical force. For each atom of air (oxygen or nitrogen) ionized, roughly 35 eV of work must be done. Thus, for example, an alpha particle moving through air loses about 35 eV for each ion pair created in that air by it. In creating, say $10^5$ ion pairs, an alpha particle would therefore expend about (35 eV $pair^{-1}$)($10^5$ pairs) = 3.5 MeV of its energy. If this quantity were the original kinetic energy of the particle, the particle would brought to rest by the production of $10^5$ ion pairs. "The" distance which an alpha particle travels before expending all of its kinetic energy in air is called the range of the particle in air. Actually, "the" range of an alpha particle is somewhat a misnomer, because the energy distribution of ion pairs created is a statistical process which averages about 35 eV per ion pair. In some cases, for example, the alpha particles will create ions which themselves are energetic enough to cause ionization of other atoms of the air. In any case, the result is a spread of ranges. This feature is called the straggling of alpha particles.

The earliest reported determinations of alpha particle ranges in air were made by W.H. Bragg[64] and by H. Geiger and J.M. Nutall.[65] Figure 2-16 illustrates the kind of curves they obtained. As Figure 2-16 shows, the ionization per unit path length (the specific ionization) increases toward a maximum near the end of the range, then drops off sharply. The shape and end-point of these curves (today known as Bragg curves) are characteristic of the source of the alpha particles.

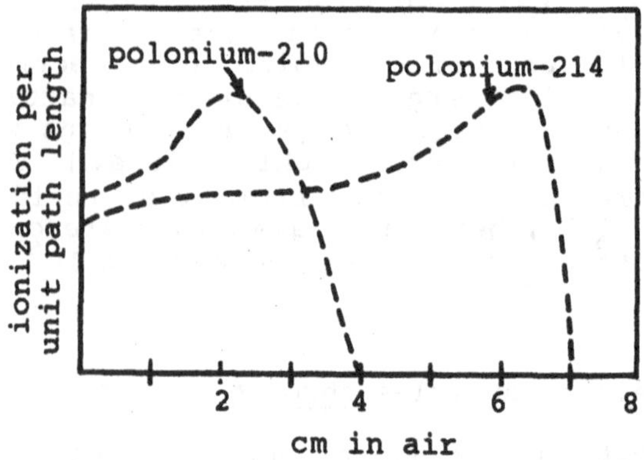


Figure 2-16. Results of Bragg's determination of alpha particles ranges in air.

What do Bragg curves say about the behavior of individual alpha particles? Early work on this problem was done by Geiger in Rutherford's laboratory.[66] Geiger first sought a relation between the range x of an alpha particle and its initial velocity V. He determined V to be approximately

$$V^3 = a(R - x) \tag{2-25}$$

where

V is the initial velocity of the particle

a is a constant

R is the distance along the particle's path at which the maximum specific ionization occurs

x is the total distance traveled by the particle before it comes to rest.

Equation 2-25 allows us a quick way to show the connection between the alpha particle's kinetic energy and its residual range. In particular, from Equation 2-25 we have

$$V^3 = a(R - x)$$

Raising both sides of Equation 2-25 to the 2/3-power, we obtain

$$V^2 = (a(R - x))^{2/3} \qquad (2\text{-}26)$$

Multiplying Equation 2-26 by (1/2)M gives

$$(1/2)MV^2 = (1/2)(Ma^{2/3})(R - x)^{2/3} \qquad (2\text{-}27)$$

or, to put the point less formally, the kinetic energy (the left-hand side of Equation 2-27) is proportional to $(R - x)^{2/3}$. If we let $T_a$ be the kinetic energy of the alpha particle and compute (using calculus) how that energy changes with absolute range x, we get

$$\frac{dT_a}{dx} = -(1/3)Ma^{2/3}(R - x)^{-1/3} \qquad (2\text{-}28)$$

Now if we extract the cube root of both sides of Equation 2-25, we get

$$V = a^{1/3}(R - x)^{1/3} \qquad (2\text{-}29)$$

Taking the reciprocal of both sides of Equation 2-29 yields

$$1/V = a^{-1/3}(R - x)^{-1/3} \qquad (2\text{-}30)$$

Notice that the rightmost factors of the right-hand sides of Equations 2-28 and 2-30 are the same and that the leftmost factors of those sides are constants. Since

$$-(1/3)Maa^{-1/3} = -(1/3)Ma^{2/3}$$

we have, by multiplying both sides of Equation 2-30 by

$-(1/3)Ma$,

$$-(1/3)Ma(1/V) = -(1/3)Ma^{2/3}(R - x)^{-1/3} \qquad (2\text{-}31)$$

But the right-hand side of Equation 2-31 is just the right-hand side of Equation 2-28. Thus, by substitution

$$-(1/3)Ma(1/V) = \frac{dT_a}{dx} \qquad (2\text{-}32)$$

This just says that the change in kinetic energy of the alpha particle per unit length of excursion along its path is inversely proportional to its initial velocity V. Thus as the velocity V decreases, Geiger argued, the loss of kinetic energy per unit path length due to ionization increases, in good agreement with the shape of the Bragg curves.

Geiger further conjectured that the shape of the Bragg curves could be derived by making the reasonable assumption that the specific ionization I produced by an alpha particle is proportional to the energy absorbed, that is

$$I = \frac{-kd(K.E.)}{dx} \qquad (2\text{-}33)$$

or

$$I = \frac{-kd((1/2)MV^2)}{dx} \qquad (2\text{-}34)$$

Substituting the value of $(1/2)MV^2$ obtained from Equation 2-27 into Equation 2-34, we obtain

$$I = \frac{-k'd(R - x)^{2/3}}{dx}$$

or

$$I = -k''(R - x)^{-1/3} \qquad (2\text{-}35)$$

where

I is the specific ionization

k, k' k'' are constants of proportionality

K.E. is the energy absorbed from the alpha particle

M is the mass of the ionizing particle

V is the velocity of the ionizing particle

R is the range corresponding to maximum specific ionization

x is the total distance traveled by the alpha particle

If the relation between I and x is plotted, a curve similar to that of Figure 2-16 (the Bragg curves) results. Thus as a first approximation, Geiger's assumption that the specific ionization is proportional to the kinetic energy given up by the alpha particle is reasonable.

We now know that Equation 2-35 holds only approximately for alpha particle ranges between 3 and 7 cm in air at STP. Below an energy corresponding to a maximum range of 3 cm, the total range is more nearly proportional to $V^{3/2}$; above an energy corresponding to a range of 7 cm, the range is approximately proportional to $V^4$. More generally, it can be shown that[67]

$$\frac{dT_a}{dx} = - \frac{4\pi e^4 z^2 z' n}{m_a V^2} \ln k_a \qquad (2\text{-}36)$$

where

x is the distance along the particle's path

e is the (magnitude of the) unit charge

n is the number of electrons per unit volume in the absorber

Z is the atomic number of the alpha particle

Z' is the atomic number of the absorber

$m_a$ is the mass of the alpha particle

V is the initial velocity of alpha particle

$k_a$ is a constant involving the ionization potential of the atoms in the absorber

$T_a$ is the kinetic energy of the alpha particle

The general picture we now have of the passage of alpha particles through matter, then, looks something like this. An alpha particle typically moves through a gas in a straight line, removing electrons from atoms along its path (assuming that it does not approach a nucleus too closely). In passing through solids or liquids, the path of an alpha particle is also relatively straight, but much shorter than in gases. Sometimes alpha particles can impart so much energy to the electrons of an environmental atom that these electrons can themselves ionize other atoms in the environment. It has been shown experimentally that alpha (and also beta) particles lose, on average, about 35 eV per ion-pair produced in air. This quantity is typically designated $W_{air}$. The first ionization potentials (i.e., the energy required to free the first electron) of oxygen and nitrogen are each about 14 eV. Thus, each ion-pair creation has about 20 eV (35 eV - 14 eV) of residual energy which can transfer kinetic energy to electrons, induce excitation of atoms, and generate thermal effects in environmental atoms near which an alpha particle passes. Below an energy of a few MeV, we now know, an alpha particle is not always doubly ionized. It can pick up one or two electrons, thus existing for part of its life as a univalent ion or a neutral atom (of helium).

Passage of beta rays through matter

The passage of beta rays through matter differs from that of alpha particles in several ways. First, because the mass of an electron is only about 1/7000 the mass of an alpha particle, we might expect a beta particle to be deflected much more easily than an alpha particle. Indeed, when the track of a "typical" beta particle in a photographic emulsion is examined, one can see that its path is extremely tortured compared to the track of an alpha particle in the same emulsion.

Second, it can be shown on theoretical grounds that the change in the kinetic energy of the beta particle, $dT_b/dx$, per unit path length is[68]

$$\frac{dT_b}{dx} = - \frac{4\pi e^4 Z' n}{m_b V^2} \ln k_b \qquad (2\text{-}37)$$

where $m_b$ is the mass of the beta particle, and the remaining quantities are as for Equation 2-36. We won't derive Equation 2-37 here, but is is worth comparing with Equation 2-36, which gives the corresponding change in kinetic energy for an alpha particle. To do this, it's instructive to rearrange the terms of Equation 2-36 a bit

$$\frac{dT_a}{dx} = - \frac{4\pi e^4 Z' n}{m_a V^2} z^2 \ln k_a \qquad (2\text{-}36')$$

If $\ln k_a$ and $\ln k_b$ were approximately the same and if the kinetic energies of the two particles were about the same, that is, if

$$m_b v^2 = m_a v^2$$

then we would have, from Equation 2-37 and Equation 2-36 that

$$\frac{dT_a}{dx} = Z^2 \frac{dT_b}{dx} \qquad (2\text{-}38)$$

The rate at which the kinetic energy of an alpha particle drops off, that is to say, is about 4 times (let Z = 2 in Equation 2-38) larger than the rate at which a beta particle loses its kinetic energy. Thus, for the same kinetic energies, Equation 2-38 says that the <u>range of beta particles is substantially longer than that of alpha particles in the same environment</u>, which is just what we observe.

There is a third difference between the passage of alpha and beta particles through matter. According to classical electromagnetic theory, a charged particle radiates energy when it is accelerated. Thus when an electron, whether it is part of a cathode ray stream or is a beta particle, enters an absorbing material and is accelerated, it radiates electromagnetic energy, typically in the form of x-rays. For example, cathode rays incident on the glass walls of a gas discharge tube give rise to x-rays. It has been determined experimentally that these x-rays have a continuous energy spectrum with an upper limit equal to the maximum electron energy. Because these x-rays arise from electron-stopping, they are known as <u>bremsstrahlung</u> (literally, "braking radiation").

We can, in fact, make a good case that the bremsstrahlung phenomenon rather sharply distinguishes the passage of beta and alpha particles through matter. The argument goes as follows. Classical electromagnetic theory says that the amplitude of the electromagnetic energy radiated by a charged particle when it is accelerated is proportional to the acceleration, that is,

$$A = ka \qquad (2\text{-}39)$$

where

A is the amplitude of the radiated energy

a is the acceleration of the charged particle

k is a constant of proportionality.

Furthermore, classical theory says that the intensity, I, is proportional to the square of the amplitude

$$I = k'A^2 \tag{2-40}$$

where k' is some constant of proportionality. By Coulomb's law, the electrostatic force F between the nuclear charge Ze and an alpha particle, or any other particle with charge E is

$$F = \frac{ZeE}{d^2} \tag{2-41}$$

where d is the separation between the two charges (considered as point-charges). Newton's second law says the centrifugal force on this particle is given by

$$F = Ma \tag{2-42}$$

where a is directed radially outward from the center of the atom. If the electron is to maintain a fixed orbit, the electrostatic force and the centrifugal force must just balance. Thus, setting the right-hand sides of Equations 2-41 and 2-42 equal, we have

$$\frac{ZeE}{d^2} = Ma,$$

or

$$a = \frac{ZeE}{d^2M} \tag{2-43}$$

where M is the mass of the charged particle. From Equations 2-39 and 2-40, we get

$$I = k'k^2a^2 \tag{2-44}$$

Substituting the value of a in Equation 2-43 into Equation 2-44, we obtain

$$I = k'k^2(ZeE/Md^2)^2$$

or

$$I = k'k^2(ZeE/d^2)^2(1/M)^2 \tag{2-45}$$

Thus, the intensity of the radiation is inversely proportional to $M^2$. Because the mass of the alpha particle is 7000 times greater than that of the beta particle, its bremsstrahlung is about $1/(7000)^2$ that of the electron, which was to be shown.

In addition to producing the bremsstrahlung phenomenon, the passage of beta particles through matter can generate x-rays which are distinctive (characteristic) of the materials with which they interact. These x-rays arise when extranuclear electrons, bumped to a higher energy state by excitation from passing beta rays, return to their ground state. Recall, in particular, Bohr's account of how line spectra are generated. If an electron is moved from a higher (potential) energy shell, say, the L shell, to a lower energy shell, say, the K shell, a photon with energy

$$hf = W_L - W_K \qquad (2\text{-}46)$$

will be emitted by the atom as a result of the transition. For hydrogen, the wavelengths of this radiation are in the visible spectrum. For heavier atoms, the binding force between the nucleus and the electron in the K shell is much larger than for hydrogen. As the atomic number Z increases, the transition energy $W_L - W_K$ increases and hence by Equation 2-46 the frequency f of the emitted radiation increases. For sufficiently heavy elements, the frequency of the energy emitted in an L-K transition is well beyond the visible light range. For these elements, characteristic x-rays arise from L-K transitions. These x-rays have energies on the order of 1 keV at Z = 11 (sodium).

During interaction with environmental atoms, the average rate of energy loss due to electromagnetic radiation, $(dT/dx)_{rad}$, from electrons having large kinetic energy $T(T >> 137\ mc^2/Z'^{1/3}$, to be precise) was calculated by H. Bethe and Heitler[69] to be given by

$$(dT/dx)_{rad} = - \frac{Z'^2Nr^2T}{137}(4 \ln (183/Z') + 2/9) \qquad (2\text{-}47)$$

where

$$r = e^2/mc^2$$

and the remaining quantities are as defined for Equation 2-36 ($137 = hc/2\pi e^2$). For lower values of T ($mc^2 << T << 137mc^2/Z^{1/3}$), Heitler and Sauter[70] determined

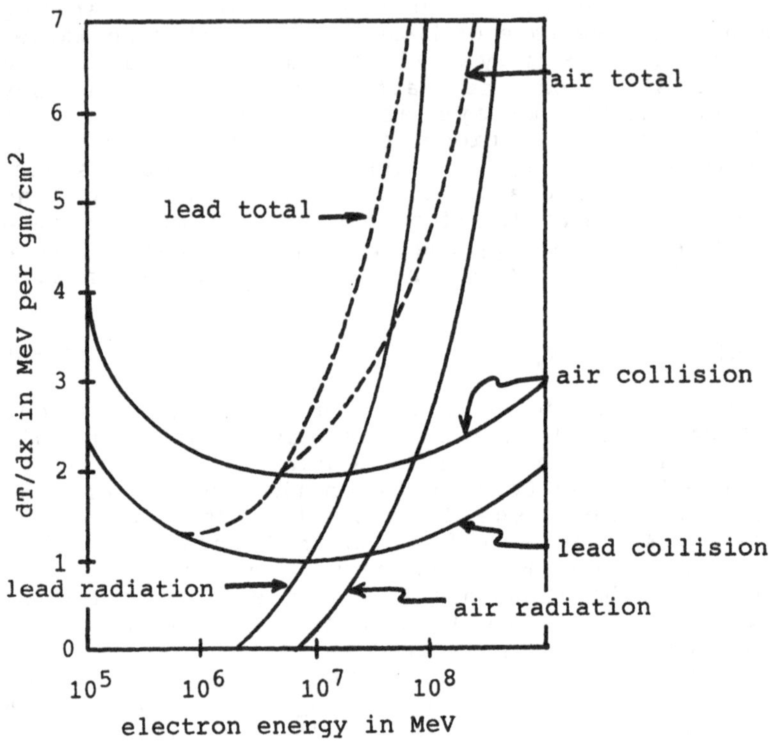


Figure 2-17. Average rates of energy loss for electrons passing through air at STP, and through lead, as a function of electron energy. Collision includes both atomic excitation and ionization.

that

$$(dT/dx)_{rad} = - \frac{Z'^2 N r^2 T}{137}(4 \ln (2T/mc^2) - 4/3) \tag{2-48}$$

Bethe and Heitler[71] were also able to derive a simple approximation for the relative rates of energy loss by the mechanisms of collision, $(dT/dx)_{coll}$, (that is, through ionization and atomic excitation) and radiation. In particular, if Z' is the atomic number of the material through which the beta particle passes, then

$$\frac{(dT/dx)_{coll}}{(dT/dx)_{rad}} = \frac{1600\ mc^2}{TZ'} \tag{2-49}$$

Thus, the energy $T_e$, at which energy losses due to the two mechanisms are equal (that is, the energy at which the right-hand side of Equation 2-49 is 1) is

$$T_e = 1600\ mc^2/Z' \tag{2-50}$$

For electrons in lead, for example, $T_e$ is about 10 MeV, MeV, while for electrons in air, $T_e$ is about 100 MeV.

Figure 2-17 shows the rate of energy losses as a function of electron energy for the passage of the electron through both lead and air. The rate of energy loss is not greatly different at lower energies for materials as different as lead and air. As a first approximation, then, we might expect that electron energy losses for lower-energy electrons passing through biological tissues (density of about 1 gm per $cm^3$) exhibit similar energy loss characteristics.

In addition to generating x-ray emission by bremsstrahlung or by exciting bound extranuclear electrons, beta particles are scattered as they move through matter.

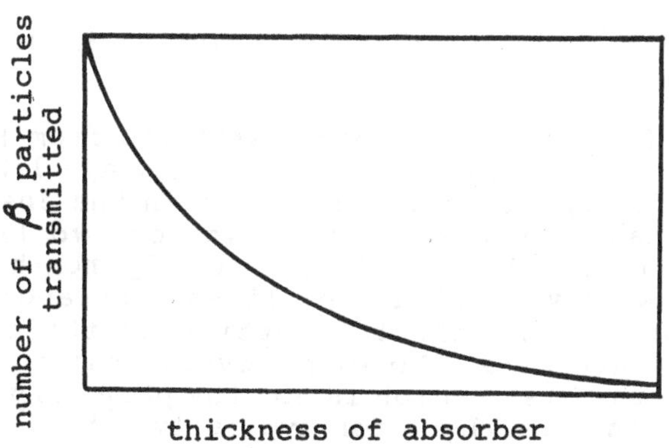


Figure 2-18. Transmission of beta particles as a function of absorber thickness (representative).

A fifth difference between the behavior of beta and alpha particles in their passage through matter derives directly from the continuity of the beta ray spectrum itself. Recall, in particular, that while almost all alpha rays from a given source have the same energy, beta rays, as Becquerel was able to show, exhibit a continuous spectrum of energies which have some upper bound $E_{b(max)}$. Thus the depth to which beta rays from a given source penetrate a given absorber varies over a wide range.

In spite of the rather rich diversity of phenomena exhibited by the passage of beta particles through matter, these phenomena combine to induce an approximately exponential decrease in the number of beta particles transmitted by the absorber as a function of absorber thickness (see Figure 2-18). If we plot the

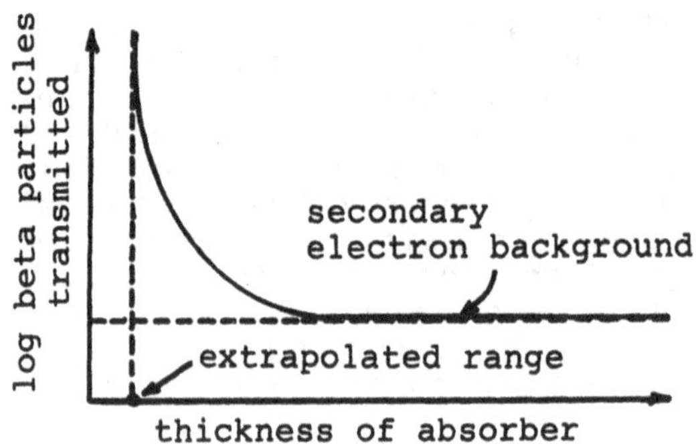


Figure 2-19. Semilogarithmic replot of Figure 2-18.

logarithm of the number of beta particles from Figure 2-18 versus absorber thickness, we get a graph like Figure 2-19. The "range" of beta particles in the absorber is typically taken as the intersection of two lines: (1) the linear extrapolation of the clearly non-horizontal part of this curve; and (2) the linear extrapolation of the more or less horizontal portion of the curve, as shown in Figure 2-19. Note, however, that the "range" determined in this fashion is not the path length of the electron in the absorber, for typically, the path length of a beta particle is much greater than its range.

Again, in spite of the diversity of phenomena which accompany the passage of electrons through matter, the absorption of beta particles is about the same for any of the light elements, provided that the amount of absorbing material is expressed in mass per unit area

(Note: the product of absorber thickness has dimensions of length; the density of an absorber has dimension of mass length$^{-3}$. Thus multiplication of absorber density by absorber thickness gives a quantity with dimension of mass length$^{-2}$). A curve of this type shown in Figure 2-20. This curve is closely approximated by the expression

$$I = kI_0 \exp(-(A/d)m_x) \qquad (2\text{-}51)$$

where

$$m_x = xd$$

where x is the depth in the absorbing material

and d is the density of the absorbing material

I is the number of beta particles transmitted to depth x

$I_0$ is the number of beta particles normally incident on the surface of the absorber (per unit area)

A is the mass absorbtion coefficient, a constant which is characteristic of the absorber

k is some constant of proportionality

(The similarity of the form of Equation 2-51 to the form of Lambert's law for the absorbtion of light by translucent materials is evident.) A is very nearly the same for the lighter elements. Furthermore, the value of A/d is nearly independent of the phase state of the absorber, for a given beta ray spectrum. With increasing beta particle energies, A/d decreases. Let $E_{b(max)}$ denote the maximum energy of the beta particles from a given source. By way of example, the range of phosphorous-32 beta particles ($E_{b(max)}$ = 1.70 MeV) is around 750 mg/cm$^2$ in aluminum and about 860 mg/cm$^2$ in lead.

In 1931, Norman Feather[72] proposed that the range R of beta particles in an absorber was approximated by

$$R = 543\ E_{b(max)} - 160 \qquad (2\text{-}52)$$

provided that $E_{b(max)} \geq 0.8$ MeV. In Equation 2-53 R is expressed in mg cm$^{-2}$ and $E_{b(max)}$ is in MeV. E.C. Widdowson and F. C. Champion[73] proposed a slightly different pair of constants for Feather's formula,

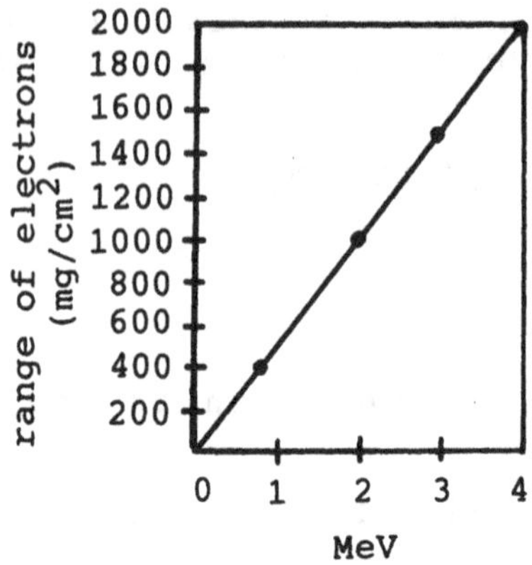


Figure 2-20. Transmission of beta particles through matter as a function of mass per unit area.

namely

$$R = 536\ E_{b(max)} - 165 \qquad (2\text{-}53)$$

which appear to fit well for $E_{b(max)} > 0.8$ MeV. W.J. Whitehouse and J.L. Putnam[74] proposed the relationship

$$R = 370\ E_{b(max)}^{3/2} \qquad (2\text{-}54)$$

which agrees to within 20% of experimental data in the range of 0.025 to 2 MeV. Equations 2-52 to 2-54 strongly suggest that the range of beta particles in matter varies linearly at "higher" energies, and to the three-halves power of maximum electron energy at "lower" energies.

## Passage of gamma radiation through matter

After emission of an alpha or a beta particle, a daughter nucleus may be left in an excited state. Gamma rays will be emitted if the nucleus passes from this excited to a lower-energy state. A gamma ray can be characterized by its quantum energy

hf

where

h is Planck's constant and

f is the frequency of the gamma ray.

Thus, for example, the emission of a beta particle by cobalt-60 ($t_{1/2}$ = 5.3 yr) is accompanied by the emission

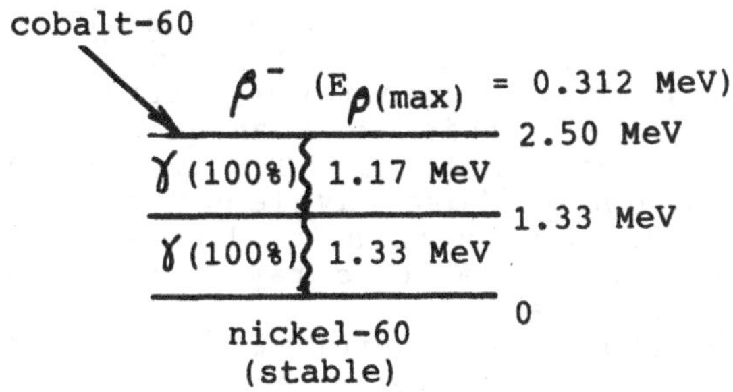


Figure 2-21. State transition diagram for cobalt-60 showing the emission of gamma rays which accompany the loss of an electron.

of a 1.17 MeV gamma ray, followed by a 1.33 MeV gamma ray. Such a transition is often represented in an energy-transition diagram of the sort shown in Figure 2-21. In Figure 2-21, the emission of the electron is shown as a line descending from upper left to lower right to suggest an increase in atomic number.

The interactions of gamma rays with matter are primarily interactions with electrons in that matter. In the following we discuss the four most important of these mechanisms.

A gamma ray can be absorbed by an environmental atom and its excess energy carried away by release of one of the extranuclear electrons. The kinetic energy of the resulting freed electron is equal to the initial energy of the gamma ray minus the energy required to free the electron from its parent atom. This mechanism is fundamentally the same as the photoelectric effect.

Second, a gamma ray coming from a decaying nucleus can transfer its energy to an electron of the same decaying atom, rather than to an electron of some atom in the environment. This process is called <u>internal conversion</u>. In such a case the excess energy of the daughter nucleus is transferred to one of the extranuclear electrons which is expelled with the (virtual) energy of the gamma ray, less the energy required to remove the electron from the atom. For example, in the decay of cobalt-60, all of the excess energy of nickel-60 (the daughter) is removed in the form of radiation. In contrast, when mercury-203 decays the excess energy is emitted in the form of gamma radiation in only 81% of the disintegrations; the remainder gives rise to internal conversion electrons.

There is a third mechanism by which a gamma ray can interact with an electron. When a gamma ray interacts with a free electron, it can be scattered by the electron in a process known as the Compton effect. The phenomenon is named after A.H. Compton, who showed that when $K_a$ x-rays from aluminum were scattered by carbon, the scattered rays consisted of an unmodified component with the same wavelength as the incident x-rays, and a modified component whose wavelength was greater than that of the incident x-rays.[75] Furthermore, Compton demonstrated, these scattered rays are less penetrating than the primary beam.

To understand the Compton effect, we need to take a brief digression into electromagnetic and relativity theory. Electromagnetic theory says that electromagnetic radiation transfers energy through space, exerting a pressure on any material surface which it contacts. With the energy density of such radiation, there must also be an associated momentum density. It can be shown that the momentum density for radiation in free space is equal to the energy density divided by the speed of light. Therefore, the momentum p of a photon of frequency f, and energy hf, is

$$p = hf/c \tag{2-55}$$

From relativity theory, we know that the momentum $p_e$ and kinetic energy $T_e$ of an electron of rest mass $m_0$ moving with velocity v is

$$p_e = \frac{m_0 v}{(1 - b^2)^{1/2}} \tag{2-56}$$

and

$$T_e = m_0 c^2((1/(1 - b^2)^{1/2} - 1) \tag{2-57}$$

where

$$b = v/c$$

and

c is the speed of light.

We can use the binomial theorem to expand $(1 - b^2)^{-1/2}$ (the multiplicative equivalent of the denominators of Equations 2-55 and 2-56) to obtain

$$(1 - (v/c)^2)^{-1/2} = 1 + \frac{1}{2}(v/c)^2 + \frac{3}{8}(v/c)^4 + \ldots \qquad (2\text{-}58)$$

Substituting the right-hand side of Equation 2-58 into the relevant parts of Equations 2-56 and 2-57, we get the following. For $v << c$, $b^2 << 1$, so that to good approximation for small v, Equations 2-56 and 2-57 become

$$p_e = m_0 v \qquad (2\text{-}59)$$

and

$$T_e = (1/2) m_0 v^2 \qquad (2\text{-}60)$$

which are familiar forms.

Let's consider now an x-ray photon of energy hf scattered by an electron at point O through an angle A with the electron recoiling at angle B, as shown in Figure 2-22. From the principle of conservation of energy, the total energy before should equal the total energy after the collision; thus

$$T_t = T_s + T_e \qquad (2\text{-}61)$$

where

$T_t$ is the energy before the scattering

$T_s$ is the quantum energy of the scattered photon

$T_e$ is the kinetic energy of the recoil photon

From this it can be shown that

$$f' = \frac{f}{1 + \dfrac{hf}{m_0 c^2}(1 - \cos A)} \qquad (2\text{-}62)$$

Note that when A is zero, f' = f in Equation 2-62 meaning that the frequency of the incident x-ray is unchanged by the scattering. Equation 2-62 shows that the fractional change in f, and hence in hf, depends on the initial energy hf: the larger the energy hf of the incident photon, the larger the fractional energy drop.

The wavelengths corresponding to f and f' are L = c/f, and L' = c/f', respectively. Substituting these values into Equation 2-62, we obtain

$$L - L' = (h/m_0 c)(1 - \cos A) \qquad (2\text{-}63)$$

This formula, derived by A.H. Compton[76] and also by P. Debye,[77] was verified by experiment by Compton. It shows, quite plainly, that the increase in wavelength, L - L', of the scattered photon, at any given scattering angle B, is completely independent of the initial wavelength of the incident x-ray photon. The constant

$$h/m_0 c \quad (\text{about } 2.43 \times 10^{-10} \text{ cm})$$

in Equation 2-63 is called the Compton wavelength.

Equation 2-63 has a further consequence of substantial interest. If the gamma ray interacts with a bound electron, then it can be shown that instead of $m_0$, a larger mass term, involving the mass of the atom binding the electron, must be substituted in the denominator of Equation 2-63. In this situation, that mass is so large that L - L' is negligible for all angles A, that is, the energy loss during scattering is negligible. Such scattering is called coherent scattering. This is just the unmodified component which Compton had originally observed. This coherent component of scattered radiation becomes decreasingly small with increasing frequency.

There is a fourth way gamma rays can interact with matter. In order to motivate our account of this phenomenon, we first need to take another brief digression into relativity theory. According to that theory, the total energy E of a particle moving with a speed v is given by

$$E = m_0 c^2/(1 - b^2)^{1/2} \qquad (2\text{-}64)$$

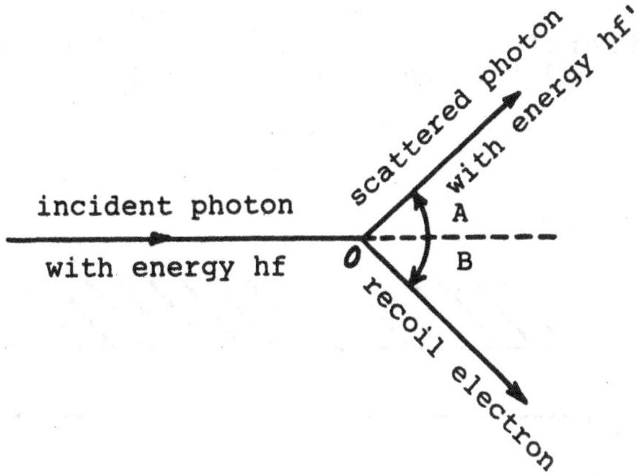


Figure 2-22. Schematic of Compton scattering of a photon by a (free) electron.

It can further be shown that

$$E^2 = m_0^2c^4 + p^2c^2 \tag{2-65}$$

If a particle is at rest, then $m = m_0$ and $p = 0$. Equation 2-65 then becomes

$$E = \pm m_0c^2 \tag{2-66}$$

the positive value of which is, of course, Einstein's well-known mass-energy equivalence expression.

In 1928 P.A.M. Dirac[78] derived a relativistic wave equation for the motion of electrons in an electromagnetic field. He pointed out that there are two sets of solutions to this this equation, one set corresponding to each of the possibilities on the right-hand side of Equation 2-66. In general, then, Dirac argued that there was a set of solutions corresponding to

$$E \geq m_0c^2 \tag{2-67}$$

and one set of solutions corresponding to

$$E \leq m_0c^2 \tag{2-68}$$

We can represent the possible sets of energy states described by Equations 2-67 and 2-68 in a diagram like that shown in Figure 2-23. According to classical electromagnetic theory, an electron cannot go from energy $-m_0c^2$ to energy $+m_0c^2$. (On the quantum theory, such a transition is possible.) Now in actual experience, all the free electrons that we observe

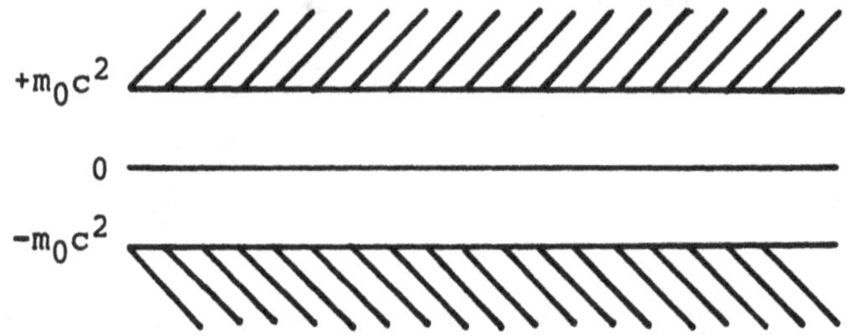


Figure 2-23. Energy state diagram for the relativistic electron.

have only positive energy. This is unexpected, to say the least. To explain this oddity Dirac first invoked the <u>Pauli exclusion principle</u>: no two electrons (this also applies to nucleons) can occupy the same energy state which is described by the same quantum descriptors (numbers). Dirac then made two very bold assumptions

1. Nearly all of the negative-energy states in the world are filled, and,

2. The normal electrification of the world (ground zero potential) corresponds to the zero level in Figure 2-23 when all negative energy states are filled.

On this view, we ordinarily have a plenum of negative energy states which are filled but are unobservable. If an energy greater than $2m_0c^2$ is imparted to a negative-energy electron, however, it can make the quantum leap from a negative-energy state to a positive-energy state and thus appear as a negative electron with positive energy. When an electron undergoes such a bumping in energy, the "hole" that is left in the plenum of negative-energy states appears as a positive charge with positive energy.

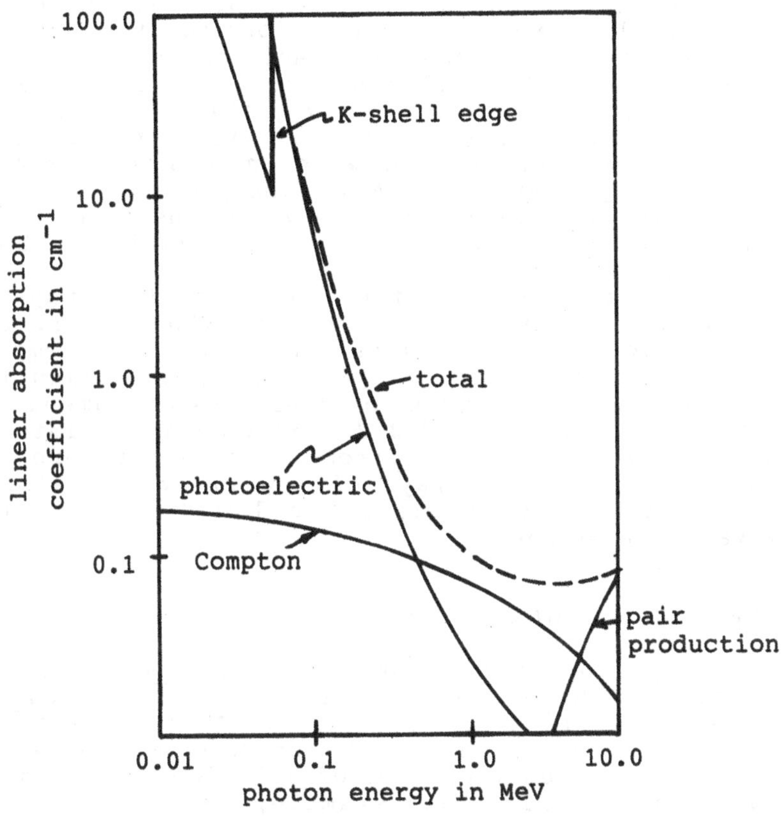


Figure 2-24. Regimes of photoelectric effect, Compton-effect, and pair-production for lead.

Initially Dirac thought that these holes were protons, although he was well aware that there was a very serious discrepancy in the mass of the holes and the then-accepted mass of protons. Finally rejecting this idea, in 1931 he proposed that the holes were a heretofore unknown kind of particle, an antielectron or positron, which has the same mass as an electron but carries a unit positive charge.[79] Dirac proposed that an encounter between two gamma rays of self-energy at least 0.511 MeV could lead to the simultaneous creation of an electron and an antielectron. Subsequently, C.D. Anderson[80] confirmed through cloud-chamber photographs of cosmic radiation that the positron existed, and P.M.S. Blackett and G.P.S. Occhialini[81] were able to observe electron-positron pair creation.

The energy which theory attributes to the positron is consistent with observation. According to the energy-mass equation, $E = mc^2$, the self-energy of an electron is 0.511 MeV. It is therefore possible, by the quantum theory, for gamma rays having energies greater than 2 x 0.511 MeV = 1.022 MeV to raise a negative-energy electron to a positive-energy state and create in the process an electron-positron pair. In fact this process is observed to occur with increasing frequency with gamma rays above 1.022 MeV. The process must take place within the Coulomb field of a nucleus so that the recoil of the nucleus can allow conservation of both momentum and energy.

We have examined the four most important ways in which gamma radiation can interact with matter: the photoelectric effect, internal conversion, the Compton effect, and pair-production. The relative regimes for the first, third and fourth of these the phenomena are shown in Figures 2-24 and 2-25 for lead and aluminum, respectively. As with electrons, the mass absorbtion coefficient for gamma rays of low energy is almost independent of the material for low Z over a wide interval of energies.

## MEASUREMENT OF RADIOACTIVITY

The <u>standardization of a radioactive source</u> is defined as the determination of the <u>specific disintegration rate</u> of the source. This quantity is usually expressed in units of disintegrations per second per gram (or per unit volume) of solution (or per some other reference unit). For example, a standardization unit which is widely use to specify natural alpha activity in water supplies is the picoCurie per liter (pCi/l). (When expressed in units of activity per unit volume, the specific disintegration rate is also called the <u>activity concentration</u>.)

One of the most direct procedures used to standardize solutions (such as drinking water) involves evaporating a weighed amount of the solution to dryness, then measuring the activity of the dried source with a suitable radiation detector/counter which subtends a fixed, known solid angle at the source. The disintegration rate, $N_0$, may then be obtained from the relation

$$N_0 = N_o \frac{4\pi}{O} \frac{1}{E} \frac{1}{1 - A_s} f(S_i) - N_b \qquad (2\text{-}69)$$

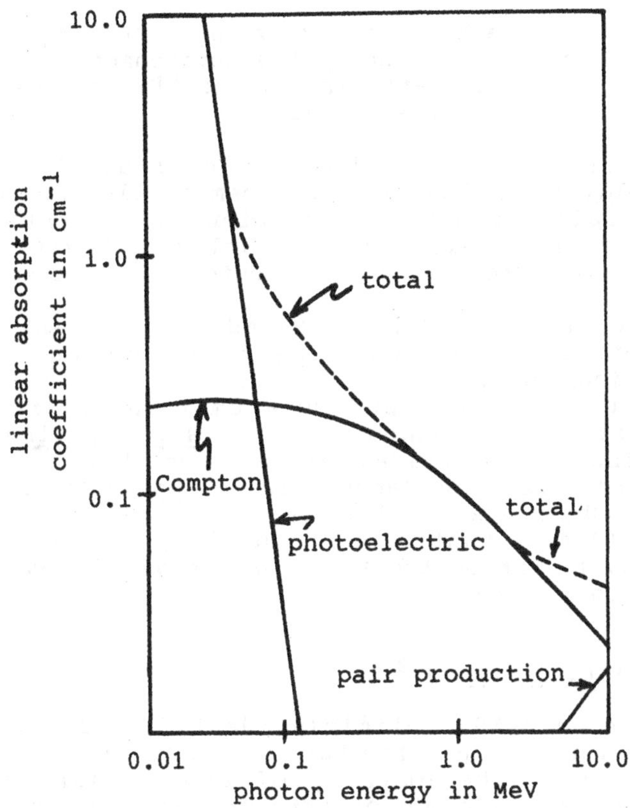


Figure 2-25. Regimes for the photoelectric effect, the Compton effect, and pair-production for aluminum.

where

$N_o$ is the observed counting rate

O is the solid angle subtended by the counter at the source (in steradians)

E is the detection efficiency of the detector for the radiation from the nuclide being measured

$A_s$ is a self-absorbtion factor which corrects for the fractional amount of radiation which is entirely absorbed in the source

$f(S_i)$ is a composite correction factor which accounts for various kinds of scattering, including

a) backscattering by the source mount
b) scattering by the environment
c) selfscattering by the finite mass of the source itself

$N_b$ is the counting rate caused by the background (primarily from cosmic radiation and the naturally occurring radiation from construction materials. This may also include activity in the materials from which the counter is constructed.)

Typically the counter is operated with a blank in the source chamber to help identify the background contributions to $N_o$.

Frequently alpha and beta emitting sources are standardized by a method known as <u>$4\pi$ counting</u>. The name of the method derives from the fact that the source to be standardized is placed between two detectors, each of which subtends a solid angle of $2\pi$ steradians at the source. (Notice that in $4\pi$ counting, $O = 4\pi$, making the second factor on the right hand side of Equation 2-69 equal to unity.)

## Statistics of counting[82]

Equation 2-69 explicitly identifies a number of factors which can influence the quality of a measurement. The quality of any measurement is determined by two kinds of errors: unknown, typically uncharacterizable but systematic errors which affect the measurement's accuracy (how close the measurement is to the theoretically predicted ("true") value for that measurement), and what we will call <u>statistically characterizable errors</u>, or more briefly, <u>statistical errors</u>, which affect the reproducibility of a result (i.e., its <u>precision</u>).

Statistical errors arise in the measurement of radioactivity measurement from two sources: the inherently random nature of the decay process itself, and variations in the technique and instrumentation used to measure that decay. We consider these in turn.

The <u>binomial probability function</u> provides the fundamental description of random events, and is therefore central to a description of radioactive decay. The principal features of the binomial probability function are as follows. Suppose we have a very large set of objects (e.g., radioactive atoms), some of which are of type A (e.g., decayed) and the rest of which are of type B (e.g., not decayed). Let p be the probability that any object selected at random from this large set will be of type A. By the axioms of probability, 1 - p is the probability that the object will be of type B. The probability P(n) that exactly n of $N_0$ objects selected from the set will be of type A is given by

$$P(n) = \frac{N_0!}{(N_0 - n)!n!} p^n (1 - p)^{N_0 - n} \qquad (2\text{-}70)$$

Equation 2-70 is called the binomial probability function.

Because decay is inherently random, over any given time interval the number of disintegrations which have occurred in two different physical samples from the same source of radioactive material will typically differ. This kind of variation is, of course, a feature of all random processes. For example, a set of one hundred flips of a coin (a random process) will typically not produce the same number of heads and tails as another set of one hundred flips. The long-term average frequency of heads (or tails), however, tends to 0.5. Thus, any set of counts of radioactive disintegrations over a finite time interval will produce only an estimate E of the true average rate of disintegration even if the counts are, for the time interval over which they are observed, free of all other errors.

It is obviously important to have a good method for determining E. Whatever else we may mean by "good," E must be computable from observational data alone, because obviously that is all we have. The rationale for choosing E also ought to show us how close E might be to the "true" average disintegration rate, which would at least require that we understand what we mean by "true" average. In general, we would intuitively expect E's closeness to the true average rate to be better with more observations than with fewer: the more we know, the smaller the error in E should be. One way to articulate this intuition is to find some measure of the true average rate, then try to show that there is a quantity which can be computed from observational data alone and which is, in some sense, the best estimate of the true average rate. In the case of the binomial distribution, it can convincingly argued that the mathematical measure of the true average is given by the mean m, which for the binomial distribution is given by

$$m = \sum_{n=0}^{N_0} nP(n) = pN_0 \qquad (2\text{-}71)$$

where $N_0$, n, P(n), and p have the same meaning as in Equation 2-70. The simple arithmetic average of a set of observations, given by

$$E = \frac{1}{N} \sum_{i=1}^{N} x_i \qquad (2\text{-}72)$$

where

N is the total number of distinct observations

$x_i$ is the value obtained on the $i^{th}$ observation

it can be shown, is the best estimator of the mean for the binomial (and several other) distribution(s).

In addition to having some handle on the true average of a distribution of outcomes of a random process such as radioactive decay, it is highly informative to know just how wildly observed values would, "on the average," differ from the mean. As a first guess, we might expect this measure to give us the average difference between the mean of the distribution and each of the individual outcomes in that distribution. Mathematicians have formalized this notion in a statistic called the variance, s. For the binomial distribution the variance is defined by

$$s^2 = \overline{(m - n)^2} = \sum_{n=0}^{N_0} (m - n)^2 P(n) \qquad (2\text{-}73)$$

where

$\overline{*}$ is the simple arithmetic average of the values of * and * is an arithmetic expression

and m, n, $N_0$, P(n) are again as defined for Equation 2-70. The positive square root of this value, called the standard deviation, can be shown to be

$$s = (m(1 - p))^{1/2} \qquad (2\text{-}74)$$

We can apply the binomial distribution to radioactive decay in the following way. Let $N_0$ be a large group of radioactive atoms. These can be divided into two types, those which do, and those which don't, decay in time period t. From von Schweidler's derivation of Equation 2-8, the probability that a given atom will not decay is exp(-kt). The probability that it will decay is thus 1 - exp(-kt). From Equation 2-70, the probability P(n) that n atoms will decay in t is therefore

$$P(n) = \frac{N_0!}{(N_0 - n)!n!} (1 - \exp(-kt))^n (\exp(-kt))^{N_0 - n} \tag{2-75}$$

The mean is, by substitution into Equation 2-71

$$m = N_0(1 - \exp(-kt)) \tag{2-76}$$

and the standard deviation is, from Equation 2-74

$$s = (m \exp(-kt))^{1/2} \tag{2-77}$$

For $kt << 1$, that is, for observation times short compared to the half-life, the standard deviation is approximately

$$s = m^{1/2} = (\bar{n})^{1/2} \tag{2-78}$$

If $p << 1$, $N_0 >> 1$, and $pN_0 << N_0^{1/2}$, the binomial probability function can be expressed as

$$P(n) = \frac{N_0 p^n}{n!} \exp(-pN_0) = \frac{m^n \exp(-m)}{n!} \tag{2-79}$$

Equation 2-79 is called the <u>Poisson probability</u> or <u>Poisson distribution function</u>. It provides a good approximation to the binomial distribution probabilities when $N_0 \geq 100$ and $p \leq 0.01$.

The binomial and Poisson probability functions describe the behavior of discontinuous quantities (e.g., being disintegrated or not) which take on integer values only. A more general approach to the theory of errors assumes that errors are <u>normally distributed</u> and can assume continuous values. The behavior of radioactivity measuring instruments and techniques is typically taken to be normally distributed. Events (errors, in this case) are normally distributed just in case their probabilities are given by the <u>normal probability</u>

<u>function</u> or

$$P(n) = \frac{(2\pi)^{1/2}}{s} \exp(-(m - n)^2/2s^2) \qquad (2\text{-}80)$$

In a sense, the normal distribution can be thought of as a limiting case of Poisson's distribution, provided that $s^2$ is about the size of m, if the absolute value of m-n is much less than m, and if m >> 1.

Typically we do not know the true mean of the population from which our observations are drawn because we can almost never afford to measure the properties of interest of each member of the population. Thus there will be some uncertainty in what we report as the mean of that population. There are a number of ways to report this uncertainty, but one of the most common is to report the observed value (say, counts) as the midpoint of some interval. The choice of this interval is a matter of convention, but ideally, the choice ought to suggest something about the probability that the true mean lies in the interval. A common approach is to report the observed value, n, bounded by $n^{1/2}$, that is, $n \pm n^{1/2}$; if the errors are normally distributed, and if n, m and $s^2$ are about the same size (frequently the case in the measurement of radioactivity), it can be shown that there are only about 33 chances out of 100 that the true mean number of counts for the interval differs from n by more than $n^{1/2}$. If n is the total number of counts observed over elapsed time t, then the average counting rate r observed is obviously given by

$$r = n/t \qquad (2\text{-}81)$$

From this it follows that

$$r \pm s_r = \frac{n}{t} \pm \frac{n^{1/2}}{t} = r \pm (r/t)^{1/2} \qquad (2\text{-}82)$$

where

$s_r$ is the standard deviation of the disintegration rates.

Stated in terms of a percentage error, Equation 2-82 is

$$r \pm \frac{100}{(rt)^{1/2}}\,\% = r \pm \frac{100}{n^{1/2}}\,\% \qquad (2\text{-}83)$$

Equation 2-83 says something quite important: the percentage error of a rate determination is determined entirely by the total number of counts accumulated. This confirms what our intuitions suggest--that the quality of an estimate improves with an increasing number of measurements.

When several normally distributed noninteracting populations or processes $Q_1, Q_2, \ldots, Q_n$ with standard deviations $s_1, s_2, \ldots, s_n$, respectively, are used together in a measurement, the standard deviation $s_s$ of the collective is given by

$$s_s = (s_1^2 + s_2^2 + \ldots + s_n^2)^{1/2} \tag{2-84}$$

For example, suppose a background counting rate were reported as $r_b \pm s_b$ and the total counting rate was reported as $r_T \pm s_T$. The counting rate $r_s$ due to the source only would then be, by Equation 2-84

$$r_s \pm s_s = (r_T \pm s_T) - (r_b \pm s_b) =$$
$$= (r_T - r_b) \pm (s_b^2 + s_T^2)^{1/2} \tag{2-85}$$

The above example is, not surprisingly, a common one in practice. The optimum allocation of time to the determination of $r_T$ and $r_b$ can be calculated. There are two experimental situations to be considered. In the first, one may be able to make a series of counts on different samples during a period in which there is little reason to expect the background to be changing. In such a case, it usually advantageous to try to make $s_b$ small, say, an order of magnitude smaller than $s_T$. (It is advisable during the counting period, of course, to make several measurements on the background to determine whether the assumption that the background is not changing is sound.) The second type of measurement situation which we might face is one in which a fixed time is available for making both a total-counting-rate and a background-counting-rate determination. If $t_b$ and $t_T$ are the times taken for the background and total activity counts, respectively, then $s_s$, the standard deviation of the net counting rate is, by Equation 2-84

$$s_s = ((r_b/t_b) + (r_T/t_T))^{1/2} \tag{2-86}$$

From this equation it can be shown (by calculus) that the condition for <u>overall</u> minimum error is

$$t_b/t_T = (r_b/r_T)^{1/2} \tag{2-87}$$

Given that statistical fluctuations are intrinsic to both the nature of the radioactive process itself and to the technology and technique of measurement, how can we determine when an observed change is real and when it is a result of variations in the measuring instruments or methodology? In practice, if measurements of a counting rate exceed an accurately determined background rate by 1.96s, the increase is typically taken to be <u>significant</u>, since it is greater than that expected due to statistical fluctuations in the background except in 5% of the cases. (More precisely, statisticians define <u>significance</u> as a measure which can assume any value from 0 to 1 depending on the nature of the distribution and how far a given observation is from the mean of that distribution.)

Consistency (reproducibility, or precision) of the behavior of measuring instruments (counters) can be checked by statistically analyzing the deviations obtained from the results of successive counts. For example, one can theoretically calculate the standard deviation of the difference between two counting rate determinations on the same source and then compare this with the observed difference between the two determinations. The standard deviation of the difference of two counting rates $r_1$ and $r_2$ taken in time intervals $t_1$ and $t_2$, respectively, is

$$s = ((r_1/t_1) + (r_2/t_2))^{1/2} \qquad (2\text{-}88)$$

The probability of observing an absolute difference $/r_1 - r_2/$ in counting rates equal to or greater than ks ($k > 0$) can be obtained from tables[83] of values for the normal distribution. If the probability of the observed differences is less than 0.05, conventional wisdom advises it is likely the deviations are due to causes other than the nature of the disintegration process itself, and probably arise from fluctuations in the instrumentation or measurement technique.

The "chi-square" test provides a second means of determining the trustworthiness of the measuring equipment or procedures. To make the test, the quantity

$$x^2 = \sum_{i=1}^{N} \frac{(\bar{n} - n_i)^2}{\bar{n}} \qquad (2\text{-}89)$$

is calculated, where

N is the number of distinct measurements

$n_i$ is the value obtained in the $i^{th}$ measurement,

$\bar{n}$ is the average value.

Notice that Equation 2-88 provides a normalized (against $\bar{n}$) measure of the scattering or dispersion of a set of measurements. Tables which give the probability P of obtaining given values of $X^2$ for various values of N are widely published.[84] If the value of $X^2$ is so large that there is only a small probability assigned to its being this high, one has a clear indication of a problem in the experiment. Conventional wisdom holds that data which give a value of P equal to 0.02 or smaller are suspect. But if P is 0.98 or larger, the corresponding data should also be questioned because the spread in the data may be too small to be credible.

## DETECTOR CONSTRUCTION[85]

All methods for detecting natural radioactivity involve its interaction with matter. This interaction produces ionization or excitation of extranuclear electrons. In this section we briefly survey how that interaction is exploited by the detection devices used today.

### Gas conduction detectors

There is a family of radiation detectors which rely on the fact that ionizing radiation passing through a gas will cause the conductivity of the gas to increase. In each of the detectors in this family, which we will generically call gas conduction detectors, an electric potential is applied across a pair of electrodes which are in contact with a gas. Gas conduction detectors are of three types: the ionization chamber, the proportional (gas) counter, and the Geiger-Mueller counter. The fundamental distinction among the three turns on differences in the ionization behavior of the gas in the chamber under different applied voltage regimes (electric field strengths).

Consider in particular an alpha or a beta particle with sufficient energy to create at least one ion pair in the gas entering a gas conduction detector. Let us call the event of the first such ion pair being created by a particle or ray the primary ionization, or the primary event. Our ultimate aim is to measure the incidence or rate of such events, and to relate these events to the activity of the sample under measurement. To understand the relation between gas conduction detector behavior and the incidence of primary events, we must evidently understand the relation between the electric field in the gas chamber and the number of ion pairs produced per primary event, as shown by Figure 2-26. The number of ion pairs produced per ionizing particle entering the chamber is seen to vary over many

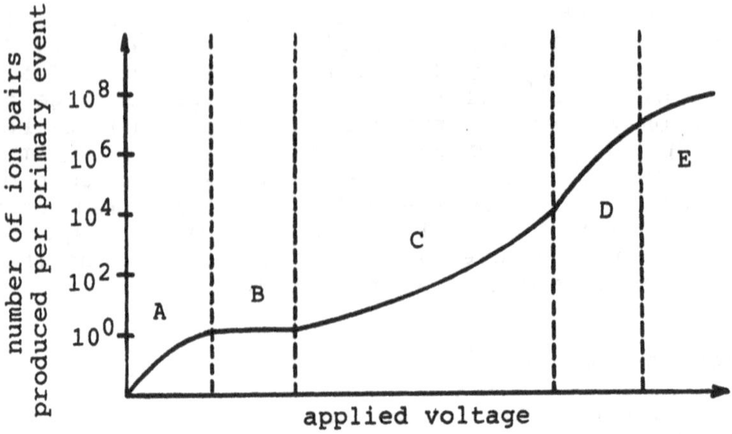


Figure 2-26. Number of ion pairs produced in an ionization chamber as a function of voltage applied across the detector electrodes.

orders of magnitude, depending on the voltage applied across the chamber electrodes. In Region B, for example, one ion pair is produced per primary event; this region is accordingly called a region of zero gas amplification. In Region C, the acceleration of electrons created by the primary ionization in the electric field is sufficient to drive these electrons to create new ion pairs themselves. Thus, in Region C more than one ion pair is produced per primary event. This phenomenon is known as gas multiplication or gas amplification.

## The ionization chamber detector

An ionization chamber is a gas conduction detector which operates in Region A of Figure 2-26. Typically an ionization chamber consists of a cylindrical cathode enclosing an anode and the gas; such a device is shown schematically in Figure 2-27. Charged particles and electromagnetic radiation undergo inelastic collisions with atoms or molecules of the gas in the chamber, ionizing them and forming positive ions and electrons. When a voltage is applied across the electrodes, ions drift along the lines of electric force, producing an ionization current. Typically these electrons drift at the speed of $10^6$ cm/sec, while positive ions drift several orders of magnitude more slowly because of their relatively greater mass. When an ionization chamber is exposed to ionizing radiation, the current measured at M increases as the potential across the electrodes of the chamber is increased, up to a certain potential P (see Figure 2-28). As the voltage is increased past P, the current levels off. At potentials below P (region R in Figure 2-28, recombination of ions occurs, that is, the number of ions collected is less than the number actually formed. At P, the electric field is strong enough to prevent recombination.

The limiting value of current in Figure 2-28 is called the <u>saturation current</u>. In general, the

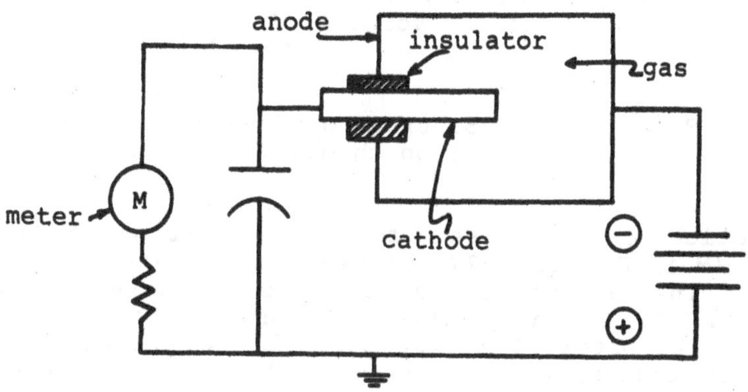


Figure 2-27. Schematic of a gas ionization chamber.

voltage required to achieve saturation for an ion chamber will depend on the rate at which ions are being produced. At saturation, the ionization current i and the number of ion pairs N produced per unit time are related by

$$i = Ne \tag{2-90}$$

where e is the electronic unit charge. Each ion pair generated results in a charge segregation of $1.602 \times 10^{-19}$ coul. Thus 1 picoampere of current, which a small current to measure directly, would require the generation of about $(10^{-12}$ coul $sec^{-1}/(1.602 \times 10^{-19}$ coul $pair^{-1})$ = $6.3 \times 10^{6}$ ion-pair $sec^{-1}$. Evidently, an ionization chamber requires rather high rates of ion pair production to produce measurable currents.

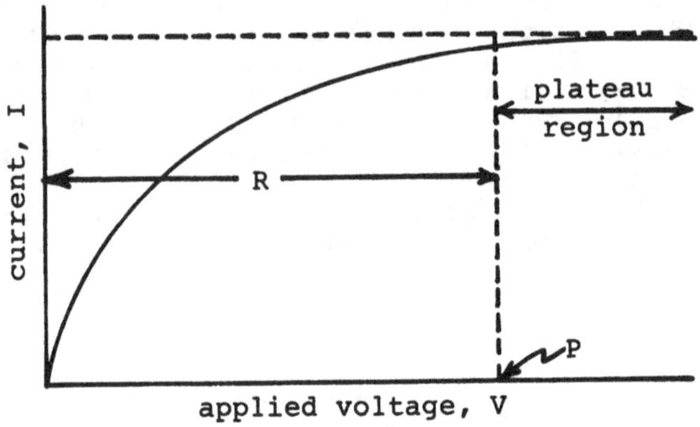


Figure 2-28. Saturation curve for an ionization chamber.

In practice, the response speed of the current-measuring device (M in Figure 2-27) is made long compared to the time of instantaneous fluctuations of current in the chamber. If the ionization current is large enough (say, on the order of a microampere), the current can be measured directly. But frequently, other means of current measurement must be employed. One way of doing this is to allow the charge to collect on the combined capacitance of the chamber and another capacitor in the circuit. The resulting rate of change of voltage can be measured with an electrometer. Since $Q = CV$, where $Q$ is the accumulated charge, $C$ is the capacitance, and $V$ is the potential across the capacitor plates, it follows that

$$i = \frac{dQ}{dt} = \frac{CdV}{dt} \qquad (2\text{-}91)$$

In some circumstances, the ionization chamber is used as a spectrometer, and in this mode, current is not measured. Rather, the amplitudes of individual voltage pulses are measured. This can be done by exploiting the fact that

$$V = Q/C \qquad (2\text{-}92)$$

Thus, the amplitude V of the pulses is directly related to amount of charge Q liberated in the gas of the ionization chamber by the complete absorbtion of individual particles. The energy of the particles can then be computed, provided that $W_{gas}$ is known.

The measurement of the effects of radiation on biological tissues is of obvious importance to the management of natural radioactivity in water supplies. For this reason, there has been extensive development of tissue-equivalent ionization chambers. These chambers contain, in both the wall of the chamber and the enclosed gas, the significant elements of tissue in the correct ratios. A reasonable approximation for this ratio for soft tissue is given by the empirical formula $(C_5H_{40}O_{18}N)_n$. A satisfactory tissue-equivalent gas consists of 38.1% hydrogen, 22.2% methane, 37.6% oxygen, and 2.1% nitrogen. For wall material, a gel of 66.2% water, 20.2% gelatin, 5.2% glycerol, and 8.4% sucrose has been used. In addition, tissue-equivalent plastics are now available.[86]

## Proportional counters

When a gas conduction detector is operated in Region C of Figure 2-26, enough of a charge can be collected as the result of a single beta or gamma ray entering the chamber to produce a 1 to 2 millivolt pulse. A gas conduction detector operating in Region C of Figure 2-26 is called a proportional (gas) counter. A crucial feature of the operation of a proportional counter is that the size of the resulting voltage pulses are exactly proportional to the amount of energy deposited by the primary ionizing event.

The center electrode of a proportional counter is typically a fine wire of diameter $10^{-3}$ to $10^{-2}$ inches. This choice of diameter is advantageous for the following reason. The electrical field strength E(r), at radius r in a cylindrical counter is given by

$$E(r) = \frac{V_0}{r \ln b/a} \qquad (2\text{-}93)$$

where

$V_0$ is the voltage imposed across the counter

a is the radius of the inner electrode

b is the radius of the outer electrode.

Thus the field strength close to the anode increases rapidly as r decreases. In the volume close to the anode, in fact, the acceleration of the electrons is greatest and hence the greatest secondary ion production occurs there. Secondary ion production is called avalanching. Provided that these secondary ions do not interact, the charge collected will be some multiple (greater than 1) of the number of primary events; thus, operation in this regime is proportional gas multiplication. This multiplication is particularly strong within 2 x $10^{-3}$ inch of the anode. Avalanching multiplication on the order of $10^3$ to $10^4$ ion-pairs per primary event is commonly used. One obviously desires a proportional counter to be designed so that the interaction between ions produced during avalanching is minimized: when avalanche- produced ions interact, the number of ions sensed may not be proportional to the number of primary events. Proportional counters can be used to measure the energy of alpha and beta particles and of low-energy gamma rays provided they are completely absorbed by the gas in the counter. Such detectors are typically connected to high-gain amplifiers.

## Geiger-Mueller counters

If the potential difference between the electrodes of a gas conduction detector is raised past the proportional region (Region C in Figure 2-26) to Region D, the number of ions produced is no longer strictly proportional to the number of primary events. Among other reasons, this loss of proportionality occurs because the charges developed locally in the chamber become large enough to distort the electric field. In addition, a discharge begins to spread along the center electrode as the potential difference is increased caused in part by the repulsion of space charges and in part by the formation in the counter of ultraviolet photons (emitted when some ions combine), which induce photoelectrons to be emitted by the cathode. When the proportionality of the counter is lost, amplification factors may depend on whether alpha or beta radiation is being measured.

When the voltage of an gas conduction detector is raised to region E of Figure 2-26, the counter becomes completely nonproportional. A gas conduction detector operating in Region E is Figure 2-26 is called a Geiger-Mueller counter. In this regime, the discharge phenomenon spreads along the entire length of the center electrode, and is independent of the magnitude of the primary event. The same size pulse is developed whether from one primary event or $10^9$ events. When a primary event occurs, electrons are accelerated highly in the region of the center electrode, and ionize more and more molecules of gas. The total discharge finally ends when an ion sheath is formed around the center electrode. The sheath reduces the field strength to the point at which further ionization is impossible. The time elapsed between the primary ionization event and the completion of the collection of all electrons is less than a microsecond. In this short time interval, the much more massive ions have, for practical purposes, remained motionless; thus the ion sheath has been motionless during this period. The sheath then begins to migrate toward the outer electrode. A few hundred microseconds later, it reaches the surface of the outer electrode. During this passage, some of the ions may be moving with a high enough velocity to eject secondary electrons.

If by this time the voltage across the tube has been able to return to its original value, these secondary electron emissions coming from the movement of positive ions toward the outer electrode can cause further avalanching. These spurious avalanches are highly undesirable because they are seen by the detector as noise spikes superposed on the signals generated by the primary events. Two approaches to controlling this

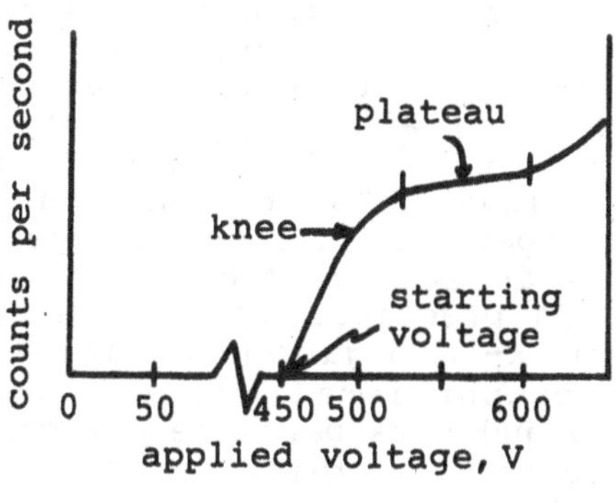


Figure 2-29. Response curve of a nominal Geiger-Mueller counter.

type of spurious avalanching are taken in practice. These approaches allow us to classify Geiger-Mueller counters into two types. In the first, called the <u>nonself-quenching</u> type, the voltage across the tube is automatically reduced following a discharge for a period large compared to the time of migration of the ion sheath. In the second, called the self-quenching type, small quantities of ethanol, ethyl formate, chlorine, or bromine are added to the gas in the chamber. These "adulterants" absorb photons emitted at the anode, thus preventing them from reaching the cathode and causing photoelectron emission there. Organic-compound-quenched counters have a finite lifetime because the organic molecules dissociate upon quenching, and cannot subsequently provide quenching for spurious photons.

A graph of the number of counts versus voltage applied across the electrodes of a Geiger-Mueller tube is shown in Figure 2-29. Below a certain voltage, called the threshold potential, no counts occur. As the voltage is increased, the count rate then rises steeply to an inflection in the curve, called the knee, after which the response curve becomes relatively flat. This flat region is called the plateau of the curve. In the plateau, the counting rate is relatively insensitive to changes in the applied voltage. Typically, a Geiger-Mueller counter is operated about 100 volts above the knee.

## Gas flow counters

Gas flow counters are gas conduction detectors in which the gas in the chamber flows continuously through the chamber from some external source. Gas flow counters can be operated as ionization chambers, Geiger-Mueller counters, or as proportional counters.

Ionization chambers, proportional gas counters, and Geiger-Mueller counters all rely on a gas-filled tube for radiation detection. They are distinguished primarily by the regions of the response curve of Figure 2-26 in which they operate. All three types of devices may be used for the detection of charged particles and gamma radiation. In practice, ionization chambers are used more often for x- and gamma radiation detection, while Geiger and proportional gas counters are used primarily for alpha and beta particle detection. In the latter case, the source is typically placed inside the counter or near a thin mica window on one end of the counter. The chief advantage of the Geiger-Mueller tube is its great sensitivity: one or two primary ion-pairs can cause a complete discharge. The output pulse of a Geiger-Mueller tube is also large compared to that of other gas conduction counters. In general, the electrical signal generated by any of these detectors is amplified electronically prior to actual counting or recording.

Scintillation counters

One of the earliest methods of detecting alpha particles (note, for example, Rutherford's alpha-scattering experiment) was a device called a spinthariscope. In this device, alpha particles were allowed to strike a zinc-sulfide screen. Upon collisions of this sort with alpha particles of sufficiently high energy, zinc sulfide emits flashes of light bright enough to be seen by a microscope. This phenomenon, in which the energy of the radioactivity is converted to light, is the basis of modern scintillation counters.

The electron-multiplier phototube is one of the more commonly used scintillation devices. The phototube is a vacuum tube, one end of which is coated by a phosphor. Incoming energy (say, alpha or beta particles) excites the phosphor atoms. When the phosphor atoms return to their ground states, they emit photons, typically in the far blue and ultraviolet portions of the spectrum (see Figure 2-30). The phosphor is optically connected to the envelope of the phototube; photons arising from the phosphor strike a photocathode which in turn ejects photoelectrons; the number of such electrons emitted is proportional to the conversion efficiency of the cathode, typically on the order of 0.10. The conversion efficiency of the tube is by definition just the ratio of the energy incident on the tube to the energy of the electrons released by the photocathode. As the photoelectrons emerge from the cathode, they are directed by an electrostatic focusing element to another electrode, called the first dynode. This electrode emits 3-5 electrons for every electron which strikes its surface. This shower of electrons then strikes yet another dynode, which emits another 5 or so electrons for each electron which strikes it. Each dynode is maintained at roughly 100 volts above the potential of its predecessor. There may be 10 - 15 dynodes in a phototube. The overall behavior of such a series of dynodes may be thought of as an ever-increasing shower of electrons--virtually, an avalanche. A single electron leaving the photocathode of a photomultiplier tube can give rise to $10^6$ - $10^8$ electrons at the anode of the tube. In practice, the anode is connected to a load resistor. In such a configuration, a burst of electrons arriving at the anode will give rise to a negative voltage pulse which can be amplified and analyzed by appropriate electronic instrumentation.

Sodium iodide containing about 0.1% thallium is the phosphor most commonly used in photomultiplier tubes for gamma ray detection in current practice. The energy output per unit energy absorbed makes this material the

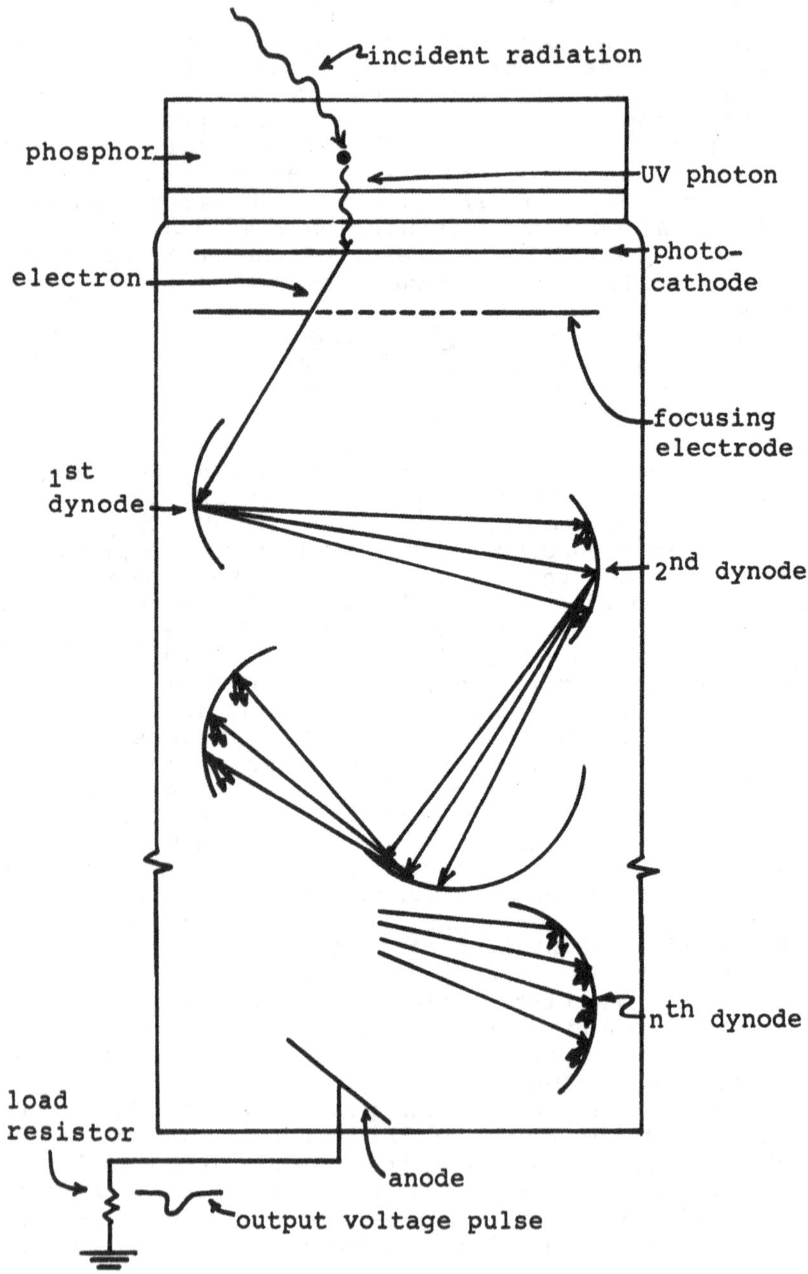


Figure 2-30. Schematic of a photomultiplier tube.

most efficient known scintillator. For beta-particle detection, anthracene, stilbene, and a number of plastic phosphors are available. The energy outputs of all the photomultiplier tube phosphors are nearly directly proportional to the absorbed energy.

A second type of scintillation detector, called a liquid scintillation counter, is also in widespread use today. In this type of counter, the radioactive sample is dissolved in a solution along with the phosphor. The solvent accepts energy from the primary radiation and transfers it to the phosphor. Solvents commonly used include toluene, xylene, and the ethers. Since the most efficient phosphors flouresce in the far ultraviolet, it is common practice to incorporate a second phosphor, called a wavelength shifter, into the solution. The wavelength shifter is excited by the ultraviolet emissions from the primary phosphor and in turn emits radiation in the more sensitive part of the detector (typically, a phototube) response spectrum. Liquid scintillation counters are used primarily for those radionuclides which decay by beta emission.

## Calorimeters

A calorimeter is a device for measuring heat. Such a device can be used to measure radioactivity because radioactive energy is absorbed by matter and is, in favorable circumstances, ultimately converted to heat. In its simplest form a calorimeter is a chamber whose heat loss to its surroundings is well understood. A source of heat is placed inside the chamber and the energy output of the source is determined by one of several possible methods.

One of the most common types of calorimeters is the concentric chamber calorimeter. Such a device consists of a cylindrical inner chamber concentrically surrounded by an outer jacket. In the adiabatic version of this type of calorimeter, as the temperature of the inner chamber rises, the temperature of the outer jacket is raised by an external heat source to keep heat from flowing from the inner to the outer jacket. The quantity of heat added to the outer jacket is thus a measure of the heat released from the source. In the isothermal concentric chamber calorimeter, the inner chamber is allowed to come to thermal equilibrium with its surroundings by removing heat in a uniform manner (say, by circulating water past it).

Yet another calorimetric method exploits the fact that a Peltier junction can remove heat from a source. In this type of calorimeter, the heat source to be measured is placed on a small disk or in a small cup and the heat output from the source is just balanced by Peltier cooling.

The twin-differential calorimeter consists of two identical chambers. In one the heat source is placed. The other contains a variable and calibrated heating element which is adjusted to maintain a zero-temperature difference between the chamber containing the source and the chamber containing the heating element. When this state is obtained, the output of the unknown heat source and the calibrated chamber are equal.

All of the calorimetric devices described above have the virtue of relative simplicity compared to other detection methods. To relate the heat output rate of a source to the disintegration rate, of course, it is necessary to know the energies of the alpha particles and/or the distribution of the energies of the beta particles because a calorimeter intrinsically integrates particle energies. Calorimetric methods are more useful for alpha and beta radiation measurements than for gamma radiation measurement, because the former are almost completely converted to heat, whereas the latter are not. Calorimeters may be less sensitive than other detection methods for low-energy sources (e.g., for activities of a few hundred pCi or less).

## The Lauritsen electroscope

The Lauritsen electroscope is a (improved) variation of the familiar gold-leaf electroscope used to detect the presence of an electric charge. The gold-leaf electroscope consists of a pair of gold foil leaves attached to a metal rod; this assembly is typically housed in an insulating case which has at least one transparent cover to permit observation of the configuration of the leaves. A portion of the rod extends outside the case to allow it to receive or transmit the charge to be measured. When a charged body is brought near to, or in contact with, the exposed part of the rod, the leaves can be observed to separate (or, under some circumstances, converge). The leaves can be discharged by bringing a body of opposite charge in contact with the exposed part of the rod. Roughly speaking, the angular separation of the leaves serves as a measure of the charge.

In the Lauritsen electroscope, designed by C.C. and Thomas Lauritsen,[87] the role of the gold leaves of the conventional electroscope is played by a metal frame to which is attached a gold-coated quartz fiber about 3 microns in diameter and about 6 mm long. Fastened perpendicular to the free end of the fiber is a very short piece of quartz fiber to facilitate tracking of the fiber movement with a microscope (see Figure 2-31). The fiber assembly and its holder are mounted inside, and electrically insulated from, a cylindrical outer can, which forms one of the electrodes of the device.

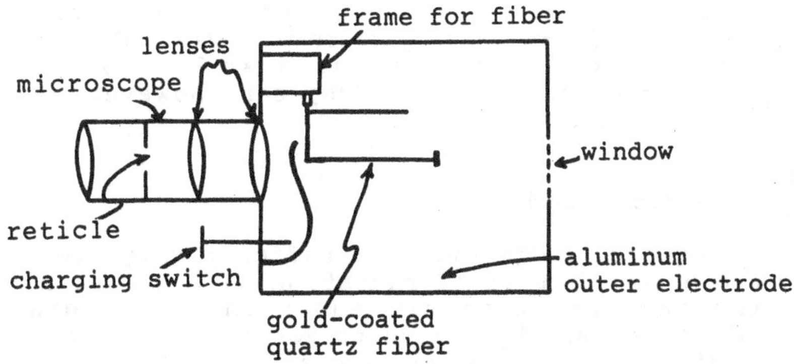


Figure 2-31. Schematic of the Lauritsen electroscope.

The fiber and its mount form the other electrode. To begin electroscope operation, a potential difference of about 100 V is applied across the electrodes. This causes the fiber to deflect away from the frame. In the presence of ionizing radiation, the negative ions and free electrons in the chamber will be attracted to the fiber, neutralizing its charge. The elastic force of the quartz will then cause the fiber to return to its uncharged position. This elastic force is large compared to the gravitational force on the fiber, thus rendering the electroscope quite insensitive to its orientation in the gravitational field. Clearly, this feature is one of the principal advantages of the Lauritsen over the conventional gold-leaf electroscope.

Cloud chambers

When air saturated with water vapor is suddenly cooled, it becomes supersaturated. Should there be any dust particles present, droplets will form immediately form on them. This mechanism is the basis of a detection device known as a cloud chamber, the invention of which is credited to C.T.R. Wilson.[88] Wilson designed a dust-free chamber connected to a piston so that the pressure in the chamber could be lowered quickly (i.e., meaning the temperature could be rapidly lowered). If an ionizing particle moves through such a chamber, the ions themselves serve as condensation nuclei and their paths can be visualized by illumination from the side of the chamber. It is relatively easy to distinguish the tracks of alpha and beta particles in such a device, because alpha particles produce about 100 times as many ions per unit path length in a cloud chamber as beta particles do. Cloud chambers have the disadvantage of producing good results for only a second or so at a time, after which they take 20 to 30 seconds to return to equilibrium. They can, however, detect single particles.

Photographic techniques

Becquerel's discovery of radioactivity resulted from the interaction of radiation with silver halide grains in a photographic emulsion. Photographic emulsions are still used to detect radiation, although the design of these emulsions has improved considerably from Becquerel's plates. Modern emulsions used for this purpose are called nuclear emulsions and contain a high concentration of silver bromide grains which are much smaller (typical diameters are on the order of 0.1 micron) than those used in ordinary photographic films. The emulsion is nearly four times as dense as water and hence can bring charged particles to rest in a few millimeters, compared to the centimeter-plus distances these particles would travel in air or in a cloud chamber. The tracks left by particles moving in nuclear emulsions are very narrow, roughly a micron in width. This allows for very fine spatial resolution of the tracks compared to cloud-chamber techniques.

Semiconductor devices

Semiconductor junction devices are widely used as radiation detectors in modern practice. They behave in

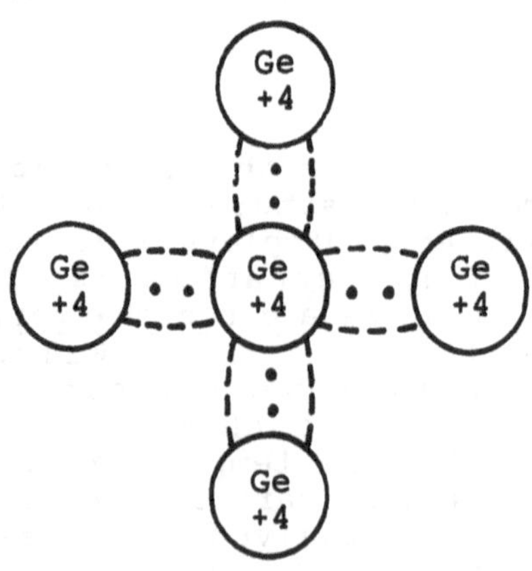


Figure 2-32. Schematic representation of the the lattice structure of pure germanium.

a way in many respects analogous to the proportional gas conduction counter, in that they produce output pulses which are directly proportional to the radiation incident on them. To get some idea of how semiconductor detectors work, it is useful to recall how semiconductor junctions in general work. Pure germanium and silicon crystal lattices have a structure in which each atom is bonded to four neighbors. Each of these elements has four valence (outer shell) electrons. These features are shown schematically in Figure 2-32. The dotted lines around each electron pair in Figure 2-32 represent

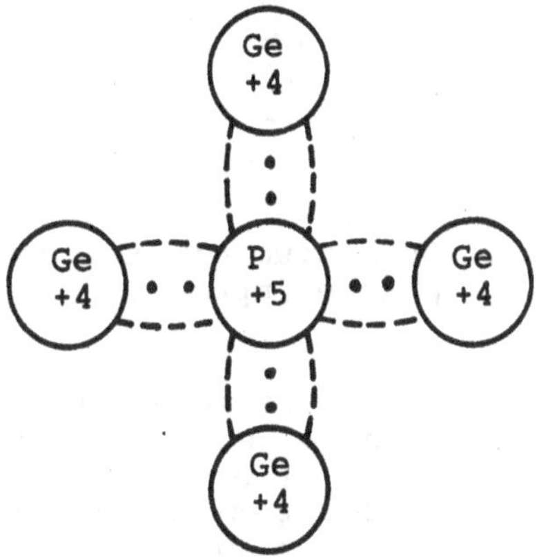


Figure 2-33. Lattice structure of germanium adulterated with phosphorous.

covalent bonds. If an electron is removed from one of these bonds, an <u>electron-hole</u> pair is said to be formed. The holes may be thought of as positive charge carriers. By carefully adulterating germanium and silicon crystals with fixed amounts of pentavalent atoms such as phosphorous or antimony, an excess of electrons is introduced into the resulting lattice (see Figure 2-33). These added impurities are called (electron) <u>donors</u> and a semiconductor adulterated in this way is said to be an <u>n-type</u> (for "excess-negative") semiconductor.

Correspondingly, if trivalent atoms such as aluminum of gallium are introduced into an otherwise pure silicon or germanium crystal, then the resulting material schematically looks like Figure 2-34. Such a material is electron-deficient, and hence is called a <u>p-type</u> (for "excess-positive") semiconductor. P-type materials are electron acceptors.

When a p- and an n-type semiconductor are physically mated, their interface is called a <u>p-n</u>

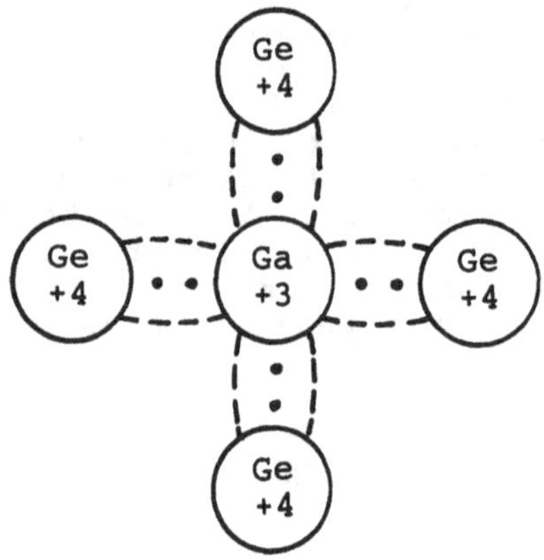


Figure 2-34. Germanium crystal containing a p-type impurity.

junction. If a potential is applied across a p-n (or a n-p) junction by an external source, a depletion region, caused by the migration of any excess charges toward their complementary electrodes, is formed to either side of the junction. Any charge carriers entering the depletion region will be rapidly swept from this region, resulting in a current pulse which is a function of the magnitude of the charge entering the depletion region. The magnitude of the pulse, it turns out, is very nearly proportional to the energy of the primary event. Thus a semiconductor junction can serve as a detector of charged particles.

A typical semiconductor detector is made by diffusing phosphorous to a depth of about one micron into a slab of p-type silicon. When ionizing radiation strikes the detector, it creates ion-pairs. If the design of the detector is sound, most of the energy deposited by the incident particle will be spent in the depletion region. The semiconductor junction detector can yield a more sensitive response to incident radiation than an ionization chamber, because only 3.5 eV are required to create an electron-hole pair, compared to about 26 eV required to produce an ion-pair in a gas like argon.

In this chapter we have surveyed the basic concepts and technology used to characterize natural (and, to a large extent, artificial) radioactivity. In the next chapter, we survey the general considerations involved in assessing the risks associated with exposure to natural radioactivity.

## NOTES

1. F. K. Richtmyer, E. H. Kennard, and T. Lauritsen, *Introduction to Modern Physics* (McGraw-Hill, 1955), p. 38. (Hereafter called "Richtmyer, *Introduction ...*")

2. Richtmyer, *Introduction ...*, p. 42.

3. Richtmyer, *Introduction ...*, p. 80.

4. This section gives only a dim intimation of the drama of late nineteenth- and early twentieth-century research into radioactivity. For a particularly thorough study of these developments, see S. G. Brush, *Statistical Physics and the Atomic Theory of Matter, from Boyle and Newton to Landau and Onsager* (Princeton, 1983).

Throughout the chapter, I have relied rather heavily on the history of this period given in W. B. Mann and S. B. Garfinkel, *Radioactivity and Its Measurement* (Van Nostrand, 1966) and Richtmyer's *Introduction ....* Neither of these works (nor, for that matter, the present volume) was intended as scholarly history, although they try to make the history accessible. Richtmyer's text cites the most important primary sources.

5. Richtmyer, *Introduction ...*, p. 345.

6. J. D. Stranathan, *The "Particles" of Modern Physics* (McGraw-Hill, 1942), p. 315. (Hereafter referred to as "Stranathan, *Particles ...*")

7. W. B. Mann and S. B. Garfinkel, *Radioactivity and Its Measurement* (Van Nostrand, 1966), p. 8. (Hereafter referred to as "Mann and Garfinkel, *Radioactivity ...*")

8. Stranathan, *Particles ...*, p. 315.

9. Mann and Garfinkel, *Radioactivity ...*, p. 11.

10. Mann and Garfinkel, *Radioactivity ...*, p. 12.

11. Mann and Garfinkel, *Radioactivity ...*, p. 13.

12.-14. Richtmyer, *Introduction ...*, p. 83.

15. Any account of a scientific concept which uses the history of that concept as an explicatory vehicle (as this volume does) will eventually have to address the question of *when* that concept was first used in its modern sense. Because the aim here is not to present an exhaustive nor even original account of the history of early twentieth-century physics, the author begs the reader's indulgence in not addressing this question with the subtlety it merits.

16.-19. Mann and Garfinkel, *Radioactivity ...*, p. 13.

20. Beta particles (electrons) in a magnetic field obey the "right-hand-screw" rule: if $\vec{v}$-perp is the component of the particle's velocity vector which is perpendicular to the magnetic induction vector $\vec{B}$, then $\vec{v}$-perp, $\vec{B}$, and the force $\vec{F}$ on the electron form a natural (intrinsic) orthogonal coordinate system. This

coordinate system has the following property: if a right-hand-threaded screw is rotated in the direction v>-perp-toward-B>, F> points in the direction that the screw would advance.

21. Richtmyer, *Introduction ...*, p. 434.

22.-23. Mann and Garfinkel, *Radioactivity ...*, p. 15.

24.-25. Mann and Garfinkel, *Radioactivity ...*, p. 17.

26. Mann and Garfinkel, *Radioactivity ...*, pp. 17-18.

27.-28. Mann and Garfinkel, *Radioactivity ...*, p. 18.

29. F. W. Sears and M. W. Zemansky, *College Physics* (Addison-Wesley, 1960), p. 487. (Hereafter referred to as "Sears and Zemansky, *College Physics*")

30. Sears and Zemansky, *College Physics*, pp. 488-489.

31. Mann and Garfinkel, *Radioactivity ...*, p. 17.

32.-33. Mann and Garfinkel, *Radioactivity ...*, p. 18.

34. Mann and Garfinkel, *Radioactivity ...*, p. 15.

35. Stranathan, *Particles ...*, p. 330.

36.-40. Mann and Garfinkel, *Radioactivity ...*, p. 19.

41. The Introduction promises that the mathematics presumed of the reader for the corpus of this volume will not exceed competence at the high school Algebra II level. To avoid extremely awkward notation, however, it is useful for this and subsequent sections to use the notion of a *derivative*, which is typically taught in most mathematics curricula one course-level past Algebra II. We define that entity as follows. *The limit of a function f(x) as x approaches a is L*, written

$$\lim_{x \to a} f(x) = L$$

if, for all $e > 0$, there exists a real number $d > 0$ such that $/f(x) - L/ < e$ whenever $0 < /x - a/ < d$. (For the purpose of this definition, /z/ denotes the absolute value of z.) We then define the *derivative of a real-valued function f(x) at $x_0$* as follows. The derivative of f with respect to x at $x_0$ denoted

$$df(x_0)/dx$$

is defined as

$$\frac{df(x_0)}{dx} = \lim_{h \to 0} \frac{f(x_0 + h) - f(x_0)}{(x_0 + h) - x_0}$$

provided that this limit exists; the derivative of f at $x_0$ is otherwise undefined. The derivative of f at $x_0$ may be intuitively (but not rigorously) thought of as the slope of a straight line tangent to the graph of f at $x_0$.

42. Mann and Garfinkel, *Radioactivity ...*, p. 20.

43. Mann and Garfinkel, *Radioactivity ...*, p. 23.

44. Mann and Garfinkel, *Radioactivity ...*, p. 24.

45. From here on, the expression

$$\exp(x)$$

will mean

$$e^x$$

where e is the base of the natural logarithms.

46. Mann and Garfinkel, *Radioactivity ...*, p. 29.

47. R. C. Weast, ed., *Handbook of Chemistry and Physics*, 60th edition (CRC Press, 1974), p. F-99.

48. The actual derivations used in Equations 2-6 through 2-12 require calculus; only the results are stated here. A full calculus-based derivation is given in Appendix A.

49. Sears and Zemansky, *College Physics*, pp. 477 - 480.

50. Sears and Zemansky, *College Physics*, p. 957.

51.-53. Sears and Zemansky, *College Physics*, p. 959.

54. Richtmyer, *Introduction ...*, p. 124.

55.-56. Richtmyer, *Introduction ...*, pp. 77-78.

57.-58. Richtmyer, *Introduction ...*, pp. 89-94.

59. N. Bohr, *On the Constitution of Atoms and Molecules: Papers of 1913 Reprinted from the Philosophical Magazine* (Munksgaard, Copenhagen, 1963).

60. N. Bohr, "The Quantum Postulate and the Recent Development of Atomic Theory," *Nature*, Vol. 121(1928), pp. 580-590.

61. W. Heisenberg, *The Physical Principles of the Quantum Theory* (Dover, 1930).

62. W. T. Scott, *E. Schroedinger: An Introduction to his Writings* (University of Massachusetts Press, 1967).

63. B. L. Van Der Waerden, ed., *Sources of Quantum Mechanics* (North-Holland, 1967). Also reprinted in a Dover edition.

64.-65. Mann and Garfinkel, *Radioactivity ...*, p. 42.

66. Mann and Garfinkel, *Radioactivity ...*, pp. 44-49.

67.-68. Mann and Garfinkel, *Radioactivity ...*, p. 50.

69.-71. Mann and Garfinkel, *Radioactivity ...*, pp. 55-56.

72.-74. Mann and Garfinkel, *Radioactivity ...*, p. 61.

75.-77. Mann and Garfinkel, Radioactivity ..., pp. 67-69.

78.-81. Mann and Garfinkel, Radioactivity ..., pp. 71-74.

82. Only an overview of the statistical issues involved in measurement is given here. For a more thorough treatment, the reader is urged to consult any college-level statistics text. See, for example, I. Guttman and S. S. Wilks, Introductory Engineering Statistics (John Wiley, 1965). For a more general, technical treatment, see H. Cramér, Mathematical Methods of Statistics (Princeton, 1946).

83. See, for example, R. C. Weast, ed., Handbook of Chemistry and Physics (CRC Press, 1974), p. A-127.

84. See, for example, H. Cramer, Mathematical Methods of Statistics (Princeton, 1946), p. 559.

85. This section provides only an overview of detector technology. For a more thorough treatment, the reader is urged to consult W. J. Price, Nuclear Radiation Detection (McGraw-Hill, 1958) (hereafter called "Price, Nuclear Radiation Detection"), or N. Tsoulfanidis, Measurement and Detection of Radiation (McGraw-Hill, 1983).

86. Price, Nuclear Radiation Detection, p. 101-102.

87. Price, Nuclear Radiation Detection , pp. 87, 94.

# 3 Mechanisms of Exposure to Ionizing Radiation

## INTRODUCTION

The overall aim of this book is to survey current views about the effects of natural radioactivity in water supplies on human well-being. Such views actually concern two distinct questions:

1. How do we describe the physical and biological effects of the exposure of humans and their environment to natural radioactivity in water supplies?

2. How do we assign significance to the biological and physical effects of natural radioactivity in water supplies on human well-being?

Modern attempts to answer both (1) and (2) are modeled on assessments of other, more well known, agents or conditions. Questions like (1) can in principle be answered by scientific theories. Scientific theories have at least one distinctive feature in this context: they can be cast in a language in which terms like "ought," "should," "right," "wrong," "obligation"--more generally, value terms or their semantic equivalents--do not occur. Questions like (2), in contrast, can be answered only by theories which must be cast in a language which contains value terms or their semantic equivalents. Just to have a brief way of referring to these distinct types of concerns, we will call those corresponding to (1) "scientific" and those corresponding to (2) "value" or "value-related." Used in this sense, both are essential to an adequate understanding of public policy concerning natural radioactivity in water supplies.

Unfortunately, sufficient attention has not been given to the distinction between scientific and value-related concerns in many of the assessment models used to answer (1) and (2). Most current radiation protection standards, in fact, implicitly presume specific answers to (2). To help maintain the distinction, this[1] and the next chapter[2] survey the scientific models which have been used to answer (1); Chapter 5 surveys the general value theories which underlie recent attempts to answer (2).

## MEASURES OF EXPOSURE

A distinctive feature of radioactivity is its power to ionize materials with which it interacts. In most current work, the effect of ionizing radiation on individuals, organs, or tissue masses is taken as fundamental, and more complex effects (say, on populations) are defined in terms of these fundamental ones. The choice of what is fundamental is to some extent a matter of convention, for in principle the effects on populations could be taken as primitive.

The choice of a fundamental measure is not entirely arbitrary, however. Traditional biology and medical practice presume that the individual cell is the fundamental system from which all other biological systems derive their properties. In contrast, some theorists have proposed that the most fundamental and important biological units are individual (or small collections of) genes. On this view, individual cells, organs, or even individuals play only a secondary role in the scheme of things, and there is no guarantee that these more complex structures have any significance. Genes instead, this view holds, rule the living world. "Selfish-gene" theories, which we will generically call such ideas, are too recently proposed for us to know just how useful they will prove in explaining biological systems. But should they persevere, some of the most fundamental concepts of traditional biology will have to be changed, and, along with them, the significance of exposure measures expressed in terms of cell-based concepts.

### Mean absorbed dose

*Mean absorbed dose* is a measure widely used to describe exposure to ionizing radiation. In this volume, it will be defined as the *mean* energy absorbed per unit mass. Unless otherwise noted, furthermore, the terms "dose" and "absorbed dose" will be used interchangeably with "mean absorbed dose." The SI unit of absorbed dose is the *Gray* (Gy), which is defined as 1 Joule per kilogram (J/kg) deposited in the mass of interest. Another unit widely used to express mean absorbed dose is the *rad* (radiation absorbed dose); one rad is defined to be 100 ergs/g deposited in the mass of interest. (Thus, 1 rad = 0.01 Gy.) The *absorbed dose* rate is defined as the time rate at which an absorbed dose is imparted to the mass of interest.

Absorbed dose measures report nothing about the kind of energy imparted to the mass of interest, nor do they say anything about what the mass of interest is. These limitations can obscure significant detail if not respected. It is well known, for example, that different kinds of radiation, even when they have the same energy, produce different biological effects. Some biological damage from radioactivity, for example, comes from ionization of specific cellular structures (primarily genetic material); it is therefore important to be able to distinguish at least the efficiency with which various types of energy sources alter the specific biological materials with which they interact. Not only do alpha, beta, and gamma radiation differ in these efficiencies for the same absorbed dose, furthermore, there is a wide range of efficiencies in a given type of radiation for a given absorbed dose in different tissues. (Details of these differences are reviewed in Chapter 4.)

The limitations of what mean absorbed dose can express can be partially overcome by defining a measure which explicitly takes into account at least some of the ionizing features of radioactivity. In order to do this, we must first appropriate some language. We will henceforth use the word "exposure" to refer to ionization, or measures of ionization, unless context clearly dictates otherwise; more specifically, when used in its technical sense, "exposure" will be defined as "charge released per unit mass of air." "Dose" will be defined as energy (or activity) absorbed per unit mass (as above). The term "biological effect" will be used to denote the biological changes, or measure of those changes, that result from the interaction of ionization radiation with biological tissues.

The roentgen (R) was the first measure of exposure to be defined. One roentgen is that quantity of X or gamma radiation which, on passing through dry air at STP, produces 1 esu of charge of either sign per cubic centimeter of that air. Notice that the roentgen is, strictly speaking, defined only for X or gamma radiation in dry air at STP. Application to any other context, such as to alpha radiation or to interaction with biological tissues, implies an extension of the definition. Care must be taken to ensure that any such extension is meaningful and unambiguous.

The SI unit of exposure is defined as 1 C/kg air (dry, at STP). No name has yet been given to this unit. Notice that the SI unit is a somewhat broader concept than the roentgen, because the SI exposure unit concerns only the number of disintegrations per unit mass air and is also indifferent to the source of the ionizing radiation, whereas the roentgen is concerned both with the magnitude of the <u>charge</u> produced during, and the <u>nature of the source</u> of, those disintegrations. The relation between these two units, <u>for passage of X or gamma radiation through dry air at STP only</u>, is

$$1\ R = 2.58 \times 10^{-4}\ C/kg\ air\ . \qquad (3\text{-}1)$$

Application of this equivalence to any context other than passage of X or gamma radiation through dry air at STP is ambiguous without further qualification.

Similarly, the relation between the roentgen and the rad, <u>for X and gamma radiation interacting with dry air at STP</u>, is

$$\begin{aligned} 1\ R &= (1\ esu)(1.293 \times 10^{-3} g) \times \\ &\quad (2.082 \times 10^{9}\ ion\ pairs\ esu^{-1}) \times \\ &\quad (34\ eV/pair) \times \\ &\quad (1.062 \times 10^{-12}\ ergs/eV) \\ &= 88\ erg/g = 0.88\ rad = 8.8 \times 10^{-3}\ Gy. \end{aligned} \qquad (3\text{-}2)$$

From this it follows that for X or gamma radiation in dry air at STP

$$D_a = 0.88 D_e \qquad (3\text{-}3)$$

where

$D_e$ is the exposure in air, <u>in rads</u>.

The concepts of absorbed dose and exposure (air) clearly do not capture all we want to say about the the interaction of ionizing radiation and tissue. Ideally, one would like to have a measure which expresses the biological effect of ionizing radiation, uniform for all types of energy. Unfortunately, no such measure is currently possible, because of the limitations of our knowledge of the biological effects which arise in the interaction of tissue and radiation. This state of affairs is hardly surprising, since even the simplest cell is an extremely complex, dynamic system of thousands of different biochemicals, each of which is relatively easily altered by radiation. Nevertheless, some headway can be made toward the ideal. We can introduce a measure which empirically relates the changes produced by radiation with known effects to those changes produced by the radiation exposure of interest. One such measure is <u>relative biological effectiveness</u> (RBE), which is defined as the ratio of the dose of 200-300 kV X- or gamma radiation producing a certain biological effect to the dose of radiation of a given type producing the same biological effect

$$RBE_i = \frac{D_x}{D_i} \qquad (3\text{-}4)$$

where

$D_x$ is the dose of 200-300 kV X or gamma radiation producing a certain biological effect in a given biological system

$D_i$ is the dose of radiation of type i producing the same biological effect in that same system.

Several features of Equation 3-4 are worth noting. First, RBE is relativized to the effects of X or gamma radiation of a specific energy. Second, a given type of radiation need not have a single RBE, and typically, it has many: RBE values depend on the energy of the radiation, the nature of the tissue and its systemic environment, the biological effect being studied, the total dose, dose rate, and the spatiotemporal distribution of the dose in the tissue mass. In practice, one RBE is often used for all biological effects of a given kind of radiation, although technically it should be used in this way only with substantial qualification.

In 1963 the International Commission on Radiological Units and Measurements (ICRU) proposed replacing RBE by a nominally more informative measure. The Commission argued that RBE does not allow as useful a basis for comparing biological effects of different types of radiation as we may be able to achieve, even though the ideal of a truly uniform measure may not currently be attainable. In place of RBE, the Commission recommended a new measure, the <u>dose equivalent</u> (DE), and correspondingly, the <u>dose equivalent rate</u> (DE rate), which explicitly identify energy deposited per unit distance in the tissue of interest (<u>linear energy transfer</u> (LET)) as a separate factor in biological effectiveness. The Commission also recommended that other factors, including the distribution of the dose, known to contribute to the biological effects of radiation, be identified in the new measure. The recommendation in its final form defined DE as the product of dose in air ($D_a$), a quality factor (Q) related to LET, a distribution factor (DF), and possibly other factors $F_1, F_2, \ldots, F_n$, if known. That is,

$$DE = Q(DF)(F_1)(F_2)\cdots(F_n)(D_a) \quad (3\text{-}5)$$

The unit of dose equivalent in the extended CGS system is the rem (radiation equivalent man), which is defined as

1 rem = the deposition in soft tissue of 1 rad of radiation with quality factor Q = 1 (e.g., 1 R of 200kV - 300kV x or gamma radiation)

(3-6)

The SI unit dose equivalent, the Sievert (Sv), is defined by

1 Sievert (Sv) = the deposition in soft tissue of 1 Gy of radiation with quality factor Q = 1.

(3-7)

Q values range from 1 for gamma rays, X-rays, beta particles, and electrons, to 20 for alpha particles ($E < 10$ MeV). (See Equation 2-37 for a closely related issue: even in interaction with elemental metals, alpha particles experience a path-length rate of change of kinetic energy four-fold greater than that experienced by beta rays.) All other factors (linear distance in tissue, absorbed dose, distribution factors, and so forth) being equal, therefore, alpha particles produce 20 times the dose equivalent (roughly, 20 times the "damage") of beta rays or X-rays per unit dose in air.

It is very important to recognize that dose equivalent is a conceptual half-way house, with all the confusion that implies. Unlike absorbed dose, dose equivalent is not strictly defined in terms of physical quantities, because it contains the quality factor Q which concerns biological effects. The latter may not be reducible to physical quantities. (In practice, one does not attempt such a reduction, and very likely such a reduction is not currently possible.) Because there is no known single measure which uniformly represents the biological effects of all types ionizing radiation on all tissues, furthermore, dose equivalents can in some cases provide only a very rough guide for comparing the biological effects of exposure. There is no guarantee, in fact, the biological effects of one exposure regime with corresponding dose equivalent $DE_A$ are less than the biological effects of another exposure regime with corresponding dose equivalent $DE_B$, even if

$DE_A < DE_B$; order of biological effects does not necessarily follow the order of dose equivalents, although in most cases of interest, each measure reflects the order of the other. This nuance is of substantial consequence because the casual assumption that DE can be used at least as a comparative measure is often made in ordinary radiometric practice. Strictly speaking, this presumption is an oversimplification, but to be charitable, it is roughly correct within a limited scope of applications; in others, it represent the best we can do, given the present state of our knowledge.

It is likely that the same dose equivalent has different probabilities of mortality when delivered to different tissues. To try to capture this fact, the International Commission on Radiation Protection (ICRP) has recommended the use of a special refinement of the dose equivalent measure, called the <u>effective dose equivalent</u>, which is defined as a probability-weighted sum which reflects (1) the mortality probability of the dose equivalent exposures delivered to each tissue in the body, and (2) the probability of hereditary effects in the first two generations from that dose. That is,

$$DE_{eff} = \sum_T w_T DE_T \tag{3-8}$$

where

$DE_{eff}$ is the effective dose equivalent

$DE_T$ is the dose equivalent in tissue T

$w_T$ is a weighting factor representing the proportion of the probability of stochastic effects resulting from irradiation of organ or tissue T to the probability when the whole body is irradiated uniformly.

The weighting factors $w_T$ are defined by international agreement, and are subject to change.

## <u>Dosimetry</u>

In actual practice, we typically know something about the field created by a radioactive source, and are faced with the task of computing the dose created by that field. This determination is the essence of radiation dosimetry. More precisely, the general problem of dosimetry is to calculate the dose rate in a region, given the power density (energy per unit time per unit area) of the radiation in that region. In typical practice, dosimetry calculations as such are rarely done; rather, the investigator determines the dose rate at a certain location based on conversion tables or graphs that relate power density to dose rate.

Suppose in particular that N(E) is the number of particles per unit area per unit time in a small region of a radiation field and that E is the energy of each of these particles; then N(E)dE is the total energy per unit area per unit time for particles in the region of interest with energy between E and E + dE. Let C(E) be the factor which converts power density to dose (equivalent) rate; in general, C will be a non-constant function of E. A tissue mass in the radiation field determined by N(E) will then receive a dose equivalent rate given by

$$\text{DE rate} = \int_0^{\infty} C(E)N(E)dE \qquad (3\text{-}9)$$

Typically, both the radiation field described by N(E) and the conversion factors C(E) are not analytic functions but are known for a discrete-energy-group distribution. That is, one has available the quantities

$$N_g = \int_{E_{g-1}}^{E_g} N(E)dE \qquad (3\text{-}10)$$

where

$N_g$ is the number particles per unit area per unit time with energy between $E_{g-1}$ and $E_g$. (The energy interval $E_g - E_{g-1}$ is called an energy group.)

and

$$C_g = \frac{1}{E_g - E_{g-1}} \sum_{E_{g-1}}^{E_g} C(E)dE \qquad (3\text{-}11)$$

If $N_g$ is in particles per square meter per second, Equation 3-7 then assumes the form

$$\text{DE rate (in Sv/s)} = \sum_{g=1}^{G} C_g N_g \qquad (3\text{-}12)$$

If the radiation is strictly a natural alpha source external to the body, the dose to an exposed individual will be effectively zero, because natural alpha rays cannot penetrate the skin-forming cells. If the external source emits electrons, beta, gamma, or X-rays, a part of the tissue deeper than the skin may be exposed, depending on the energy of the radiation.

If the source of charged particles is inside the body, in contrast, the dose equivalent rate can be obtained from the simpler relation

$$\text{DE rate} = \frac{\text{energy released/s}}{\text{mass}} Q \qquad (3\text{-}13)$$

because the range of charged particles in tissue is deposited within a very small volume (on the order of, or less than, cellular dimensions). If the source strength is

$$S(E)dE$$

where

$S(E)$ is the number of particles emitted per second at energy E

$S(E)dE$ is the number of particles emitted per second with kinetic energy between E and E + dE

and

$$S_g = \int_{E_{g-1}}^{E_g} S(E)dE$$

where

$S_g$ is the number of particles emitted per second with energy between $E_{g-1}$ and $E_g$

then the equation for dose rate becomes

$$\text{DE rate} = \frac{\left(\int_0^{\infty} dES(E)\right) EQ(E)}{M(t)} =$$

$$= \frac{\left(\sum_{g=1}^{G} S_g E_g Q_g\right)}{M(t)}$$

(3-14)

The mass in Equation 3-14 depends on the distribution of the source. If the source has a uniform spatial and temporal distribution, then the mass of that organ should be used. If the distribution is not uniform, one must explicitly modify Equation 3-14 to reflect the distribution of the dose. Note also that the mass of the organ itself may be a function of time. The mass of bone affected by internal alpha emitters, for example, changes rapidly in growing children. For monoenergetic sources, Equation 3-14 consists of one term, with the energy $E_g$ being the energy of the source particles. In the case of a beta source, the energy $E_g$ can to good approximation be taken as the average energy of the beta spectrum, which to a first approximation is equal to $E_{max}/3$.

Biological half-life. A radioisotope which has entered the body is biologically eliminated at a rate that depends only on the chemical nature of the isotope, since all isotopes of an element are chemically identical and the body's elimination mechanisms discriminate on chemical properties only. For most radioisotopes, the rate of elimination is proportional to the amount of the isotope in the body. This is, of course, just an exponential law of the very sort discussed in Chapter 2. Specifically, that law takes the form

$$dN/dt = -k_B N \tag{3-15}$$

where $k_B$ is the biological decay constant, and N is the quantity of the material in the body at time t, on strict analogy with Equation 2-7. The <u>biological half-life</u> $T_B$ is, therefore, by analogy to Equation 2-8

$$T_B = \frac{\ln 2}{k_B} \tag{3-16}$$

If we let

N(t) be the number of radioactive atoms attime t

and

$k_R$ be the radioactive decay constant

then the rate of change dN/dt due to both radioactive decay and biological elimination can readily be computed to be

$$dN/dt = -(k_R + k_B)N \tag{3-17}$$

On analogy with Equation 2-7, Equation 3-17 has the solution

$$N(t) = N(0)\exp(-(k_R + k_B)t = N(0)\exp(-k_e)t \tag{3-18}$$

where

N(t) is the number of atoms which have not decayed at at time t

$k_e$ is $k_R + k_B$.

The <u>effective half-life</u> $T_e$ can then be computed to be

$$T_e = \frac{T_B T_R}{T_B + T_R} = \frac{\ln 2}{k_e} \qquad (3\text{-}19)$$

The rate of elimination of an element from the body is not necessarily the same for the whole body as for a particular organ. For example, the biological half-life of iodine is 138 days for rejection from the thyroid, 7 days for the kidneys, 14 days for the bones, and 138 days for the whole body. For this reason, biological half-lives should be referenced to specific organ or tissue masses.

If the biological half-life is large compared to the radioactive half-life, that is, if $T_B >> T_R$, radioactive decay removes the material much faster than the body eliminates it. For example, for $^{131}I$ $T_B$ = 138 days (thyroid), $T_R$ = 8 days, and $T_e$ = 7.6 days. If the biological half-life is small compared to the radioactive half-life, that is, if $T_B << T_R$, then biological elimination is mainly responsible for the removal of the isotope. For tritium, for example, $T_B$ = 12 days, $T_R$ = 12 years, and $T_e$ = 12 days.

Because both radioactive decay and the body's elimination mechanisms combine to remove an isotope from the body, the dose rate to the body or organ is not constant after a single ingestion. Suppose, in particular, that the ingestion of an amount of a radioactive isotope contributes a dose DE(0) at the time of ingestion. If the effective half-life of the isotope is $T_e$, the total dose, $DE_T$, delivered to the body, or organ over time T is[3]

$$DE_T = \int_0^T DE(0)\exp(-k_e t)dt =$$

$$= \frac{DE(0)}{k_e} (1 - \exp(-k_e T)$$

$$= \frac{DE(0)}{\ln 2} T_e \left[ 1 - \exp\left( -\ln 2 \frac{T}{T_e} \right) \right] \tag{3-20}$$

If $T >> T_e$, then

$$DE_{total} = \frac{DE(0)}{\ln 2} T_e \tag{3-21}$$

Equations 3-8 through 3-21 describe the dose equivalents on individuals or organs. Frequently, however, one does not have enough of the sort of information required by these equations. There are a number of reasons for this. First, it is reasonably likely that in a heterogeneous population of all ages and states of health, there is a nonuniform distribution of factors which may conspire with radiation insults to enhance or diminish the biological effects of exposure for certain individuals; the effects of these synergies or antagonisms are typically unknown. Second, even apart from such phenomena, there may be a nonuniform distribution of sensitivities to ionizing radiation, and we typically do not know this distribution. Third, the very distribution of the dose, even to as coarse a level as whole-body measure, is often not known. These limitations need to be given high visibility in any technique used to estimate exposure and risk.

*Collective dose rate, collective dose, and dose commitment.* Radiation sources typically give rise to a nonuniform distribution of per capita dose rates in an exposed population. There is no single quantity (parameter) which can represent a nonuniform distribution. One could argue, however, that the effective sum of the dose rates to all affected individuals provides at least a crude measure of the exposure of the population as a whole. Formally, this measure is called the *collective dose rate in a population from a source of radiation at time t*, S', defined by

$$S' = \int D'N(D')dD' \tag{3-22}$$

where

N(D') is the number of individuals in the populationwho are individually exposed at dose rate D' (i.e., N = N(D') is the population dose-rate spectrum curve at time t)

N(D')dD' is the number of individuals who are exposed at D' to D' + dD'.

(It should be noted that S' collapses the very information required to compute it: in practice we would need to know the dose-rate spectrum to find S'.)

Intuitively speaking, the collective dose S in a population over a specified time T would be just the sum of all the particular collective doses during the time interval T. For a continuous function such as collective dose rate, this sum is given by an integral. Since dose is just dose rate times time, the collective dose can thus be defined as the time integral of S', that is

$$S = \int_0^T S'dt \tag{3-23}$$

Sometimes we know virtually nothing about the distribution of the total dose, but nevertheless want to say something meaningful, if only statistical, about the dose to individuals. All other ignorance being the same, the average dose $\bar{D}$ to an individual, called the per caput dose, answers this need. The average dose is just the total dose divided by the number of individuals N in the exposed population. Formally

$$\bar{D} = \frac{\int DN(D)dD}{N} \tag{3-24}$$

Particular care must be taken in using the per caput dose measure: $\bar{D}$, is, after all, just an average. Equation 3-24 accordingly provides reasonably good information about the dose distribution when the distribution is more or less uniform. If the dose distribution in a population is highly nonuniform, however, Equation 3-24 provides relatively little information about the actual dose to any individual.

In some cases, enough is known about the distribution of the dose rate that we can subdivide the population into several groups, each member of which can be presumed to first approximation to be exposed at the same known rate. Let us denote these groups and their dosimetric attributes by the subscripts j = 1, 2, 3, ... The collective dose rate may then be expressed as

$$S' = \sum_j \int D' N_j(D') dD' = \sum_j S'_j \qquad (3\text{-}25)$$

where

$N_j(D')$ is the number of individuals in group j who are exposed at rate D'.

If there are multiple sources of exposure, denoted by the subscripts k = 1, 2, 3, ..., then the collective dose rate from source k can be expressed as

$$S'_k = \sum_j \bar{D}^T_{kj} N_j \qquad (3\text{-}26)$$

where

$\bar{D}^T_{kj}$ is the <u>per caput</u> dose rate contributed by source k in group j, and $N_j$ is the number of individuals in group j. All these quantities are functions of time.

The <u>collective dose commitment</u> $S^C_k$ due to a given source is defined as

$$S^C_k = \int_0^T S'_k(t)\,dt \qquad (3\text{-}27)$$

Notice that Equation 3-27 is just a trivial variant of Equation 3-23.

It is often useful to be able to succinctly express what the total dose to an individual from a given event or incident would be, <u>assuming</u> that that individual were exposed at the average per caput dose rate $\bar{D}^T$. Intuitively, this would be just the average (per caput) dose rate times the time to which the individual was exposed at that rate. This concept is called the <u>dose commitment</u> $D^C_k$ and is formally defined to be the infinite time-integral of the *per caput* dose rate due to a given source k:

$$D_k^C = \int_0^\infty \bar{D}_k^T(t)\,dt \qquad (3\text{-}28)$$

where

$\bar{D}_k^T$ is the quotient of the collective dose rate and the population size at t, that is, $\bar{D}_k^T = S_k'(t)/N(t)$.

*Dose commitment in continued practices.* The collective dose commitment and the dose commitment (Equations 3-27 and 3-28, respectively) are frequently proportional to the size (activity) of the source. For example, if the event under consideration is the release of a quantity of a radionuclide into the environment, the dose commitment and the collective dose commitment are proportional to the activity of the release, all other factors being the same. Under such conditions, we can considerably simplify the above equations. In particular, if the dose commitment and the collective dose commitment are proportional to the activity of the source, then the *per caput* dose rate at given times after the originating event is also proportional to the size of that event. Given this, it is meaningful to define a *per caput dose rate per unit of activity of the originating event*, $\bar{D}_1^T$ which is a function of time. The time integral of this value is then the *dose commitment per unit activity in the originating event*, $D_1^C$

$$D_1^C = \int_0^\infty \bar{D}_1^T(t)\,dt \qquad (3\text{-}29)$$

Similarly, we may define the *collective dose commitment per unit originating event*, $S_1^C$, as

$$S_1^C = \int_0^\infty \bar{D}_1^T(t)N(t)\,dt \qquad (3\text{-}30)$$

where N(t) is the population size. If the population size is constant over the period of interest, then Equation 3-30 reduces to $S_1^C = ND_1^C$. Over longer periods, the calculation of collective dose commitments requires the use of a population growth model. For short periods, we may assume this model to approximately exponential. For extremely long periods ($\geq$ 10 generations), we can assume sigmoidal growth approaching a world population of nominally $10^{10}$ individuals.

A continued exposure at a given dose rate can be treated as a sequence of separate exposures of the sort discussed above. If this repeated exposure results from a repeated practice at a known an constant rate R, the dose commitment per unit activity of the originating event can be used to gain some information on the per caput dose rate. In the case of short-lived nuclides, in particular, the per caput dose rate will eventually reach a steady-state value $\bar{D}^T_{eq}$, which can be determined by

$$\bar{D}^T_{eq} = \int R\bar{D}^T{}_1(T - t)dt$$

$$= R \int \bar{D}^T_1 dt = R\bar{D}^C_1 \quad (3\text{-}31)$$

If the population is constant, $S^C_1 = ND^C_1$, and thus $\bar{D}^T_{eq} = (R/N)S^C_1$. Often, furthermore, it is possible to make a rough estimate of the <u>practice rate per caput</u>, R/N. It is then possible to estimate the maximum per caput dose rate that will be experienced due to the continuing exposure.

It is not possible to model a dose delivered over a very long time after single originating event--such as that arising from a single ingestion of $^{238}U$--as a continued practice. We can determine the maximum per caput dose rate, $\bar{D}^T_{max}$, caused in the future by such a practice, however, by other means. If the practice continues to a time x at rate R and then is discontinued, then the <u>per caput</u> dose at any time T > x is given by

$$\bar{D}^T(T) = \int RD^{\prime}_1(T - t)dt = R \int D^{\prime}_1(t)dt \quad (3\text{-}32)$$

It can easily be shown that if the function $\bar{D}^T_1$ has a single maximum at time m and then declines monotonically, the maximum <u>per caput</u> dose rate occurs at a time $T_m$ such that $x < T_m < x + m$. Since in most practical cases $x \gg m$ (implying that $T_m$ is relatively the size of x) the maximum <u>per caput</u> dose rate is approximately given by

$$\bar{D}^T_{max} \doteq R \int \bar{D}^T_1(t)dt \quad (3\text{-}33)$$

*Genetically significant dose*. The concept of *genetically significant dose* is often used to describe the long-term genetic impact of ionizing radiation. Intuitively, it is that child-producing-weighted dose, which, if administered to the gonads of *each* member of a population, would give the same child-producing-weighted dose as the dose to the gonads of those individuals actually exposed. Formally, the genetically significant dose (GSD) is defined by

$$GSD = \frac{\sum D_i N_i P_i}{\sum N_i P_i} \tag{3-34}$$

where

$D_i$ is the average gonad dose to persons age i who are exposed

$N_i$ is the number of persons in the population of age i who are exposed

$P_i$ is the expected number of children for persons of age i

$N_i$ is the number of persons in the population of age i

## THE NATURAL IONIZING RADIATION BACKGROUND

Most current radiation protection standards or proposals explicitly take existing sources of ionizing radiation into consideration. This only makes sense, because an exposure ceiling more restrictive than, say, the exposure produced by the natural background, would be impractical to enforce. Regulatory agencies, furthermore, must often attempt to determine whether a proposed or actual practice will "significantly" increase the radioactive burden already present in the environment; making such a determination typically requires knowing something about the magnitude of both natural background radiation field and the consequences of human redistribution of, and contributions to, that field. This section surveys the contributions known to be made by each of those phenomena.

### *External irradiation from the natural background*

*External irradiation* is defined as exposure of man

to radioactive sources which lie outside of the human body. The natural background of ionizing radiation contributes significantly to the total external irradiation dose experienced by man. External irradiation comes comes from the earth itself and from high-energy cosmic radiation entering the earth's atmosphere from space.

External irradiation from cosmogenic sources. The high-energy radiation which enters the earth's atmosphere from space is called primary cosmic radiation. This radiation consists largely of high-energy protons from interstellar regions, together with $^{4}He$ ions (which constitute about 10% of the total primary particle flux) and very small proportions of heavier particles, electrons, photons, and neutrinos. The primary cosmic radiation flux density is strongly affected by the earth's magnetic field, which deflects low-energy charged particles back into space; this deflection is dependent on geomagnetic latitude, being strongest at the poles and weakest in equatorial regions.

The origin of primary cosmic rays is still not completely known. Calculations show that most of the primary cosmic radiation in the upper atmosphere could have originated in our own galaxy. Those primary cosmic rays which do arise within our galaxy are called galactic cosmic rays; the sun also produces primary cosmic rays, called solar cosmic rays, which are associated with solar flares.

When primary cosmic rays interact with molecules of the atmosphere, secondary particles and electromagnetic radiation are produced: these secondary products are called secondary cosmic rays. Secondary cosmic rays include neutrons, protons, pions, kaons, and a variety of nuclides, including $^{3}H$, $^{7}Be$, $^{10}Be$, $^{22}Na$, and $^{24}Na$. The high-energy protons, neutrons, and pions formed in the primary reactions react further with nuclei in the air to form more secondary particles; this cascade continues until sufficient energy no longer remains to propagate the chain, or until the particles intercept the earth. The pions formed in these reactions subsequently decay into muons or photons, initiating further cascades.

Primary cosmic ray protons and neutrons contribute significantly to the absorbed dose index rate (defined as the maximum absorbed dose rate that would occur in a 30-cm diameter tissue-equivalent sphere located with its center at the point of interest) in the upper atmosphere. The neutrons, furthermore, lose energy by inelastic collisions, and when thermalized, are captured by $^{14}N$ to form $^{14}C$. The primary nucleonic fluence rate is substantially attenuated by nucleonic energy losses through ionization and collisions with air molecules. Thus, primary nucleons directly account for at most a few percent of the absorbed dose index rate from cosmic rays at sea level. The major contribution to the cosmogenic fluence rate at sea level comes instead from muons produced by the decay of charged pions at higher altitude, and by the electrons which result from secondary ionization, from muon decay, and from other secondary processes.

The ion production rate per unit volume in free air is an indirect measure of the fluence rate of the total charged-particle component of the cosmic-ray field. That rate is about 2.1 $cm^{-3}$ $s^{-1}$ at ground level. Assuming that an average of 33.7 eV must be imparted to a neutral air molecule to produce an ion pair, the resulting absorbed dose rate in air is $3.2 \times 10^{-8}$ Gy $h^{-1}$. It is reasonable to assume that this value is numerically equal to the absorbed dose index rate. About 75% of the dose is from muon collision electrons, 15% from muon decay electrons, and 10% from other sources. The (unshielded outdoor) annual absorbed dose index at sea level which results from the charged-particle component of cosmogenic radiation is then about $2.8 \times 10^{-4}$ Gy.

The neutron fluence rate at sea level is about $8 \times 10^{-3}$ $cm^{-2}$ $s^{-1}$. If we assume that the conversion factor from neutron fluence rate to absorbed dose is $5 \times 10^{-8}$ Gy $h^{-1}$ $cm^2$ s, the absorbed dose index rate at sea level from the neutron component arising from cosmic radiation is about $4 \times 10^{-10}$ Gy $h^{-1}$, yielding an annual absorbed dose index of $3.5 \times 10^{-6}$ Gy at sea level. The neutron absorbed dose index rate increases rapidly with altitude, reaching a maximum at 10 to 20 km above sea level.

For both the charged-particle and the neutron components of cosmogenic radiation, it turns out that the absorbed dose index reasonably approximates the absorbed dose in all organs and tissues of the body.

The quality factor (Q in Equation 3-5) of the charged-particle component of cosmogenic radiation is taken to be one, giving an annual effective dose equivalent for that component of 280 microSievert at ground level. The quality factor for the neutron component is taken to be 6, giving an annual effective dose equivalent for this component of about 21 microSievert at sea level. Roughly speaking, then, the cosmogenic neutron flux contributes less than 10% of the total external cosmogenic exposure at sea level.

External irradiation from terrestrial sources. The main primordial (of terrestrial origin) contributors to external exposure are the decay series and subseries headed by $^{40}K$, $^{87}Rb$, $^{238}U$, and $^{232}Th$. Each of these are long-lived nuclides which have existed in the earth's crust throughout most of its history. The concentration of the primordial radionuclides in soil is determined primarily by the activity of the source from which the soil is made. In igneous rocks, the concentration of radionuclides is related to the quantity of silicates, being highest in acidic rocks and lowest in the ultrabasic rocks. Igneous rocks as a whole generally exhibit higher radioactivity than sedimentary rocks, while metamorphic rocks have concentrations about the same as those of the rocks from which they derive; notable exceptions to this are certain shales and phosphate rocks (particularly rock phosphate fertilizer sources) which are sometimes highly radioactive.

It is estimated that the average external outdoor absorbed dose rate in air from gamma radiation arising from natural terrestrial sources is about $4.4 \times 10^{-8}$ Gy $h^{-1}$, although the dose rate can be as high as 10 times this value in certain regions.

The annual gonad absorbed dose D due to external irradiation from primordial sources can be as the time-weighted sum of indoor and outdoor exposures, that is,

$$D = (cqD_a')_{outdoor} + (cqD_a')_{indoor} \qquad (3\text{-}35)$$

where

c is the conversion factor from absorbed dose rate in air to gonad annual absorbed dose

q is the occupancy factor, i.e., proportion of time spent indoors and outdoors

$D_a'$ is the dose rate in air

UNSCEAR has estimated that the appropriate coefficients for 3-35 are

$$c_{outdoor} = 7.2 \text{ microrad}^{-1} \text{ h}$$

$$c_{indoor} = 6.0 \text{ microrad}^{-1} \text{ h}$$

$$q_{outdoor} = 0.2$$

$$q_{indoor} = 0.8.$$

If we assume $D_a'$ = 4.5 microrad $h^{-1}$ outdoors and 5.3 microrad $h^{-1}$ indoors, the annual external gonad absorbed dose rate from terrestrial radiation is about 30 mrad. People living in areas of high external terrestrial radiation, for example in certain parts of Brazil and India, can experience much greater doses: rates as high as 400 - 500 mrad/yr exist in these areas.

## Internal irradiation from the natural background

Internal cosmogenic irradiation. Internal exposure is defined as exposure to ionizing radiation whose sources lie within the body. These sources typically enter the body through inhalation or ingestion, although they may be injected directly into the blood in certain radiological practices.

Very little of the internal irradiation dose comes from the cosmogenic background. Of the the many nuclides produced by cosmic rays, only $^{3}H$, $^{7}Be$, $^{14}C$, and $^{22}Na$ contribute significantly to the internal primordial dose.

The principal source of cosmogenic tritium is the atmosphere, where it is formed from the interaction of cosmic-ray neutrons with nitrogen and oxygen. Before atmospheric nuclear testing began, tritium activity concentrations in surface waters were 100 - 900 Bq $m^{-3}$; this value thus represents a nominal estimate of the natural background tritium activity concentration. Tritium enters food crops in the form of HTO (hydrogen-tritium-oxygen, the analog of ordinary water) and is there partly incorporated into organic matter; from there, it enters man through ingestion. Best estimates of the annual internal dose arising from cosmogenic tritium are on the order of $10^{-8}$ Gy in all tissues.

Concentrations of cosmogenic $^{7}Be$ in the temperate zones are about 3000 microBequerels (microBq) $m^{-3}$ in air and 700 Bq $m^{-3}$ in rainwater. The principal pathway to man is through leafy vegetables; annual per capita intake probably averages about 50 Bq. The resulting soft tissue dose per year is 12 microGy in the walls of the lower large intestine and somewhat lower in the other tissues.

Cosmogenic carbon-14 is produced in the upper atmosphere by the reaction $^{14}N(n,p)^{14}C$, which is induced by slow cosmic-ray neutrons. $^{14}C$ is incorporated into the food chain by green plants. The specific activity in biological carbon in trees is on the order of 225 Bq $kg^{-1}$, and the consensus view holds that this activity is approximately the same in all cosmogenic biological carbon, resulting in an annual internal dose of 5 to 25 microGy from this source. From this, one can estimate that the annual average effective dose equivalent from internal irradiation by cosmogenic $^{14}C$ is about 12 microSv, assuming an "average" diet.

The annual absorbed dose from cosmogenic $^{22}Na$ has been calculated assuming an annual ingestion of 50 Bq. This implies an annual effective dose equivalent of about 0.2 microSv. Although the the production rate and the atmospheric concentration of cosmogenic $^{22}Na$ is very small compared to cosmogenic $^{3}H$, the estimated annual absorbed dose arising from the ingestion of $^{22}Na$ is larger than that of $^{3}H$ because of differences in the metabolism of the two nuclides and because of the decay properties of $^{22}Na$.

Internal irradiation from naturally occurring terrestrial radionuclides. Primordial nuclides are the principal contributors to internal irradiation arising from the natural background. The most significant of these are the members of the $^{235}U$, $^{238}U$, and $^{232}Th$ series, together with $^{40}K$ and $^{87}Rb$.

In the average adult male, the concentration of potassium is about 2 g $kg^{-1}$ body weight. This corresponds to a concentration of $^{40}K$ in the body of about 60 Bq $kg^{-1}$. Red bone marrow receives the highest annual absorbed dose from this source at 270 microGy; the annual effective dose equivalent is estimated to be about 180 microSv.

Rubidium-87 contributes a minor amount to the internal exposure. Assuming that the concentrations of rubidium are the same throughout the body, the annual absorbed dose in all tissues is from 0.4 to 0.9 mrad from this source. The annual effective dose equivalent from this source is estimated to be about 6 microSv.

Uranium-238 and thorium-232 each head a series of major contributors to internal irradiation from primordial sources. For the sake of describing the dose to man from these sources, it is convenient to identify certain subseries within each of these terrestrial series. These subseries are natural in the sense that, by themselves, they contribute significant parts of the total exposure in the whole series, and are terminated by relatively more stable nuclides. In the $^{238}U$ series, these subseries include (a) $^{238}U$, two short-lived nuclides and $^{234}U$, (b) $^{230}Th$, (c) $^{226}Ra$, (d) $^{222}Rn$ and its short-lived decay products through $^{214}Po$, a subseries which is important both for external radiation because of the energetic gamma rays emitted through the decay of $^{214}Bi$, and for internal radiation, and (e) the long-lived $^{222}Rn$ decay products $^{210}Pb$, $^{210}Bi$, and $^{210}Po$. The $^{232}Th$ subseries of interest are (a) $^{232}Th$ itself, (b) the subseries $^{228}Ra$, $^{228}Ac$, $^{228}Th$, and $^{224}Ra$, (c) $^{220}Rn$ and its decay products.

The Uranium-238 subseries ($^{238}U$, $^{234}Th$, $^{234m}Pa$, $^{234}U$). It is reasonable to assume that most human exposure arising from this subseries from natural sources comes from $^{238}U$ in radioactive equilibrium with with $^{234}Th$, $^{234}Pa$, and $^{234}U$; thus 1 kg of uranium contains 12 MBq of each of those four nuclides. The contribution of $^{235}U$ and its decay products to the total dose from the natural background is several orders of magnitude lower than that from the natural background, and hence will not be examined here.

The main natural source of uranium in the atmosphere appears to be the resuspension of dust particles from the earth. If we assume that the atmospheric dust concentration is about 50 micrograms $m^{-3}$ and that the average concentration of $^{238}U$ in soil is 25 Bq $m^{-3}$, the activity concentration in ground level air from this source is about 1.2 microBq $m^{-3}$. Actual measurements tend to confirm this estimate. The corresponding intake through inhalation is approximately 0.01 Bq.

The annual dietary intake of $^{238}U$ has been found to be about 5 Bq in areas of "normal" activity, and the contribution to that intake from drinking water is generally negligible in comparison. There are some rather dramatic exceptions to this generalization, however; in certain parts of France, Finland, the USSR, and Colorado in the United States, activity concentrations as high as 1 - 2.6 kBq $m^{-3}$ have been found. In Helsinki, for example, activity concentrations on the order of 100 kBq $m^{-3}$ $^{238}U$ alpha have been found in several wells. These very high concentrations are probably caused by small, localized, uranium-rich deposits.

Average activity concentrations of $^{238}U$ in the bone of adults are on the order of 100 to 200 mBq $kg^{-1}$. In soft tissues, the concentrations cover a much greater range, from about 1 - 10 mBq $kg^{-1}$ in most of the soft tissues to 10 - 80 mBq $kg^{-1}$ in the kidneys. The resulting annual absorbed does range from 2 x $10^{-8}$ Gy in the lungs to 3 x $10^{-6}$ Gy in the bone lining cells.

Thorium-230. The annual activity intake of $^{230}Th$ through inhalation is about 0.01 Bq. There is no systematic information on the dietary intake of $^{230}Th$, however. Thorium, like uranium, is a bone-seeker, and is assumed to reside for relatively long periods on bone surfaces. Current research suggests that the typical activity of $^{230}Th$ in bone is about 140 mBq, resulting in activity concentrations in bone of approximately 20 mBq $kg^{-1}$ in cortical bone and 70 mBq $kg^{-1}$ in trabecular bone. The concentrations in the soft tissues range from 300 mBq $kg^{-1}$ for the lymph nodes to 0.3 mBq $kg^{-1}$ in muscle. The annual effective dose equivalent from internal irradiation by $^{230}Th$ is estimated to be about 7 microSv.

Radium-226. The principal natural source of radium in the air at ground level is, like $^{238}U$, probably resuspension of soil particles. If we assume that the concentration of dust in air is 50 micrograms $m^{-3}$, the calculated activity intake of radium inhaled is about 0.01 Bq per annum per capita.

Food is a much more important source of radium for intake and blood uptake than is inhalation. The average annual dietary intake of $^{226}Ra$ in areas of normal radiation background is about 15 Bq. Drinking water typically contributes little to the dose to man if the drinking water source is surface water. Activity concentrations of 0.1 Bq $l^{-1}$ are not infrequent in well water, however. The annual effective dose equivalent resulting from $^{226}Ra$ intake in "normal" areas is about 7 microSv.

Radon-222 and its short-lived decay products ($^{218}Po$, $^{214}Pb$, $^{214}Bi$, and $^{214}Po$). Humans are exposed to $^{222}Rn$ (radon) mainly through inhalation. Mean annual values of radon concentration in outdoor air vary between 0.1 and 10 Bq $m^{-3}$. (The equilibrium equivalent concentration of a species is that concentration of the species, which, if in equilibrium, would produce the same activity concentration as that observed.) In the temperate latitudes, the average equilibrium equivalent concentration of radon in air is several times higher indoors than outdoors; the mean values of the indoor equilibrium equivalent radon concentrations range from 5 to 60 Bq $m^{-3}$. 15 Bq $m^{-3}$ is regarded as a reasonable approximation of the mean indoor equilibrium equivalent concentration for the world population. These concentrations result in an annual effective dose equivalent of 0.92 mSv from indoor exposure and about 0.06 mSv from outdoor exposure, resulting in a weighted average of about 1 mSv for inhaled $^{222}Rn$ daughters.

Long-lived decay products of radon-222 ($^{210}Pb$, $^{210}Bi$, and $^{210}Po$). The main source of inhaled $^{210}Pb$ and $^{210}Po$ is $^{222}Rn$ emanation from the ground. In the mid latitudes of the northern hemisphere, the average concentration of $^{210}Pb$ in surface air has been estimated to be 0.5 mBq $m^{-3}$. The activity concentration of $^{210}Po$ in air is about 0.1 mBq $m^{-3}$. If we make the reasonable assumption that the concentration of these two products are the same indoors as out, the annual per capita intake is about 4 Bq of $^{210}Pb$ and 0.8 Bq of $^{210}Po$.

Most of the $^{210}Pb$ and $^{210}Po$ ingested by humans comes through food. Annual per capita uptake through this route is estimated to be about 40 Bq. Values lower than this are consistently obtained for the United States, however. Populations which have a high proportion of seafood in their diet are likely to have higher intakes, because the edible portions of marine organisms can contain on the order of 10 Bq $kg^{-1}$.

Lead is a bone-seeker in man, and it is found incorporated into bone tissue. The principal mechanism for biological elimination of this element appears to be long-term skeletal remodeling. In continental areas of the northern latitudes, a typical activity concentration of $^{210}Pb$ in bone is about 3 Bq $kg^{-1}$; typical total skeletal activity at those latitudes is about 15 Bq. The total per capita $^{210}Pb$ activity in soft tissues, more or less distributed uniformly, is about 6.4 Bq per capita.

Polonium, in contrast to all other natural alpha emitters, is not a bone seeker but is distributed in the soft tissues after intake. Therefore the greatest part of the $^{210}Po$ bone activity arises from the decay of deposited $^{210}Pb$. This assumption leads to a concentration of $^{210}Po$ in bone of about 2.4 Bq $kg^{-1}$ for populations in temperature regions of the northern hemisphere. In soft tissues, the activity of $^{210}Po$ is about the same as that of $^{210}Pb$, averaging about 0.2 Bq $kg^{-1}$.

The absorbed doses from the $^{210}Pb$ subseries depend primarily on the highly energetic alpha particles of $^{210}Po$; the contribution from beta emissions amounts to about 10% of the total.

$^{210}Bi$ has a relatively short half-life ($t_{1/2}$ = 5 days). Given this, and the fact that $^{210}Bi$ is the immediate successor of $^{210}Pb$ in the series, the activity intakes of the two nuclides may be taken, to first approximation, to be the same. $^{210}Bi$ may be assumed to be in radioactive equilibrium with $^{210}Pb$ in body tissues, so that the absorbed doses from $^{210}Bi$ mainly arise from the intake of $^{210}Pb$ and not from the intake of $^{210}Bi$ itself.

Thorium-232. The average activity concentration of $^{232}Th$ in soil is about the same as that of $^{238}U$ and its decay product $^{230}Th$--about 25 Bq $kg^{-1}$. The principal natural source of $^{232}Th$ in air at ground level is generally taken to be resuspension of soil particles. Given this, the estimated annual intake of $^{232}Th$ through inhalation is about 0.01 Bq per capita. Since the digestive tract cannot readily absorb thorium, the principal source for internal exposures is probably through inhalation, rather than through ingestion. Recent studies indicate that the total average body content of $^{232}Th$ is about 80 mBq, about 60% of which is in the skeleton. The annual effective dose equivalent from this source has been calculated to be about 3 microSv.

Radium-228 subseries ($^{228}Ra$, $^{228}Ac$, $^{228}Th$, and $^{224}Ra$). Radium is much more readily available to plants and animals than $^{232}Th$; thus, the activity concentrations of $^{228}Ra$ in man are primarily due to the dietary intake of $^{228}Ra$ directly, rather than through the decay of ingested or inhaled $^{232}Th$. The most important contributors to the dose from this subseries are the alpha emitters $^{228}Th$ ($t_{1/2}$ = 1.9 years) and $^{224}Ra$ ($t_{1/2}$ = 3.6 days).

The annual activity intake of $^{228}Ra$ arising from inhalation of resuspended soil particles is estimated to be about 0.01 Bq. Food intake of $^{228}Ra$ is a more important source of internal exposure to this nuclide than inhalation, however: in areas with normal concentrations of the isotope, the annual ingested contribution per capita is about 15 Bq.

Radium and thorium are bone seekers. The estimated average activity concentration of these nuclides in bone is about 90 mBq $kg^{-1}$; in soft tissue, it is about 4 mBq kg$^{-1}$. The corresponding annual effective dose equivalent for the subseries headed by $^{228}Ra$ is about 13 microSv.

Radon-220 and its decay products ($^{216}Po$, $^{212}Pb$, $^{212}Bi$, $^{212}Po$, and $^{208}Tl$). Like Rn-222, inhalation is the principal route for internal exposure to $^{220}Rn$ and to its short-lived decay products. In outdoor air, the equilibrium equivalent concentration of $^{220}Rn$ is about 0.2 Bq m$^{-3}$. For the total population in temperate regions of the world a mean value of about 0.7 Bq m$^{-3}$ appears to be reasonable, although very little field data is available. These concentrations correspond to an average indoor equilibrium equivalent concentration of 0.2 mSv and an average outdoor equilibrium equivalent concentration of 0.02 mSv, giving a weighted equilibrium equivalent total of 0.22 mSv for inhaled $^{220}Rn$ daughters.

## Summary of exposures from the natural background

Among the wide variety of contributors to internal irradiation arising from the natural background, the short-lived decay products of $^{222}Rn$ are by far the most important; they contribute almost 60% of the effective per annum dose equivalent from natural internal emitters. Then follow, in decreasing order of importance, $^{40}K$ (about 13%), the short-lived decay products of $^{220}Rn$, and $^{210}Pb$/$^{210}Po$ (8%). The total per annum effective dose equivalent from internal emitters which come from the natural background is about 1350 microSv.

The effective external dose equivalent irradiation from cosmic rays is slightly larger than that from terrestrial radiation. The total per annum per capita effective dose equivalent from external emitters from all natural sources is about 650 microSv.

Overall, it appears that the natural sources of ionizing radiation (including the cosmogenic neutron component) produce a median annual effective per caput dose equivalent of about 2 - 3 mSv. The available data suggest that on a global scale the distribution function of natural radiation exposure can be approximated in its central part by a normal or log-normal distribution function with a standard deviation of about 0.3 - 0.6 mSv.

# CONTRIBUTIONS TO AMBIENT EXPOSURE BY HUMAN ACTIVITIES

In addition to the natural background, there are a number of human activities which contribute to the current ambient exposure to ionizing radiation. In this section, we survey what is known the most important contributions from these activities.

## Medical practices

Use of radiation in medical practices is the largest man-made contributor to the collective radiation dose rate in most of the industrialized countries. These uses include medical diagnostic radiology, clinical nuclear medicine, and radiation therapy.

Diagnostic radiology. A number of studies have indicated that despite substantial improvements in X-ray techniques and detector sensitivities, the typical dose to the patient per examination has not decreased since 1970. Several factors contribute to this state of affairs, including the fact that newer, more sensitive detectors have not rapidly replaced older instruments, nor have newer examination techniques significantly replaced existing ones.

Dental radiography. Dental radiography is the most common type of diagnostic x-ray examination. Exposures are highly dependent on the type of film used, beam collimation, and patient shielding. Absorbed doses from full-mouth radiographic examinations can result in about 15 mGy to the lens of the eye, 5 mGy to the thyroid, and 260 mGy to the skin. A number of studies have indicated that these quantities can vary as much as 10- to 100-fold.

Mammography. Mammography is another source of exposure to medical diagnostic radiation. The use of mammography is steadily increasing. It is difficult to estimate the exposure from this practice, but rough calculations can be made. The consensus view holds that the most sensitive indicator of risk from mammography is the energy imparted to the gland tissue of the breast. In theory, determining risk to this tissue would require knowing at least the spatiotemporal distribution of the dose to that tissue. Very little, unfortunately, is known about this distribution. A crude estimate can nevertheless be made. If one assumes that the linear density of the gland tissue is 35 g/cm and its total mass is 175 g on average, the average absorbed dose to that tissue from typical practices per examination is between 1 and 4 mGy.

Tomography. The introduction of computer-assisted tomographic (from the Greek tomos, meaning "a piece cut off") x-ray methods in 1971 revolutionized medical diagnostics. In this procedure, a finely collimated x-ray beam scans the plane of interest at various discrete angles. The attenuation of the transmitted beam is recorded by a detector and the resulting data are processed by computer to generate a cross-sectional image of the body, expressed in terms of relative attenuation coefficients in the layer examined. Because a computer plays a major role in this procedure, the method has become known as "computed tomography (CT)." The eye can receive as much as 50 mGy in a complete CT examination of the head. In normal clinical CT applications, the total dose to skin ranges from 50 to 500 mGy.

Summary of exposures from diagnostic irradiation. The reported organ doses for all types of diagnostic x-ray examinations range from less than 0.01 to about 50 mGy per examination. Although it has been difficult to collect complete information on the total dose from medical diagnostic practices, the consensus view suggests a round figure of 1000 man Sv per million population as the annual collective dose equivalent for the industrialized countries.

Diagnostic nuclear radiology. The use of radionuclides in the field of diagnostic nuclear medicine has rapidly increased all over the world. Estimates of from 10 to 150 man Sv per million population per year for the collective effective dose equivalent from this source have appeared in the literature. The growth of the use of these materials suggests that radiopharmaceuticals could contribute as much as 15% of the total somatic man-made dose by the year 2000.

Radiation therapy. High energy radiation is also widely used in the industrialized countries for therapeutic purposes, primarily in the treatment of certain cancers. Doses to the gonads ranging from 50 to 300 mGy for $^{60}Co$ have been reported; the differences depend on the location of primary irradiated region. Mean bone marrow doses under the same conditions of about 500 to 1000 microSv have been reported. The genetically significant dose equivalent due to radiotherapeutic procedures has been estimated at between 0.7 and 23 microSv per year per capita.

Summary of absorbed dose from medical procedures. In developed countries the genetically significant dose equivalent from medical procedures has been estimated to be on the order of 0.15 mSv per year. Because the distribution of this exposure is far from uniform, however, this value should be taken for what it is--a technical statistic.

Nuclear power

Until quite recently, the nuclear power industry in the United States grew rapidly; growth rates in Europe and Asia remain large. The total installed generating capacity in the world in 1981 was about 145 GW; an additional 210 GW was under construction in that same year. Estimates of installed capacity for the year 2000 range from 1000 to 1700 GW.

The ionizing radiation burden which will placed on man by the activities associated with the nuclear power industry depends very heavily the growth of this industry. Accurate projections of that growth have been very difficult to make for the U.S. because of a lack of understanding of the effect regulatory practices. In an inflationary economy, for example, regulatory delays can easily double the cost of building a nuclear power plant while drastically eroding the real value of capital available to construct it. The cost-effectiveness of nuclear power, furthermore, is a function of the cost of conventional power-generation technology; this cost is itself driven in part by a maze of regulations, by the long-term cost of capital, by the total demand for electrical energy, and by the cost of foreign oil. In the last five years, the US electrical power industry has tended to believe that conventional fuel technology is less expensive than nuclear power generation. Estimates of the ionizing radiation burden imposed by the nuclear power industry therefore have to be considered in light of a complex of economic factors which are often difficult to predict.

Uranium mining. Uranium for power reactors is extracted from ore mined throughout the world. The major producers are Canada, France, South Africa, the United States, and Australia. Annual world production was about 40,000 tons (metric, as pure uranium) in 1979; it is predicted to rise to about 120,000 t by 1990. Roughly the same quantity of uranium is currently mined from open pit and deep sources.

Uranium mining involves the removal of large quantities of ore containing uranium and its daughter products at concentrations up to several thousand times the typical concentration of these nuclides in the natural terrestrial background: the concentration of uranium in mined ores is between 0.1% and 3% $U_3O_8$ by weight.

$^{222}Rn$ is the principal nuclide released from underground uranium mines; it enters the biosphere through mine ventilation air. $^{222}Rn$ emissions from 0.1 to 8 GBq $t^{-1}$ (metric) ore have been reported in the literature. The major airborne radioactive component in effluents from open pit mining is also thought to be

$^{222}Rn$. About 0.1 - 0.2 GBq of $^{222}Rn$ is released per ton of ore from surface mines.

Radiocontaminated liquid wastes from both surface and underground mines result from mine drainage and process feed water; this water is typically discharged to ponds for settling of solids. The settling pond water is either allowed to evaporate or is released into the environment. Mine drainage water can also be used as process feed for a mill, or may be diluted, treated, and discharged into sources available to the public. Leaching of mine tailings may also be a source of radiocontaminated liquid waste.

The activity content of the mined ore is primarily due to $^{238}U$ and its daughter products; there is almost no thorium in uranium ores. One metric ton of ore containing about 2 kg of $U_3O_8$ has an activity of 21 MBq from each of the 14 principal members of the $^{238}U$ decay chain, for a total of about 0.29 GBq per metric ton ore.

The mining of uranium increases the amount of uranium and its radioactive decay products in the environment, and hence increases the exposure of man. Deep mines <u>in general</u> produce negligible contamination outside the immediate vicinity of the mine.

Substantial local contamination may occur, however: North Table Mountain Reservoir, once a major drinking water supply for the city of Golden, Colorado, was contaminated by the Schwartzwalder Uranium Mine, located several kilometers from the Reservoir. When the Schwartzwalder ceased operations, dewatering pumps were switched off, allowing water to fill the mine. This water then found its way through unknown mechanisms into the Reservoir. Uranium activity in the Reservoir rose to over 400 pCi/l, eight-fold above the highest proposed ceiling for that contaminant.

<u>Uranium mills.</u> From the mine, uranium ore is sent to a uranium mill. Mills are frequently located adjacent to mine sites, because the cost of transporting large quantities of what will ultimately be waste material (mill tailings) is prohibitive. The mill processes the ore (nominally, 2000 metric tons per day) to extract partially refined uranium. The uranium concentrate which results from the milling process is called <u>yellowcake</u>.

About 14% of the total radioactivity in the ore feed to a mill appears in the uranium concentrate; the mill achieves better than 90% uranium extraction. In the the resulting solid wastes some 70% of the original activity remains and is due primarily to $^{230}Th$ and its daughters. These solid wastes, called tailings, are discharged from the mill in a slurry of about 50% solids to an impoundment area. Nominally, mill tailings are composed of about 70% sand and 30% slimes; about 85% of the total activity in the tailings is contained in the slimes. Overflows from tailings are treated with barium chloride to co-precipitate the radium as $Ba(Ra)SO_4$. Further barium chloride treatment and settling may be be used before effluents are discharged into waters available to the public. In typical practice, the dissolved $^{226}Ra$ concentration remaining after such treatment is usually less than 0.4 Bq $l^{-1}$; suspended $^{226}Ra$ concentrations in the same effluents are roughly 0.2 - 7 Bq $l^{-1}$.

Some 120 million tons of tailings are currently stored at active mill sites, mainly in the United States and Canada. Current projections suggest that this quantity will increase to $5 \times 10^8$ t by the year 2000.

In general, in dry areas, there are essentially no liquid discharges from uranium mills, provided all of the water which does not evaporate in holding ponds is recycled in the mill operation. It is not clear, however, just how successful mill operations have been in this respect: there is some evidence, for example, that despite the best impoundment practices, uranium apparently from the Cotter Uranium Mill in Canon City, Colorado, migrates into the Arkansas River through unknown groundwater mechanisms.

Mill operations can be a source of atmospheric contamination by radionuclides, also. The ore crushing and grinding circuits, the yellowcake drying and packaging operations, and the tailings themselves are sources of such emissions. For a mill processing nominally 2000 t of ore per day, the major sources of atmospheric dust emissions are the yellowcake drying and packaging operations; typical emissions from these operations have been reported to be 1 - 4 GBq per year for $^{238}U$, 0.2 - 2 GBq per year for $^{230}Th$, $^{226}Ra$ and $^{210}Pb$, and 1 - 7 TBq per year for $^{222}Rn$ per mill site. The most modern mills may able to restrict atmospheric particulate releases from the crushing and storage processes to about 0.04 - 0.16 GBq per year per mill site.

Atmospheric emissions from the mill tailings areas range from 7 to 500 MBq $^{238}U$ and $^{234}U$, 0.1 - 8 GBq for $^{230}Th$, $^{226}Ra$, and $^{210}Pb$, and 500 GBq to 300 TBq per year for $^{222}Rn$ per mill site. The amount of particulate airborne emissions from tailing areas depends on the size of the tailings dry beach areas and local meteorological conditions; the radon emissions, in contrast, depend on diffusion from the ground. The radon exhalation rate appears to be about 1 Bq $m^{-2}$ $s^{-1}$ of $^{226}Ra$ in typical uncovered tailings areas; this figure can vary by as much as an order of magnitude, however. Tailings impoundment areas almost completely covered by water will have very low radionuclide emissions to the air.

Tailings remain after the mill has ceased to operate and can become a long-term source of radioactive contamination due to wind and water erosion, leaching, and radon emanation. Erosion can be alleviated by covering the tailings with native soils, clays, asphalt, or artificial materials such as polyvinyl chloride. The major source for long lived activity (for about 5 x $10^5$ years) in the tailings is $^{230}Th$, which continues to produce $^{226}Ra$ and corresponding radon releases.

<u>Summary of exposures from uranium mines and mills</u>. Most uranium mine and mill are currently located in areas of low population density, a practice which substantially restricts the collective exposure of man to environmental radiocontamination by these sites. In arid areas, there is ordinarily negligible release of radionuclides to the aquatic environment, provided that releases to water are well controlled. Under such conditions, the main route to man would be through airborne effluents. In areas of high precipitation, the liquid effluent doses are dominated by $^{226}Ra$ in drinking water and aquatic foodstuffs. Even in such cases, it appears that the airborne component of exposure is still significantly greater than the liquid effluent as a route to man. The estimated doses to most exposed members of the public are highly dependent on the characteristics of the particular location of the mine and mill. Annual effective dose equivalents from a few hundred microSv to several milliSievert have been estimated for typical emissions from mines and mills.

<u>Uranium fuel fabrication</u>. Uranium yellowcake produced at a mill is transported to fuel fabrication plants, where it is further processed and purified, and frequently enriched in the isotope $^{235}U$, before being converted into uranium oxide or uranium metal and then fabricated into fuel elements. Graphite or heavy water moderated reactors (HWRs) can use natural uranium which contains about 0.7% $^{235}U$. Light-water reactors (LWRs) and advanced gas-cooled reactors (AGRs) require enriched fuel containing about 1 to 4% $^{235}U$. Uranium metal without enrichment is clad in a magnesium alloy can for use in Magnox reactors. For HWRs, unenriched uranium dioxide is normally used.

Before uranium can be enriched, it must be converted from the oxide form $U_3O_8$ to uranium hexafluoride ($UF_6$). There are two processes used to produce uranium fuel: the hydrofluor process and the solvent extraction process.

The hydrofluor process consists of reduction of the yellowcake uranium, followed by hydrofluorination, and subsequent fluorination to produce crude $UF_6$. The crude $UF_6$ is then distilled to produce pure $UF_6$.

The solvent extraction process employs a wet chemical solvent extraction step prior to the reduction, hydrofluorination, and fluorination steps; otherwise, it is similar to the hydrofluor process.

$^{235}U$ enrichment usually takes place at a gaseous diffusion plant, although centrifuge separation technology is also used. In the gaseous diffusion process, the purified $UF_6$ from the hydrofluor or solvent extraction process is pumped through a series of porous membranes which discriminate against the passage of the heavier isotope of uranium by a factor of about 1.0043 at each stage. Some 1700 stages are required to produce an enrichment of 4%. The centrifuge process, in contrast, uses high speed centrifuges to separate the isotopes of uranium. Centrifuge technology uses only about 10% as much electricity as the diffusion process.

In the final fuel fabrication step, $UF_6$ is chemically converted to $UO_2$ or to uranium metal for use in fuel elements. For use in LWRs or AGRs, the $^{235}U$-enriched oxide powder is sintered into pellets and loaded into zircalloy or stainless steel claddings to produce fuel pins which are filled with helium and welded with end caps.

After the enrichment process, large quantities of depleted uranium remains at the fuel enrichment plant. The $^{235}U$ content in this depleted uranium is 0.3% or more. This uranium may become a source of public exposure if it is discharged into the environment; at present it is stored for possible use in breeder reactors and for other as-yet-unspecified purposes.

Emissions of radionuclides from the conversion, enrichment, and fuel fabrication processes are small. Residual amounts of $^{230}Th$ and $^{226}Ra$ are removed from the uranium ore concentrate in the conversion process and small amounts of these nuclides appear in effluent streams. Atmospheric annual discharges from typical conversion plants in the United States are reported at about 3 GBq for $^{238}U$, $^{234}U$, and $^{234}Th$; 33 MBq for $^{230}Th$; 3 MBq for $^{226}Ra$; and 74 MBq for $^{235}U$.

Summary of exposures from uranium fuel fabrication. The collective effective dose equivalent commitment due to uranium fuel fabrication is estimated to be $2 \times 10^{-3}$ man Sv per GW electricity produced per year. The main contribution to man is thought to be through inhalation of the isotopes of uranium. Radon releases contribute about 20% of the total.

Power generating plants. Most of the electrical energy generated by nuclear power is produced by thermal reactors in which the fast neutrons produced by the fission process are slowed to thermal energies by the use of a moderator. The smaller the atomic mass of the moderator, the more efficiently it removes energy from the neutrons. The most common materials which have been employed as moderators in thermal power reactors are light water, heavy water, and graphite.

During the production of power by a nuclear reactor, fission products are formed within the fuel; in addition, neutron activation produces radioactive components in the reactor's structure and in the fuel pin cladding materials. Radionuclides are also generated in the coolant bath because the coolant itself is activated, because fission products diffuse into the coolant from fuel with defective cladding, and because activated structural and cladding materials corrode into the coolant. All reactors have treatment systems for the removal of radionuclides from gaseous and liquified wastes which arise from such sources.

The principal radionuclides present in today's reactor effluents are $^{3}H$, $^{58}Co$, $^{60}Co$, $^{85}Kr$, $^{89}Sr$, $^{90}Sr$, $^{131}I$, $^{131}Xe$, $^{133}Xe$, $^{134}Cs$, $^{137}Cs$, and $^{140}Ba$. The release rates depend heavily on reactor design. Gaseous and volatile nuclides such as $^{85}Kr$, $^{131}Xe$, and $^{133}Xe$ contribute to the external gamma dose as a result of immersion; the rest contribute externally through deposition or internally through the food chain. The principal stack discharges are the noble gases, tritium in water vapor, and iodine in both elemental and organic compounds. Principal radiocontaminants in the liquid effluents of reactors are tritium as HTO and activation nuclides of iron, cobalt, nickel, and zinc; in addition to occurring in elemental form, these metal nuclides may occur as oxides or in complexes.

If liquid metal fast breeder reactors (LMFBR) become common, other radionuclides, including $^{239}Pu$, $^{238}Pu$, $^{241}Pu$, $^{24}Na$, and $^{22}Na$ will become principal radiocontaminants in plant effluents. Normal (non-accident) release rates are expected to be low for LMFBRs.

External dose rates from power reactor effluents have been estimated by computer models assuming whole-body gamma doses of 5 mrem at the reactor boundary. Under this assumption the annual average dose the U.S. population was estimated to be 0.002 mrem for 1970 and 0.17 mrem for the year 2000 (assuming accident-free performance).

Under non-accident conditions, the consensus view holds that the principal radiocontaminant contributed by nuclear power plant effluents will be $^{131}I$; in general, internal doses will be much lower than the overall external gamma radiation doses.

Accident conditions. Nuclides escaping from reactors under design-basis accident conditions will consist of the release of volatile and non-volatile components. The former include the noble gases, iodine, and tritium, as described above. The latter include all fission products, many activation products, uranium, and plutonium. Most design-basis accident scenarios suggest that inhalation of radiocontaminated particulates is likely to be a more important source of exposure from an accident than other mechanisms. In an LMFBR design-basis accident the initial inhalation problem will come from plutonium; in the long run, ingestion of $^{22}Na$ is likely to be the greater problem.

The most serious accident which could occur to a light water thermal power reactor would begin with a rapid loss of coolant. Although the absence of the water moderator would stop the fission process, the decay heat from the radioactive materials inventory could cause the meltdown of the reactor core. Most power reactors have backup cooling systems to respond to this kind of accident. The effectiveness of the backup cooling system, and the speed with which it is activated will determine whether the core melts. If the core does melt, the molten core material could break through the containment vessel, and in such an event, substantial quantities of volatile and nonvolatile radioactive materials would escape from the site in the form of a radioactive cloud. There is not a little dispute about the frequency with which such accidents will occur, about the severity of effects which would result, and what could be done to help prevent them. A very large problem in making such estimates involves assessing the contribution of human error to the probability of a meltdown scenario.

Fuel reprocessing plants. Spent fuel elements are removed from the reactor and stored under water while waiting for reprocessing. Fuel elements are usually stored until all the short-lived isotope $^{131}I$ has decayed to insignificant amounts (usually about 120 days). The most important gaseous radioactive effluents from fuel reprocessing plants include tritium in the form of HTO, noble gases (particularly $^{85}Kr$), elemental iodine, organic iodides, and probably NOI and HIO. Short-cooled fuels may contain $^{131}I$; and all fuels will contain $^{129}I$. Liquid effluents from reprocessing plants contain radionuclides of cesium and strontium, tritium in HTO, and long-lived fission products including some rare earth nuclides.

Dose estimations for exposure to nuclides released by reprocessing plants are heavily dependent on the assumptions made. The EPA has estimated that the average per capita exposure of the U.S. population from fuel reprocessing activities was about 0.0008 mrem in 1970 and is estimated to be 0.2 mrem in the year 2000. The primary tissues involved in this exposure will be the respiratory lymph nodes, the thyroid, and the skin.

Storage of radioactive waste. Although substantial concern has been expressed about the safety of long-term nuclear waste storage, at present such storage does not contribute significantly to the ionizing radiation burden placed on man. The scale of hot waste storage is currently orders of magnitude lower than what it could be if nuclear generation of power were to replace conventional means, however. Estimates of the exposure of man from storage practices depend heavily on, among other factors, how thorough we believe our knowledge of the geology of the storage sites is and on the role of human error in storage. Needless to say, there is little consensus on these issues.

Summary of contributions from nuclear power activities. For releases during the operational phase in the nuclear fuel cycle, that is, excluding spent fuel and waste disposal, the collective effective dose equivalent commitment is estimated to be 5.7 man Sv per GW (electricity generated) per year; most of this is delivered within five years of the time the wastes resulting from the generation of a unit of electrical power are discharged. The commitment due to global dispersion of nuclides released during fuel cycle operations is estimated to be 670 man Sv per GW (electricity generated) per year, which is delivered to the environment $10^4$ to $10^8$ years after release from the fuel reprocessing plant. The contribution from tailings is estimated to be 30 man Sv per GW per year; this amount is highly dependent on local site characteristics, however, and may easily be two orders of magnitude larger or smaller.

## Global dispersion of radionuclides released by human activities

Of all the radionuclides redistributed by human activities under ordinary conditions, tritium and krypton-85 by far contribute most to the overall burden of ionizing radiation to man.

Nuclear reactors have contributed on the order of 1 MCi (Megacurie) of tritium to the current world environmental burden; in comparison, weapons testing contributes about $10^3$ MCi; in contrast, the natural background contributes about 10 to 100 MCi. The principal mechanism of exposure is through contaminated water. The annual dose from worldwide tritium is estimated to be 0.04 mrem/person.

Krypton-85 is distributed throughout the atmosphere and is a source of exposure to man both externally and through inhalation. The gas is produced naturally by cosmic rays and artificially by weapons testing; both these sources contribute very little, however, relative to the contribution of nuclear power activities. In particular, weapons testing has cumulatively contributed about 3 MCi to the world burden; nuclear power plants contribute at least 10 MCi/yr. The estimated annual whole-body dose to the U.S. population from worldwide distribution of $^{85}Kr$ lies between 0.0004 and 0.04 mrem. Skin doses are calculated to be about 50 times greater and lung doses about twice as great as whole body doses.

Weapons testing. Calculations have been made for exposures in the vicinity of the Nevada Test Site for the period 15 September 1961 to 15 September 1962, a period of resumption of atmospheric nuclear tests following the moratorium of 1958. The exposures were estimated as 47 mrem per capita external gamma dose to a population of 18,000; 10 mrem whole body dose from $^{137}Cs$ to a population of 792,000; 9 mrem per thyroid to the same population. Some exposure also occurred from underground testing in the early 1960s. For example, some venting occurred to produce a mainly gaseous effluent from the "Gnome" test on 10 December 1961. External gamma dose from the cloud gave a total of 30 person-rem to a population of about 45,000; there were no internal radiation exposures detected.

Weapons testing also contributes to the environmental radioactive load through worldwide global fallout. Most of the contribution from this mechanism has come primarily from large-scale, high-yield atmospheric tests conducted by the U.S. and U.S.S.R. prior to 1963; in the last several years small tests conducted by the French and Chinese have made relatively insignificant contributions in comparison. The total annual whole-body doses from global fallout averages about 5 mrem/person.

## Miscellaneous sources

Television, consumer products, air transport, and other miscellaneous sources contribute to the radioactive burden; current estimates suggest an average of 1-2 mrem per person per year for all such sources.

## Summary of ambient exposure

Figure 3-1 summarizes the estimated average annual whole-body radiation doses from known sources. This figure shows that the major contributors to the whole body dose are the natural background and medical applications. Medical diagnostic procedures account for 90% of the man-made dose and for 35% of the annual average per capita dose from all sources including the natural background. It should be emphasized that these figures are rough estimates only, although are typically the best available. It should also be noted that the average dose rates reported in this Figure may provide

| Source | Annual effective dose equivalent ($\mu$Sv) External | Internal | Total |
|---|---|---|---|
| COSMIC RAYS | | | |
| Ionizing | 280 | | 280 |
| Neutron | 21 | | 21 |
| COSMOGENIC RADIONUCLIDES | | 15 | 15 |
| PRIMORDIAL RADIONUCLIDES | | | |
| $^{40}K$ | 120 | 180 | 300 |
| $^{87}Rb$ | | 6 | 6 |
| $^{238}U$ series | 90 | 954 | 1044 |
| $^{232}Th$ series | 140 | 186 | 326 |
| MEDICAL | 1500 | 20 | 1500 |
| TOTAL | 2150 | 1360 | 3500 |

Figure 3-1. Estimated average annual doses from known ionizing radiation sources.

very little information about the distribution of that rate; subsets of the population may be exposed at a much higher rate (orders of magnitude) in localized events or practices than the population as a whole.

## ENVIRONMENTAL TRANSPORT OF RADIONUCLIDES

Measures such as average whole body dose rates often provide little insight into the distribution of global effects resulting from release of radionuclides into the environment. A sound grasp of this distribution, however, is essential to an understanding of the resulting long-term risk to man; being able to predict that distribution, in turn, requires knowing at least how the environment transports radioactive materials and energy. With few exceptions, we know very little about the fate of radioactive materials in the environment. This unfortunate state of affairs is not likely to change in the near future, because reliable, system-level field data is very costly and difficult to collect.

General features of radionuclide transport

The chain of events which leads from the release of radioactive substances into the environment to the actual irradiation of human tissue can be abstractly represented by a system of compartments through which radioactive materials and energy flow. In such a model, the real-world transfer of radioactive material and energy from one part of the environment to another, or from one part of the body to another, is functionally represented by transfer from one compartment to another. The rates of transfer of radioactivity between compartments can be specified by time functions. More specifically, steps in the transfer of materials and energy from its entry into the biosphere to the dose to man can be described in terms of quotients of the appropriate quantity ( e.g., alpha activity concentration) in a given compartment j to to the same quantity in the preceding compartment i. These quotients define dynamic transfer factors, $P_{ij}$, in the pathway from input of radionuclides into the environment to the subsequent radiation dose in man, that is,

$$P_{ij}(t) = M_j(t) / M_i(t) \qquad (3\text{-}36)$$

where $P_{ij}$ is the transfer factor to compartment j from the preceding compartment i, and $M_i(t)$ and $M_j(t)$ are the appropriate mass or energy quantities in the respective compartments at time t. Thus, each node (compartment interface) can be uniquely and exhaustively described by a transfer factor.

The network linking the release of radioactive materials into the environment to the dose to man consists of series and parallel paths. The total transfer factor of a series branch is the product of the transfer factors associated with that branch; the total transfer factor of a set of parallel branches is the sum of transfer factors of the members of the set. The final dose to man, therefore, can be expressed as the following function of the input N(0) and the transfer factors:

$$D(t) = N(0) \sum_{parallel} \prod_{series} P \qquad (3\text{-}37)$$

Equations 2-36 and 2-37 are merely an abstract though theoretically adequate descriptive scheme. Very little is known about the actual values of the transfer functions identified in that scheme: at best, we have only rough generalizations at our command. The following sections survey what we know about those factors.

## Atmospheric transport

Transfer by air. The atmosphere redistributes radionuclides through the processes of transport and diffusion. For nominally instantaneous releases such as explosions or short ventings, contaminated material is transported from the source with a velocity in the short-run determined largely by the velocity of the wind at the moment of release. Surface air concentrations from such events tend to decrease with time downwind, primarily due to horizontal dispersion. For continuously emitting sources, in contrast, meteorological conditions for times on the order of days must be considered. Reasonably good short-term estimates of average local concentrations or exposures can be made from routine local meteorological observations and the use of appropriate diffusion equations. In contrast, current understanding of long-term global dispersion is limited; a number of calculational models, in the literature for over a decade, can just now be implemented on very large computers (CRAYs, for example).

There are two known general ways that radionuclides can be removed from the air: dry deposition and precipitation scavenging.

Transfer from air by dry deposition. Dry deposition on surfaces can result from gravitational settling (fallout), surface impaction, electrostatic attraction, adsorption, and chemical interaction. Two approaches have been used to study dry deposition. The first defines deposition velocity as the ratio of the deposition rate to the air concentration immediately above ground level. The second derives deposition velocities from a material balance involving the mass flux of material through a plane normal to the mean wind direction. Both these approaches suggest that for particles less than 10-15 microns in diameter, the relative effects of impaction, diffusion and adsorption are more important; as particle size increases, gravitational effects begin to dominate.

Transfer from air by precipitation scavenging. There are two general ways in which radionuclides can be transferred from the air by water: in-, and out-of-cloud precipitation scavenging. Because nuclides released from the ground tend to be transported by low-level winds, then diffused upward by eddies, removal from the air at short distances (that is, at low altitude) from the release site will occur primarily from washout (out-of-cloud precipitation scavenging); at greater distances from rainout or snowout will dominate the scavenging activity. Releases of nuclides in the stratosphere, in contrast, will be dispersed by the general circulation and tropospheric exchange processes will control subsequent removal.

Washout by rain is generally insignificant for particles smaller than about 1 micron in diameter. Snow is apparently as effective as, and in some cases, several times more effective than, a scavenger than rain at the same precipitation rate for all sizes of particles.

*Transfer to air from soil.* Most of the natural radioactivity in the atmosphere which comes from natural terrestrial sources consists of radon and its daughter products--$^{218}Po$, $^{214}Pb$, $^{214}Bi$, and $^{213}Po$. These nuclides become attached to submicron aerosols and are borne aloft by local winds and eddies.

Atmospheric radon concentrations at the ground level usually range from 10 to 1000 $pCi/m^3$ for continental areas; the higher concentrations occur for brief periods during stagnant weather conditions. Higher concentrations may also occur over areas exposed to uranium ore tailings or natural uranium outcroppings. Concentrations over oceans may be lower than those over ground by two orders of magnitude.

*Transfer to air from nuclear power activities.* Stack effluents from nuclear power and reprocessing plants also contribute to the radionuclide load dispersed by atmospheric processes. The effluents from nuclear processing sites consist almost exclusively of gaseous species or particulate components less than 1 micron in diameter. Stack releases to good approximation can be treated as point emissions dispersed by ordinary Gaussian diffusion; the approximation must be augmented, however, by consideration of atypical local weather conditions. Reactors sited along seacoasts or near large bodies of water, for example, present special problems of this sort. Quantitative information concerning the movement of airborne material over ocean and shoreline complexes must take into account the land-sea breeze phenomenon, and such information for individual sites is often not available. Time of day, season, relative water and air temperatures, and local topography are evidently major influences on concentrations. In general, the diffusion regime of each site must be analyzed individually.

*Transport to air from atmospheric testing.* Atmospheric testing of nuclear devices also contributes to the radioactive load dispersed by the air. For fallout within about 100 miles of the test site, local wind appears to be the most important short-term determinant of dispersion; for greater distances and longer times, Fickian diffusion theory augmented by consideration of general wind conditions is required.

Aquatic transport

We have relatively little detailed knowledge of the distribution of radionuclides in the aquatic environment; what data we do have is highly particular, and generalizations from it must be made conservatively.

Transfer from water to aquatic plants and animals. Estimates for dose rates from alpha activity to the total oceanic phytoplankton mass range from 230 to 2800 mR/yr. Tissues doses in cod *Gadus callarias* range from 8 to 27 mR/yr per individual. These values may be taken as indicative of the nominal range of ambient alpha dose rates for extreme ends of the oceanic food chain.

More is known about the effects of short-term lethal doses of radiation to aquatic organisms than is known about low-level chronic exposures of those organisms. Lethal response to acute radiation varies among all organisms because of physiological differences; these responses are complicated in the aquatic environment because additional factors such as temperature, dissolved oxygen, and salinity, among others, affect response to exposure. Some generalizations can nevertheless be made. First, with the exception of the eggs and larvae of invertebrates and fish, most of the aquatic organisms which have been studied are relatively radioresistant, in the sense that the $LD_{50}$ is much higher for these organisms than for land animals and plants. The level of exposure required to produce genetic damage, however, may differ little among all living organisms. Second, marine and freshwater species are similar in (LD)radiation resistance. Third, primitive forms are more (LD)resistant than complex vertebrates, and older organisms are more resistant than young. Bacteria and algae, for example, may survive thousands of roentgens, whereas the $LD_{50}$ for the adult rainbow trout ranges from about 300 to 3000 R; for the most sensitive state of the developing trout egg, in contrast, the $LD_{50}$ is as low as about 15 R.

Data on the chronic exposure of marine organisms to low-level radiation are sketchy. Long-term experiments on chinook salmon indicate that irradiation at 500 mR/day per individual from the fertilization through the feeding stage did not reduce the reproductive capability over a period of just over one generation. Frequent abnormalities in the young fish were noted, but the number of adults returning to spawn was not decreased.

Practically all workers have seen effects from chronic exposure to activity concentrations in the range of $10^{-4}$ Ci/l for a wide variety of fish species; below $10^{-4}$ Ci/l, results are ambiguous. There is some, but conflicting evidence, on whether activity concentrations as low as $10^{-10}$ Ci/l can produce detectable effects in fish.

## Soil-plant transport

The fate of radionuclides in the soil is complex. Depending on the chemical nature of the nuclide, the soil can be a sink which the individual radionuclide enters, leaving only with great difficulty, or it can be a reservoir through which the nuclide moves with ease. Soil exposure can result from direct nuclide uptake, from entry into surface, ground, or irrigation waters, or from direct irradiation.

Transport to soil from natural terrestrial sources. The most commonly occurring radionuclides in the soil come from the following and their decay products: potassium-40, rubidium-87, thorium-232, uranium-235, and uranium-238; these have been described above. The total amount of natural radioactivity arising from these sources in one square meter of soil to normal cultivation depth (i.e., per 200 kg soil) is about 5 to 10 microCuries.

Among the decay products of thorium-232 and uranium-238 are the gases radon-220 and radon-222, respectively. A small fraction of these gases escapes into the atmosphere and decays to form radionuclides of polonium and lead; these gradually return to the soil in natural fallout. Polonium and lead radionuclides do not contribute appreciably to soil levels as such but do constitute a significant source for plant uptake of lead-214, lead-212, lead-210 and polonium-210.

Transport to soil from nuclear device testing. Atmospheric nuclear explosions also contribute to the deposition of radionuclides in soils. Soil levels of the long-lived man-made fission products $^{90}$Sr and $^{137}$Cs from these sources are widely documented. Average levels of $^{90}$Sr in soils increased from about 0.015 microCi per square meter in 1958 to a maximum of 0.065 in 1967, for example; these years mark the beginning and end of large-scale atmospheric weapons testing. The highest $^{90}$Sr level found in the United States outside the Nevada Test Site was 0.160 microCi per square meter in a high rainfall area of western Washington in 1967. Maximum accumulations of moderate-lived fission products occurred in 1959: soil levels of $^{144}$Cs, $^{106}$Ru, and $^{95}$Zr ranged from 0.40 to 0.74 microCi per square meter.

There are no direct data on accumulation of short-lived nuclides in soil but back-calculations from their concentrations in cow's milk suggest that deposits of $^{140}Ba$ and $^{131}I$ were about 1 microCi per square meter in some areas of Utah in 1962. Short-lived radionuclides by definition do not persist in the environment, and thus have no long-term relevance for human exposures.

Atmospheric testing of nuclear devices has also contributed poorly characterized amounts of carbon-14 and tritium to soils; levels of carbon-14 and tritium added to soil from fallout can be estimated only indirectly from air and rainfall concentrations. The average concentration of carbon-14 contributed to soils by atmospheric testing has been estimated to be 0.014 microCi per square meter. Tritium from the same sources soils may have reached 6 microCi per square meter in wet soils in 1963.

Transport to soil from nuclear power activities. Soil radiocontamination can occur from reactor effluents, reprocessing wastes, mining and milling operations, and accidental releases. The pathway to soil can be via the atmosphere, by irrigation or flooding of streams, or by seepage of contaminated ground water from radioactive waste disposal sites. Relatively little is known about these mechanisms. Data are available for levels of radionuclides in farm produce originating from irrigation water from the Columbia River below the Hanford reactors, however, although this data may be atypical. From 1958 to 1965, the following average radionuclide concentrations were observed in that water: $^{24}Na$ (2000 pCi/l); $^{32}P$ (200 pCi/l); $^{51}Cr$ (5000 pCi/l); $^{65}Zn$ (200 pCi/l); $^{76}As$ (1000 pCi/l); $^{239}Np$ (2000 pCi/l). Only $^{32}P$ (700 pCi/l) and $^{65}Zn$ (500 pCi/l) were found in dairy milk. Groundwater flow and radionuclide movement in groundwater from the Hanford site have been studied, also. The groundwater flow rate from the reactor site was about 1.5 kilometer per year. Most radionuclides moved only a few feet from the disposal site in the seven-year study period; only tritium, technetium-99, and ruthenium-106 moved nearly as fast as the groundwater.

Transport of radioactive wastes from uranium mining and milling to soil depends greatly on local geography and climate. In some instances, the principal redistribution mechanism is the wind; in others, solution in, and transport of particles by, streams dominates.

Transport to plants by irrigation. Irrigation may be an important process in the recycling of radionuclides from water to terrestrial food chains. In furrow irrigation, plants can become contaminated by uptake of radionuclides added to the soil; in sprinkler irrigation, there is the additional direct contamination from the wetting of the foliage. The little experimental evidence we have suggests that sprinkler irrigation produces vegetation with about the same concentrations of $^{90}Sr$ and $^{137}Cs$ as in the irrigation water.

Extended use of contaminated irrigation water can result in the accumulation of long-lived radionuclides in the soil. It has been estimated, for example, that the concentration of $^{90}Sr$ in a green crop on a fresh weight basis could reach 20 times the concentration in irrigation water after a few decades of irrigation with contaminated water.

Transport to and from soil by natural geological processes. Radionuclides in or on the soil can be redistributed by the natural geological processes of erosion, sedimentation, desorption, and leaching.

Redistribution through erosion and sedimentation may be very great in sloping areas, depending on average slope and amount of runoff. For example, more than half of the total $^{90}Sr$ fallout on experimental cultivated one-acre watersheds plots with from 10 to 15% slope had been eroded away in 1960. Individual storm runoff has been shown to carry a few percent of fallout being deposited in the current storm; radionuclide loss from this mechanism is correlated with water runoff. Estimates of $^{90}Sr$ movement from major river basins have been calculated from flow data and $^{90}Sr$ concentrations in many locations; as a rough generalization, per year, from 5 to 10 percent of the $^{90}Sr$ fallout was removed in areas with the greatest runoff (mountain and coastal regions); less than 5 percent in mid-sections of the country; and less than 1% in arid sections.

Deposition of radioactive sediment occurs in reservoirs and quiet stretches of streams, and the extent depends on particle size, stream velocity, and holding times. It has been estimated that in the United States about one-fourth of the radioactive sediment produced is trapped in man-made reservoirs (see Chapter 7 for an example of this mechanism).

Desorption of radionuclides from soil is an ion exchange phenomenon; thus the primary factors affecting desorption rates are the ion exchange properties of the radionuclide, the composition of the displacing solution, and the exchange capacity of the soil. Generally, monovalent ions are most easily displaced; divalent and trivalent ions, less easily so. Anions are more easily displaced than cations because the charges on soil particles are predominantly negative.

Downward movement of radionuclides through soils can occur by leaching or particle movement. Field studies in 1966, for example, showed that 95% or more of the $^{90}Sr$ and $^{137}Cs$ were in the top six inches of soil except where there had been mechanical movement of particles. Only tritium, technetium-99 and ruthenium-106 have been observed to move with ground water. Iodide also moves relatively freely (i.e., at roughly groundwater migration speed) through soils that are low in organic matter.

Transport to organisms living on or in the soil. Gamma-emitting radionuclides which have been spread on the soil surface irradiate any organisms living the surface. Organisms living directly on the surface of the soil (i.e., within the first centimeter or so), may experience a gamma dose 10 times greater than organisms that do not live wholly in this region. The radiation dose decreases sharply with depth unless the radionuclides are mixed in the soil. The gamma dose from fresh fission products on the soil surface, for example, decreases 100 fold for each meter depth in the soil. Beta rays penetrate much shorter distances, being appreciably absorbed by even a leaf or plant stem and almost completely by 5 mm of soil, depending on their energy. Thus, the distribution of beta-emitting nuclides on soil or plant surfaces is critical in determining radiation exposure from this source. Insects that feed or nest in leaves or flower cups and some sensitive plant tissues such as actively growing meristems may be exposed to much more beta radiation than would be expected from the general intensity of gamma radiation in the area. Alpha particles in the soil present little external radiation hazard to higher plants and animals; specific hazards to single-celled organisms are largely unevaluated.

Lower forms of life in the soil are generally more (LD-)resistant to radiation than are the higher forms. For example, 10,000 rads kills many higher plants or animals on or within the soil. Doses greater than 50,000 rads reduce the microbial population and could result in selective killing of different bacterial groups. Doses less than 1000 rads probably would not affect any of the simpler organisms in the soil, at least in the short run.

Plants may become contaminated by absorption through roots or through above-ground parts including leaves, stems, branches, flowers, and fruit. Root absorption depends largely on soil processes involving ionic form of the contaminant, pH, exchange capacity of the soil and roots, moisture, and temperature. Some elements are strongly concentrated by plants (K, Rb, P, Na); some slightly concentrated (Ca, Sr, Mn, Zn); some not concentrated (Ba, Ra, Co); and some almost excluded (Cs, Fe, Ru, Sc, Y, Ce, Pb, U, Th, and Pu). Lowering pH values generally increases cation uptake and decreases anion uptake. Increasing the soil's exchange capacity tends to decrease both cation and anion uptake. Highly organic soils permit increased Cs uptake as compared to mineral soil. Flooding of soils tends to increase the uptake of Cs and I. Legumes have tendency to absorb more alkaline earths than alkali cations but the reverse is true for grasses. Brazil nuts, curiously, are effective accumulators of barium and rare earths containing relatively high proportions of alpha emitting nuclides.

Direct contamination of foliage contributes more to the plant's radionuclide content than uptake through roots does when the fallout rate is high. Three mechanisms of direct contamination have been recognized; foliar contamination, which results from retention and absorption through the leaves; floral contamination, which involves entrapment and absorption in flower parts; and plant-base absorption, which results from entry in to the basal tissues of shoots or superficial roots by material initially deposited on them or washed down by rain from the foliage.

Material is deposited on plants by dust or other particulate matter, by precipitation, or by sprays. Retention depends on intensity and amount of precipitation, wind speed, particle size and density, wettability of leaves, leaf type and angle, and thickness and continuity of the leaf cuticle. To the extent that radionuclides are water-soluble, they may be absorbed through the leaves or basal tissues, by much the same mechanisms described above for root absorption.

Once radionuclides are absorbed, the processes of translocation influence their distribution within the plant. Metabolites accumulate in certain plant parts depending on the metabolism of the substance and the state of development. For example, calcium and strontium are found in cell wall materials and are not readily retranslocated to other parts. Phosphorous is accumulated in areas of high metabolic activity such as root tips, buds, flowers and developing leaves. Carbohydrates are transferred from leaves, where they are manufactured, to areas of active growth and metabolism. Elements such as potassium, phosphorous, sodium, rubidium, and cesium, in contrast, are freely soluble, and can be found more or less uniformly distributed throughout the plant. In general, with the exception of the highly soluble elements, substances are preferentially translocated to the plant organ or part that is developing at the time.

Radioactive substances in plants are returned to soil by death and decay, by leaching by rain or dew, by exudation, and by volatilization.

Both direct contamination and absorption from the soil are important mechanisms of radionuclide transport to plants in pasture lands. Most pasture grasses obtain their nutrients from the top few inches of soil, and in humid regions have shallow root systems; thus the concentration of contaminants in the rooting zone may decrease with time if the soil exposure is the result of a single event or practice. It appears that the rate of reduction of uptake of $^{90}Sr$ by plants from pasture land in the U.S., for example, is about 13 - 14% annually.

There are little or no data on the effects of low level chronic irradiation of plants. But in general, there appears to be an extreme range of sensitivities to acute radiation. Lilium and Tradescantia are affected by 30 to 40 R per day; gladiolus requires 9000 R per day before effects are noted. Conifers such as pines and Taxus are affected at about 2 R per day.

In general, flower parts and meristematic areas are much more sensitive to irradiation than are leaves, stems, and roots. Plants with low chromosome numbers and small nuclear volumes are in general more radiosensitive than are plants with high chromosome numbers and larger nuclear volumes.

## Transport to animals other than man

It is difficult to obtain information on the effects of environmental transport of radiocontamination to terrestrial animal populations in their natural habitats. Some generalizations are nevertheless possible. Invertebrates are more (LD) resistant than vertebrates; many insects, for example, are able to survive kilorad exposures. Mammals are somewhat more radiosensitive than birds, fish, amphibia, or reptiles. Published values for the $LD_{50}$ of mammals ranges from about 150 rad for sheep and burros, to 1500 rad for desert mice. Both the male and female germ cells of all mammals are relatively radiosensitive.

Animal populations are exposed to radionuclide contamination through many routes. Releases to the atmosphere may be of the greatest consequence. External irradiation from an airborne cloud and subsequent inhalation can present transient hazards, but the main source of external exposure is likely to be materials deposited on or in soil from atmospheric releases. In the process, food sources become contaminated by the mechanisms discussed above.

Ruminants may be particularly susceptible to radiocontamination. In their natural habitat, they either graze grass or lichen, or browse on trees or shrubs. As a result, they are highly efficient gatherers of surface contamination. For example, a cow at pasture daily consumes an airborne contaminant equivalent to that deposited on 20 square meters of ground. The unabsorbed ingesta thus constitutes an important source of internal exposure, particularly to the female gonads. It is estimated that the whole body internal exposure from mixed fission products ingested by a cow grazing from a single above-ground nuclear detonation would approximate the external exposure from material on the ground surface.

When a contaminant passes through the food chain, its concentration sometimes changes several orders of magnitude. The concentration of most elements decreases as they pass through the plant-herbivore-carnivore trophic levels. A few elements are concentrated, notably sodium in invertebrates and cesium in mammals. Ninefold increases in $^{137}$Cs have been reported the the plant-mule deer-cougar chain, fourfold increases in the lichen-caribou-wolf chain, and threefold increases going from food to the human body.

## Transport from the environment to humans

Exposure of man from environmental radiocontamination (apart from nuclear war or accidents) comes mainly from contaminated food. In the western world, milk has in general proven to be the most important conveyor of this contamination. Meat, poultry, and eggs are a potentially significant source, too.

The bulk of the work on transfer of radionuclides to man through ingestion of milk has been confined to isotopes of iodine, strontium, and cesium. Average transfer coefficients for $^{131}$I determined for cows under laboratory conditions range from 0.5 to 1.0 percent of the cow's daily intake per liter of milk. Values from field trials have range from 0.12 to 2.4 per cent of the cow's daily intake per liter of milk. Rough calculations suggest that cows continuously grazing on pasture carrying 1 microCi of $^{131}I/m^2$ would produce milk with a concentration of about 0.2 microCi/l. The transfer of $^{131}$I to muscle tissue in terms of percentage of daily intake per kilogram has been reported as 0.15% for the cow and 3% for the sheep.

A number of studies have been done on $^{90}$Sr contamination. Under laboratory conditions, the transfer coefficient from the cow's food to milk has averaged 8% of the cow's daily contaminant intake per liter of milk produced.

In parts of Florida, a combination of high milk and beef levels of $^{137}$Cs have led to animal body burdens three times higher than those reported elsewhere.

## Summary of environmental transport of radionuclides

Man's welfare depends on an equilibrium reached with the environment over millions of years of evolution. In the last forty years, dozens of radioactive waste products from electrical, industrial, military, and medical activities have been added to that burden. The exposure thus produced now roughly equals that produced by natural processes. The effects of these new, human-induced exposures on man's equilibrium with his environment are largely unknown. This state of affairs is, to be blunt, a poorly considered bet.

The next chapter surveys what is known about the consequences of losing that bet--the particular effects of natural radioactivity on human health.

## NOTES

1.-2. The data in these two chapters are derived from UNSCEAR reports (United Nations Scientific Committee on the Effects of Atomic Radiation, Ionizing Radiation: Sources and Biological Effects, United Nations, New York, 1982). The reports represent the world scientific consensus on the title topic. UNSCEAR reports are issued every five years and contain a wealth of information; they also provide the best general bibliography available on the subject.

3. More apologies. In the introduction, I promised that the concepts of calculus were not presumed of the reader, and I backslid a little there. Here, I use the prime notation to denote the time derivative of a function of time, that is

$$f'(t) = df(t)/dt$$

where t is time.

The notion of an integral, denoted by

$$\int_a^b f(x)dx$$

is used here only for the sake of notational convenience no deep knowledge of its meaning is assumed. Intuitively, the integral of a function f(x) between x = a and x = b, denoted as shown above, may be thought of as the area bounded by the lines y = f(x), the x-axis, and the lines x = a and x = b. The symbol

$$\int$$

is intended to suggest a large "S" for "sum;" more particularly, f(x) (for a specific value of x) can be thought of as the length of a rectangle which lies "between" the graph of f(x) and the x-axis; dx ("deltax") is intended to suggest the width of this rectangle. Thus

$$\int_a^b f(x)dx$$

is intuitively the sum of the areas of a large collection of small rectangles, whose collective area comprises the area bounded by $y = 0$, $x = a$, $x = b$, and $y = f(x)$. Formally, we may define

$$\int_a^b f(x)dx = \lim_{N \to \infty} \sum_{i=1}^{N} f(m_i)(x_i - x_{i-1})$$

where

$$a = x_0 < x_1 < x_2 < \cdots < x_N = b$$

and

$$x_{i-1} < m_i < x_i$$

provided that the limit exists; with trivial exceptions, the integral is otherwise undefined.

It can be shown, furthermore, that if the integral exists, there is a function $F(x)$ with the following property

$$\int_a^b f(x)dx = F(b) - F(a)$$

where

$$F'(x) = f(x)/dx.$$

Thus, integration (evaluating the integral) of a function $f(x)$ "undoes" (inverts) the differentiation of that function.

In the case of physical or biological quantities which are functions of time, the derivative of that function represents the instantaneous time rate at which that quantity is changing; the integral of that time rate over a given time interval represents the total change in that quantity over that time interval.

# 4
# Biological Effects of Ionizing Radiation

## INTRODUCTION

The previous chapter surveyed various mechanisms by which humans are exposed to ionizing radiation, including how they are exposed to natural radioactivity in water supplies, without identifying precisely what the biological consequences of that exposure are. This chapter is an introduction to what is known about those consequences.[1]

Because the damage caused by ionizing radiation is often irreversible and can sometimes be transmitted from one generation to another, we have relatively little controlled direct data on the effects of chronic (long-term) low-level exposure regimes for humans. There is substantial data on exposures from animal studies, however, and these have been used as a basis for radiation protection standards for man. Sadly, we do have data on entire human populations for large short-term exposures.

## SOME FEATURES OF HUMAN GENETICS

At the concentrations likely to be found in water supplies, ionizing radiation is not likely to cause death directly. Rather, the most serious biological consequence of exposure from such a source would involve genetic damage. Such damage is particularly insidious if it occurs in germ cells (gametes), because it has the power to propagate through reproduction, affecting generation after generation, long after the energy which caused the original damage has ceased to be measurable. This section briefly reviews[2] some of the genetic terminology necessary to describe those effects.

## Genetic structures

According to genetic theory, the biological information which can be transmitted from one generation to the next resides in the nucleus of those cells which contain such structures (red blood cells, for example, do not have nuclei). More particularly, this inheritable information is contained in a network of dark-staining material called chromatin. Chromatin contains DNA, whose chemical structure, properly speaking, embodies the genetic code. The DNA is arranged along fine threadlike structures called chromosomes. Chromosomes are composed of relatively stable, identifiable subunits called genes, which are strung more or less end-to-end. In nongerm cells, chromosomes occur in pairs, and the members of each pair are called homologs (literally, "similar structures"); alternately, each member of such a chromosome pair is said to be homologous to the other member of the pair. Ordinarily, the gene sequence of each member of a homologous pair is identical to the sequence of the other member. There is a slight, but very important wrinkle, however: although the gene sequence on homologous chromosomes may be identical, corresponding genes in those sequences may have very slightly different chemical structures. Corresponding genes on homologous chromosomes which differ in this way are called alleles ("alternate forms") of the same gene. Genes, then, are not unique molecular arrangements, but tightly related families of such structures.

The total genetic complement of a cell is called its genome. Allelic recombinations and variations which arise during the genesis of germ cells and during reproduction often give rise to variations in the genomes of offspring, and hence, give rise to variations in the structure and chemistry of the tissues and fluids in the body; these are the principal mechanisms for normal variation.

In each of the normal nongerm cells of all mammalian species, there is a fixed number of chromosomes. This number is called the diploid (literally, "two-fold") number. For any individual, only half the diploid number, or haploid number, of these chromosomes are distinct in non-allelic ways, because the chromosomes occur as homologous pairs. In man, the diploid number is 46. Half this number, 23, thus represents the number of "non-allelically" distinct human homologous chromosomes in any given normal individual.

## Mitosis

The growth of multicellular organisms occurs primarily through a type of cell reproduction called mitosis. A cell from which others arise by mitosis is called a parent cell (relative to its offspring); cells which result from a mitotic division are called the daughter cells (relative to the parent from which they come). A fundamental feature of normal mitosis is that it replicates, cell-by-cell, the genetic complement of the parent cell.

Cell growth in mammals (and most other animals) consists of a sequence of phases. In the first of these, called interphase, the DNA in the nucleus replicates. At the same time, a small ellipsoidal structure adjacent to the nucleus called a centriole replicates. The resulting centrioles then migrate to diametrically opposed positions on the outside of the nucleus. As the centrioles separate, a set of filamentous microtubules appears, joining them.

The first phase of phase of mitosis, called prophase, then begins. During prophase, the chromatin condenses and the chromosomes become more and more distinct. At this point, each chromosome consists of two intertwined parallel longitudinal subunits called chromatids. As prophase continues, the chromosomes condense into shorter units. In the process, a small, less condensed feature called a centromere becomes visible on each. The set of filamentous structures which connected the centrioles in interphase begins to evolve into a network of spindle fibers, some of which are connected to the chromosomes. The centrioles evolve into a distinct pair of diametrically opposed termini, called poles, in the spindle net. Eventually the spindle net consists of just two types of spindle structures. One set of spindle fibers joins the poles directly; each member of the remainder connects one pole to the centromere of a chromosome, and thence, to the other pole. Concurrent with the evolution of the spindle net, the nucleus becomes a less and less distinct structure, finally vanishing altogether.

This development continues and the cell enters the second phase of mitosis, called metaphase. During metaphase, the chromosomes align in a plane, called the equatorial plane, midway between the spindle poles.

With the equatorial alignment complete, the next phase of mitosis, called anaphase, then begins. During anaphase, the sister chromatids separate, and move toward opposite poles of the spindle net. The net itself becomes less and less distinct.

When the polar movement of the chromosomes is complete, the next phase of mitosis, telophase, begins. During telophase, a new nuclear membrane forms around each set of chromosomes, and the cell body itself separates to form two new daughter cells.

The result of mitosis is a pair of cells, each of which in normal circumstances has an exact duplicate of the original genetic material of the parent cell. The cell is again in interphase, and the mitotic cycle begins again.

## Gametogenesis

The generation of germ cells (gametes), called gametogenesis, is improbably elaborate in mammals. It consists of several complex stages and phases. The development of the mature germ cell is demarcated by two general stages called meiotic divisions, by means of which the number of chromosomes in the mature germ cell becomes haploid. Prior to the first meiotic division, the diploid number of chromosomes is doubled; the first meiotic division reduces this number to the diploid number. The second meiotic division, in turn, reduces the diploid number to the haploid number. The events culminating in the completion of both the first, and the second, meiotic divisions themselves have recognizable phases (prophase, metaphase, anaphase, and telophase), much akin to those of mitosis. To distinguish these phases in each of the two meiotic stages, the names of each of the phases are appended to the roman numeral of the meiotic stage. Thus, for example, "prophase I" designates prophase of the first meiotic stage; "anaphase II" designates anaphase of the second meiotic stage.

In prophase I, the chromatin threads that were present in the interphase primordial germ cell condense and thicken. DNA replication completes; each chromosome then exists as a pair of identical partners, called sister chromatids, attached to a common centromere. Each chromosome and its homologous partner, furthermore, then configure so that they lie more or less parallel to one another. At this point, the homologous pair is known as a bivalent. Each bivalent therefore contains four chromatids, and is accordingly often called a (chromatid) tetrad. The chromosome pairs then shorten and thicken and twist about one another; this process is called synapsis. At this point a piece of one chromatid of the paternally derived chromosome is exchanged for a homologous piece of one chromatid of the maternally derived chromosome. This process thus produces chromosomes that contain a mixture of paternally and maternally derived genes. The homologous chromosomes then partially separate, remaining attached at specific points called chiasmata. The chromosomes then thicken and the chiasmata vanish; the nuclear membrane and the nucleolus also vanish and the tetrads move into an equatorial plane.

The completion of these processes mark the beginning of metaphase I. During metaphase I, a spindle network forms with the bivalents arranged at random on the equatorial plane, and the centromeres of the two homologous chromosomes orient toward opposite poles of the spindle net.

Anaphase I then begins. One maternal pair of the chromatids goes to one pole of the spindle and a paternal pair goes to the opposite pole. This process is known as segregation. Segregation occurs in a random fashion; maternal chromatids from one bivalent may migrate on the spindle with maternal or paternal chromatids from other bivalents. Thus some maternal and some paternal pairs of chromatids may go to one pole, and those of opposite origin to the other pole in the spindle net.

Telophase I then begins. It ends in the same general way that telophase in mitosis does, with new nuclear membranes forming and the cell walls of the two daughter cells separating.

The phases of the second meiotic division are similar, but not identical, to those of the first. In this division, however, a spindle again forms and the chromosomes align along an equatorial plane. The centromeres divide, and the sister chromatids separate, moving as distinct chromosome to opposite poles of the cell. The cytoplasm then divides and the chromosomes elongate.

The result of the two meiotic divisions is the formation of four cells, each containing the haploid number of chromosomes.

Gametogenesis in males, more particularly known as spermatogenesis, occurs in mammals in the following way. The (diploid number of) chromosomes of a parent germ cell called a spermatogonium duplicate within that cell. The resulting cell, called a primary spermatocyte, thus contains twice the diploid number of chromosomes. The primary spermatocyte then divides into two cells, passing a diploid number of chromosomes to each of its daughters, which are called secondary spermatocytes; this division constitutes the first meiotic division. Each of the secondary spermatocytes then divides, passing a haploid number of chromosomes to each of its daughters, called spermatids; this division constitutes the second meiotic division. The spermatids then metamorphose into mature sperm cells, or spermatozoa.

Gametogenesis in females, more particularly known as oogenesis, is somewhat different in mammals than spermatogenesis. As in spermatogenesis, chromosomes first duplicate within the parent cell, called the oogonium, to form a primary oocyte containing twice the diploid number of chromosomes. The primary oocyte then divides (the first meiotic division) into two cells of unequal size, each containing a diploid number of chromosomes. The smaller of these two cells, called the first polar body, subsequently undergoes mitosis, but its daughters do not mature into female gametes. The larger of the two cells which results from the first meiotic division, the secondary oocyte, undergoes division (the second meiotic division), again producing two cells of unequal size. Each of these cells contains a haploid number of chromosomes. The smaller of these two cells is called the second polar body, and it does not mature into a female gamete. The larger of the two cells, called an ootid, then metamorphoses into mature female gamete, called an ovum.

## Development of the zygote

When a male gamete unites with a female gamete, a fertilized cell, or zygote, is produced. The zygote contains genetic information from each parent. In normal zygotes, half of the chromosome complement comes from each of the ovum and sperm.

The chromosomal makeup of male and female humans is somewhat different. In human females there are exactly 23 pairs of chromosomes that make up the genetic complement. In human males, however, there are 22 pairs, plus two non-paired chromosomes of unequal size. Furthermore, the larger of the two unpaired chromosomes in human males is identical to a member of the 23rd pair of human female chromosomes. The members of the 23rd pair of human female chromosomes, identical to the larger unpaired human male chromosome, are called "X" chromosomes. The smaller of the unpaired set of human male chromosomes is called the "Y" chromosome. Thus a female human will have two "X" chromosomes forming the 23rd pair, whereas a male human will have one unpaired "X" and one unpaired "Y" chromosome corresponding to the female "XX" pair. Because individuals who have an "XX" combination are invariably females and individuals who have an "XY" chromosome combination invariably males, the "X" and "Y" chromosomes are called the sex chromosomes. The chromosomes of the remaining 22 pairs of non-sex chromosomes (in humans) are called autosomes.

Normal meiosis separates the members of every homologous chromosome pair in the spermatogonia or oogonia which head the meiotic cycle. In the female, this means that all gametes developing from the oogonia will contain only the "X" sex chromosome, whereas in males, meiosis produces sperm, half of which have only the "X", and half of which have only the "Y," sex chromosomes. If a human female gamete is fertilized by a "X"-type sperm cell, then the resulting individual will be female; if a "Y"-type sperm fertilizes the ovum, the resulting individual will be male.

Once formed, a normal zygote undergoes a series of mitotic divisions. The first few of these divisions, collectively called cleavage, are synchronous in the sense that all cells in the cell mass undergo division at about the same rate. The number of cells in the growing mass during cleavage thus increases by a factor of two with each mitotic cycle. Cleavage, oddly enough, does not increase the size of the growing mass: even at the 16-cell stage, the cell mass is about the same size as the original zygote. However, after the 32-cell stage or so, cell division is no longer synchronous. Instead, a hollow ball of cells, called the blastula, develops; cells from the original synchronous phase partially fill the blastular cavity, or blastocoel.

A nonuniformly distributed mass of cells begins to form on the outer surface of the blastocoel. A single layer of cells, called the trophoblast, forms in the blastula. When initial trophoblast development is complete (in humans), the resulting structure migrates and then implants in the uterine lining. Differentiation of the various tissues and structures in the body then begins.

## EXISTING FREQUENCY OF GENETIC AND CHROMOSOMAL DISORDERS

Chromosomal and genetic damage are the most important of all the known biological effects of ionizing radiation. For the sake of convenience, we will say that a defect is chromosomal if it arises from a feature of an entire chromosome; we will say it is genetic if the defect is caused only by a defect in a single gene. (This distinction is a little fuzzy, because in a sense, genes compose chromosomes; thus a genetic defect is a kind of chromosomal defect.)

In order to characterize the chromosomal and genetic effects of ionizing radiation in humans, it is useful to know the existing frequency of such disorders. Knowing that frequency, furthermore, helps place the consequences of instituting a proposed radiation exposure ceiling in perspective.

Some part of the existing frequency of chromosomal or genetic disorders, it is reasonable to believe, is inherited, and some part of it is environmentally induced. Evidence for the existing incidence of genetic and chromosomal defects is both indirect and direct.

### Existing frequency of malformations

Malformations present at birth provide strong, but indirect evidence on the incidence of genetic and chromosomal disorders. The evidence is indirect in at least two ways. First, in a given instance, a malformation may not be of genetic or chromosomal origin even when all of its clinical ("easily observed by conventional clinical examination techniques") features have been associated with known genetic or chromosomal defects in other cases. Second, not all genetic or chromosomal defects immediately manifest themselves at birth; some of the more insidious produce obvious clinical indications only several years later. Nevertheless, probably the most common serious malformations present at birth are of genetic origin. With few notable exceptions, the specific mechanisms underlying malformations are not currently known, and progress toward such understanding has been relatively recent. For example, one of the most common serious chromosomal disorders, trisomy-21, was not accepted as a cause of Down's syndrome until 1959.

Roughly 10% of all babies are born with some malformation. This figure is nominal, however: various studies report a malformation incidence ranging from 4% to 16%. The higher of these values occurred only in a study which observed children from birth through age five; the lower of the figures comes from a study in which children were observed only for a few days after birth. If anything, therefore, the higher incidence value may provide a more nearly complete picture of the malformation rate. The range of severity of malformations is great, from defects which are of little more than cosmetic significance to lethal disorders. Major malformations by definition include anencephaly, microcephaly, congenital dislocation of the hip, cataract, cleft palate, cleft lip, pyloric stenosis, hypospadias, cystic kidney, and others. About half of the of children born malformed have at least one of these major malformations. Minor malformations, such as polydactyly (extra digits), syndactyly (fused or "webbed" digits), low-set ears, supernumerary nipples, postural foot problems, and so on, are present in about half of children born malformed. A combination of major and minor malformations is present in about 1%; of the major malformations, about 42% are multiple; 25% of the minor are multiple. The cardiovascular system has the highest frequency of multiple malformations (about 80%). As a group, children who die within the first year of life have a higher incidence of malformations than any other group of newborns. Males have significantly more malformations than females; this is due entirely to a larger frequency of major malformations among males. There appears to be no significant difference in the overall frequency of major malformations between so-called "whites" and "non-whites," although "non-whites" have a significantly larger frequency of minor malformations than "whites," primarily polydactyly, branchial cleft anomalies, and supernumerary nipples.

The frequency of multiple malformations is significantly higher in "whites" than in "non-whites."

Some evidence has been collected on the transmissibility (inheritability) of the most common malformations by comparing the frequency of malformations in siblings with the frequency of malformations in the offspring of those siblings. Such data consistently shows a frequency of malformation of about 2% to 5%, regardless of generation. This result suggests that at least some malformations have an environmental component, because strict Mendelian theory predicts a substantially higher frequency among offspring of malformed parents than is observed.

## Structural and numerical chromosome abnormalities

Improper segregation of chromosomes during meiosis or during the early zygote cleavage stages is thought to be the most common cause of chromosomal defects in newborns.

There are a number of different kinds of chromosomal abnormalities. There may be, for example, too many or too few chromosomes per cell; if such a defect occurs in the autosomes, it is called a numerical autosomal anomaly. If one set of parental chromosomes is missing, the defect is called haploidy; if the diploid set is doubled or tripled, the disorder is called polyploidy. In general, the condition of not having the rightfold number of chromosomes is called aneuploidy.

In addition, individual chromosomes may be damaged or parts of them may be rearranged. For example, a piece of a chromosome may break off (a deficiency). A chromosome may be "read" twice at meiotic or mitotic duplication; if this occurs, it is called a duplication. Sections between two or more nonhomologous chromosomes may exchange; if this occurs, the defect is called a translocation. If a single break occurs in each of two nonhomologous chromosomes and the parts recombine, with one piece taken from each of the nonhomologous pair, the result is called a reciprocal translocation; such aberrations are rare in nature but occur with elevated frequency in radiation damage. An entire section of a chromosome may be rotated 180$^{o}$, reversing the order of genes on that chromosome; such a defect is called an inversion. Entire chromosomes may also be lost or gained; a cell which is short exactly one chromosome is called monosomic. A polysomic genome is one that is short more than one chromosome; in polysomic genomes, one kind of chromosome may be represented multiple times.

A substantially rarer chromosomal defect arises when one of the cells in the developing undifferentiated embryo retains an extra chromosome of one of the pairs, while the other cell receives no member of that pair and dies. Thus, some of the cells in the organism will contain too many chromosomes; others will contain the normal number. The existence of different (numerically or otherwise) genomes in different cells of the same individual is known as mosaicism. The earlier in development mosaicism occurs, the greater the number of cells which will contain the aberrant chromosome mix, and, presumably, the greater the impairment (if there is any) of the individual will be.

Sometimes mistakes in the meiotic process produce anomalies in the sex-chromosome makeup of the individual. Defects in the sex-chromosomes are called sex-chromosomal anomalies. If an ootid which had no sex chromosome happens to be produced, for example, the resulting offspring can possess a so-called "XO" chromosomal makeup. This disorder is the underlying cause of Turner's syndrome. It is not fatal, and apparently occurs in about 1 in every 2,500 births.

If a meiotic mistake occurring during oogenesis allows both X chromosomes to continue to reside in an ovum, and the resulting ovum is fertilized by a Y-type sperm, the zygote will possess an XXY complement of 47 chromosomes. The sex of an individual with this complement will be male. This phenomenon is the cause of Klinefelter's syndrome; it appears to occur at the rate of 1 in 450 males. Other, similar aberrations are known, and include males with XXXY, XXXXY, and XXXXXY complements. There is also an XYY complement which occurs at the rate of about one in every 300 male births.

For the reasons given above, it is not entirely clear what the relation between perinatal (at, or near, birth) malformations and chromosomal abnormalities is, although a large part (at least a fifth and possibly most) of perinatal malformations are thought to be caused by chromosomal or genetic defects of some sort. We do in any case have some, but limited, direct evidence on the existing frequency of chromosomal anomalies. This evidence underestimates the incidence of existing chromosomal defects among newborns, because it is obviously limited to known chromosome defects. These known defects occur at the rate of between 0.4% and 1.0% of newborns. Of this group, about one-third are sex-chromosomal abnormalities, about one-quarter are numerical autosomal anomalies, about one-third are balanced structural anomalies, and one-tenth are unbalanced structural anomalies.

It is very likely that not all chromosomal anomalies are apparent during ordinary clinical examinations. Some obviously are, and a number of attempts have been made to determine the proportion of genetic defects that are "clinically significant."

Exactly what we should mean by "clinically significant" is no doubt open to dispute, but a conservative approach would be to limit the term to cases in which the correlation between known genetic defects and clinical indications is very strong. Such an approach would at least provide a plausible lower bound on the proportion of chromosomal defects which are "clinically significant" when that term is taken in its broadest sense. At least one study has accordingly taken "clinically significant" to mean all the non-mosaic XXY and 45,X instances, XYY and (n)XY (male) genotypes, all autosomal trisomies, and all unbalanced structural rearrangements reported to have been associated with congenital malformations at birth, as a basis for correlation. If this conservative approach is used, one can conclude that about half of the chromosomal anomalies (2.91 per 1000 newborns) are "clinically significant."

## Chromosome anomalies and perinatal deaths

There is some weak evidence linking chromosome anomalies with perinatal deaths. The frequency of chromosomal anomalies among stillbirths, in particular, is estimated to be of the order of 5-6%. Among the anomalies thus recorded, trisomies predominate, particularly those involving chromosomes of group E, followed by chromosomal structural anomalies, triploidy, and others.

## Chromosome anomalies and spontaneous abortions

A stronger correlation appears to exist between chromosome anomalies and spontaneous abortions: the overall frequency of chromosomal anomalies among spontaneous abortions may be as high as 50%. Trisomies as a group constitute the most common type of chromosomal anomaly among spontaneous abortions, accounting for about 50%, followed by monosomy-X (18%), triploidy (17%), tetraploidy (6%), and others (7%, including double trisomies, mosaics and structural rearrangements).

## Mutations

Mutations also contribute to the existing incidence of genetic or chromosomal defects in a population. Most mutations which have been generated in the laboratory are deleterious, and there is little reason to believe that nature is any more beneficent on this count. The data we have suggests that the following estimates of existing mutation rates are reasonable: 8.3 x $10^{-4}$ per gamete per generation for sex-chromosome errors, and 6.9 x $10^{-4}$ per gamete per generation for autosomal errors. For balanced structural rearrangements of autosomes and for unbalanced rearrangements, the rates are probably about 1.8 x $10^{-4}$ mutations per gamete per generation, and 0.45 x $10^{-4}$ mutations per gamete per generation, respectively. The order-of-magnitude agreement among these various types of mutation suggests a similarity in mutation dynamics among the types. Collectively, these suggest an existing mutation rate for the types indicated of about 16 x $10^{-4}$ mutations per gamete per $10^{-2}$ generation.

## Genetic defects

Genetic defects are classified according to the way they are inherited. They may be inherited through recessive non-sex-linked alleles, through dominant alleles, or through sex-linked alleles. Some genetic disorders are multifactorial, that is, they are caused by the joint action of several different alleles of several genes. The majority of known genetic defects are transmitted through a recessive allele; an individual must therefore be homozygous in this allele to have the disorder. Known recessive genetic defects include phenylketouria, sickle-cell anemia, Cooley's anemia, Tay-Sachs disease, and cystic fibrosis. Dominant genetic defects, in contrast, are expressed merely if the individual is heterozygous in the defective allele.

In most sex-linked genetic defects, the female parent carries the recessive allele that results in the defect on an X-chromosome of the offspring. A female offspring of this parent has a high probability of carrying a dominant allele on the homologous partner of the chromosome containing the defective gene, and thus will be protected from the expression of the recessive gene. A male offspring, however, will not be so lucky: since he does not have another X-chromosome which could carry a corresponding protective dominant allele, he will have the disorder. Sex-linked genetic disorders range from red-green color blindness to hemophilia and other severe defects.

## The cancer connection

A large number of single gene traits are known to predispose to, or are complicated by, neoplasia (tumorous growths). Among these traits, about half are autosomal dominant, one-third are autosomal recessive, and one-sixth are X-linked. The better known cancerous neoplasias which are predisposed by single gene traits, or are complicated by them, are xeroderma pigmentosum (XP), ataxia telangiectasia (AT), Fanconi's anemia (FA), Bloom's syndrome (BS), and retinoblastoma. All these except retinoblastoma are inherited as autosomal recessives. In AT, FA, and BS, furthermore, spontaneous breakage of the chromosomes can be observed in peripheral blood lymphocytes and in cultured fibroblasts.

Xeroderma pigmentosum (XP). The principal feature of XP is a marked sensitivity of the skin to sunlight-induced damage, manifested as sunburns, freckling, hyperpigmentation and keratoses, eventually leading to multiple skin carcinomas and melanomas which are the final cause of death, usually before age 30. The basis for predisposition to cancer is a metabolic abnormality in the repair of UV-induced damage to DNA. XP individuals have been found in all populations studied. The frequency of the disorder in North America and Europe is about 1 in 250,000, whereas in Japan the frequency may be as high as 1 in 40,000.

Ataxia telangiectasia (AT). The clinical indications of ataxia telangiectasia are progressive cerebellar ataxia (inability to coordinate voluntary body movements due to disorders of the cerebellum), conjunctival and cutaneous telangiectasia (chronic dilation of the small arteries and capillaries, causing reddish tumors, particularly on the skin), frequent sinopulmonary infections, subnormal immunity, a generally underdeveloped lymphoid system and a predisposition to cancer. Most of the cancers reported in AT patients involve the lympho-reticular system; less frequently, epithelial (of or relating to tissue that covers body surfaces, forms glands, and lines body cavities) tumors and leukemia become involved. Death often occurs before age 20 either from sinopulmonary infections or from malignancies. In many, but not all AT patients, there are increased chromosome fibroblasts. Homozygotes for the AT gene may occur as often as 1 in 40,000 births. Assuming Hardy-Weinberg equilibrium, the frequency of heterozygotes implied by this homozygote frequency is 1%; there is some evidence which suggests that the Hardy-Weinberg assumption overestimates the AT heterozygote frequency, however. AT heterozygotes are about twice as likely to die of malignancies as others who have malignancies.

Fanconi's anemia (FA). Fanconi's anemia is a chromosome instability syndrome associated with progressive marrow failure. The clinical features of FA include progressive underproduction of red cells, white cells, and platelets. This leads to anemia, leukopaenia (a decrease in the number of leucocytes in the blood), and thrombocytopaenia (a decrease in the number of thrombocytes), hypoplasia (underdevelopment) or aplasia (failure of development) of the radius and thumb, growth retardation, and brownish pigmentation of the skin. Skeletal malformations and anomalies of the heart and kidney may also occur. Death generally occurs in the early years, but those who survive longer have a substantially increased risk of acute leukemia. Affected patients are also at a greater risk of developing squamous cell carcinoma of the mucocutaneous junctions (that is, where mucosal and skin surfaces join, such as in the mouth or at the anus). FA victims show increased chromosome breakage and rearrangement, most evident in fresh bone marrow preparations and in lymphocytes. The breaks are typically seen in the chromatids, and the interchanges among them are mainly between non-homologous sites. The incidence at birth of FA is about 1 in 350,000 in the North American population and may be as high as 1 in 70,000 in mid-Europe.

Bloom's syndrome (BS). The main clinical features of BS are severe growth retardation, a telangiectatic erythema (a reddening of the skin caused by capillary congestion) in exposed areas and sun-sensitivity. Many BS victims have serious respiratory and intestinal infections, and their immune system is impaired. The risk of cancer is greatly increased in BS patients: primary cancer develops in about one out of six of them. About half the cancers are leukemia of the nonlymphocytic type. The cells from BS patients contain highly elevated frequencies of sister chromatid exchanges (roughly 90 per cell in leucocytes, fibroblasts, and bone marrow).

Retinoblastoma. Retinoblastoma is a tumor of the precursors of the rod and cone cells in the retina of embryos. It develops into a malignant eye tumor in children. Mortality is associated with the direct extension of the tumor into the cranial cavity, into the membranes which cover the brain and finally, into the brain itself. Estimates of the incidence of retinoblastoma range between 1 in 30,000 to 1 in 15,000. Retinoblastoma is often considered a classical example of a dominantly inherited tumor. About 60% of all retinoblastomas are unilateral and non-heriditary, however; 15% are unilateral and hereditary, and 25% are bilateral and hereditary. The penetrance is of the order of 90 to 95% for hereditary cases. Patients with bilateral, and possibly all patients with hereditary retinoblastoma, run an increased risk of other cancers, particularly bone sarcomas.

Aniridia-Wilms' tumor-urogenital abnormalities association. Aniridia is the absence or defect of the iris, and more specifically, a congenital underdevelopment of the iris. It is usually bilateral and is transmitted as an autosomal dominant trait. Approximately 1 in 50,000 of the general population is affected by it. About 30% of the cases are sporadic (i.e., not traceable to hereditary causes), and are thus presumed to represent new mutations.

Wilms' tumor is a tumor of the kidney which develops in the embryo. The incidence of this disorder has been estimated to be on the order of 1 in 10,000 live births. Wilms' tumor is most often discovered between the ages of 3 and 4, by which time it is extremely malignant.

The presence of aniridia somehow renders an affected child prone to the development of Wilms' tumor. The sporadic cases seem more at risk, because about one-third of these develop Wilms' tumor. The risk of Wilms' tumor seems highest when aniridia is accompanied by genito-urinary tract malformations and mental retardation. It now appears that the aniridia-Wilms' tumor association with mental retardation and genito-urinary abnormalities in males is caused by an interstitial deletion of the short arm of chromosome 11.

## GENETIC EFFECTS OF IONIZING RADIATION ON EXPERIMENTAL MAMMALS

There are obvious ethical problems in intentionally subjecting humans to elevated levels of ionizing radiation. Most of the information we have about such effects therefore comes from controlled studies on other animals. The genetic and chromosomal effects of these experiments can be classified into several types: interference with reproductive capacity, gene translocations, loss or addition of chromosomes, point mutations, effects of internal emitters, and miscellaneous effects.

### Dominant lethals and reproductive capacity

Short of causing death directly, one of the most serious effects of ionizing radiation is its power to profoundly interfere with reproduction by inducing lethal mutations in germ cells.

A technique frequently used to measure the induction of dominant lethals involves irradiating male mice, then mating them to females and observing the frequency of live and dead uterine implants. By imposing various delays between irradiation and mating, one can isolate the effects on various stages of germ-cell development. Immediate matings sample spermatozoa or spermatids, more delayed mating sample spermatocytes, and the most delayed sample spermatogonia. Such a technique provides relatively little detailed information on the mechanisms of genetic damage, but it does establish a coarse bound on those irradiation regimes which can produce drastic and immediate genetic effects.

Irradiation of males and mid-pregnancy lethality. For example, three different sets of 100-120 day old $B6CF_1$ hybrid male mice were respectively exposed to alpha, gamma, and neutron radiation. Each of the irradiated males was then mated twice a week with previously unmated, unirradiated 100-150 day old $B6CF_1$ females. This procedure continued for up to 45 weeks. The females were dissected at mid-pregnancy (10 to 17 days after conception) and the number of live and dead uterine implants was counted. These two numbers, when compared to controls, provided a measure of the dominant-lethal induction effect of the irradiation. (All doses reported in this series were midline-tissue-absorbed doses (MDL) as measured in a tissue-equivalent phantom. There is some indication that the dose to the testes may have been as much as 10% higher than the MDL.)

In the first of these series, male mice were injected with $^{239}Pu$ (a strong alpha particle source) at the rate of 0.19 and 0.37 MBq $kg^{-1}$ body mass and mated as described above. (A microassay showed that 0.05% of the injected dose reached the testes, and was retained there at that level over a 420-day period.) A dose/implant-lethality relation was easily detected in the dissected females. Assuming that the dose to the male gametes was uniformly distributed among the gametes present at mating, the alpha irradiation regime in this series produced about $64 \times 10^{-4}$ dominant lethals per male gamete per $10^{-2}$ Gy greater than was produced in the controls.

In the gamma-irradiation series, pre- and post-meiotic cells in male mice were exposed to gamma radiation from $^{60}Co$. The males were then mated with unirradiated females, as described above. Single gamma doses in the range of 0.45 to 5.7 Gy were delivered to one group of mice at the rate of 0.3 to 0.4 Gy $min^{-1}$. Weekly gamma doses of between 0.08 and 2.1 Gy were delivered to a second group of mice in a 45 minute period each week; the dose rates for this series were $2 \times 10^{-3}$ to $4.7 \times 10^{-2}$ Gy $min^{-1}$, respectively. A third group of mice was exposed continuously to gamma irradiation (a "chronic" gamma irradiation regime) for 22 hours per day at the rate of 0.03 and 0.06 Gy $d^{-1}$, delivered at $2.5 \times 10^{-5}$ and $4.5 \times 10^{-5}$ Gy $min^{-1}$, respectively. For each regime, dominant lethality was determined by counting live and dead implants resulting from the mating. For post-meiotic chronic gamma exposures (at the rate of $3.3 \times 10^{-2}$ Gy per day and $5.98 \times 10^{-2}$ Gy per day with total accumulated doses of between 0.59 Gy and 1.06 Gy, and between 1.05 and 1.89 Gy, respectively), the rate of induction of dominant lethals appears to be about $5 \times 10^{-4}$ per gamete per $10^{-2}$ Gy greater than the controls. The single or weekly exposure regimes produced a mutation rate approximately twice that of the chronic regime for postmeiotic irradiation. Mutation rates for premeiotic exposure regimes were approximately an order of magnitude smaller than for the postmeiotic regimes.

In the third series in this experiment, pre- and post-meiotic male mice cells were exposed to 0.8 MeV neutrons from a research reactor. Single exposures for both pre- and post-meiotic cells in the range of 0.1 to 1.6 Gy were administered to one group of mice at rates of 0.04 to 0.12 Gy per minute. Another group of males was irradiated once a week for 45 minutes for a total of 6 to 24 weeks; weekly exposures were in the range of 0.008 to 0.13 Gy. The dose rates ranged from $2 \times 10^{-4}$ Gy $min^{-1}$ to $3 \times 10^{-3}$ Gy $min^{-1}$. The males were then mated with unexposed females and the frequency of dominant lethals was determined by counting the frequency of live and dead uterine implants. The induction of dominant lethals by postmeiotic irradiation ranged from $54 \times 10^{-4}$ per gamete per $10^{-2}$ Gy for single exposure to $88 \times 10^{-4}$ per gamete per $10^{-2}$ Gy for the weekly exposure regime (both values are the quantity by which the irradiated animal rates exceed the controls). For premeiotic exposures, the rates were about 5% of the magnitudes of the postmeiotic regimes.

The results of this series of experiments produced some comparative data on the dominant-lethal induction effects of the radiation regimes used. First, there were no statistically significant differences in dominant lethal rates for gamma-irradiation delivered either singly or weekly; there were also no differences in the effects of post-meiotic neutron irradiation delivered singly or weekly. Furthermore, the effects of single or weekly neutron exposures of pre-meiotic stages were not observably different. Single acute gamma-irradiation exposures were more efficient than weekly or chronic exposures in inducing dominant lethality in pre-meiotic stages. The experiments also showed that alpha irradiation (from injected $^{239}Pu$) of post-meiotic stages is about as efficient as post-meiotic fission neutron irradiation in inducing dominant lethals, for the regimes studied. For pre-meiotic stages, fission neutrons were 19 times more effective (RBE) than chronic gamma-radiation.

In a related series of experiments, male mice were exposed to low single doses of neutrons and chronic low doses gamma rays. The mice were then mated to unirradiated females, and the uterine contents analyzed as described above. Low single doses of neutrons (2 to 40 x $10^{-2}$ Gy) and gamma rays (0.23 to 1.45 Gy) to males appear to produce a slight but measurable increase in post-implantation mortality.

Similar experiments have also been performed using relatively high-energy neutrons. 60- to 70-day old CBA male mice were administered a 1.5- or a 2.5-Gy dose of 14 MeV neutrons. Each of the irradiated males was mated to three unirradiated females of the same strain for three consecutive weeks. The pregnant females were dissected on the 17th day (late mid-pregnancy) after the matings. At 1.5 Gy the frequency of dead implants was about twice that of the controls, rising to almost four times that of the controls in the 2.5 Gy regime. The results suggest that 14 MeV neutrons (1.5 and 2.5 Gy) may be between one and two times as effective (RBE) as x rays in inducing dominant lethals in post-meiotic male germ cell stages for the same dose.

The method described above for the alpha, gamma, and neutron series was also used to evaluate the effect of x-irradiation of spermatids on reproductive capacity. This experiment showed that the frequency of dominant lethals induced in spermatids by acute x-irradiation (2 Gy) is 1.7 times that induced by protracted (0.002 Gy $min^{-1}$) gamma irradiation, for the same total dose.

Irradiation of males and late-pregnancy lethality. All of the dominant-lethal experiments described above terminated at mid-pregnancy. To determine whether some dominant lethals are expressed at later stages of pregnancy, 10-week old $B6D2F_1$ hybrid male mice were irradiated with 600 R of x rays and then mated to females of the same genotype and age. Each male was mated to two previously unirradiated, unmated females per week for seven days. The females were checked daily for the birth of new litters and the number of live an dead at birth were recorded. The progeny were weaned at three weeks of age; the parental females were then dissected and their uterine horns examined for live and dead implantation scars. The data showed that the irradiation regime caused a significant increase in the number of late-pregnancy deaths (defined as the difference between the number of "live scars" and the number of "live born"): the percentage of late deaths in the irradiated group was four times that of the controls.

Irradiation of males and early-pregnancy lethality. Similar experiments have been conducted to determine the effect on the early stages of gestation of x-irradiation at various stages of spermatogenesis. The development stages examined ranged from cleavage to the early trophoblast outgrowth stages. Random-bred 10-12 week old male mice of the Dub:(ICR) strain were administered a 4.5 Gy x-ray dose at 0.6 Gy $min^{-1}$. They were then mated to females of the same strain; each male was mated to one female twice a week for the first four weeks after irradiation. The females were killed on the second day of pregnancy and their oviducts were flushed to obtain uncleaved ova and two-celled embryos. The ratio of uncleaved ova to two-celled embryos provided a measure of fertilization. Data from this experiment showed that in the irradiated group there is a higher incidence of developmental failures in early cleavage, at the late morula state, and at the late blastocyst stage than in the controls. Furthermore, the experiment showed that dominant lethals are induced more frequently when the male germ cells are exposed as early spermatids and spermatocytes than when the cells are exposed as spermatozoa: of the two-cell embryos which derived from irradiated spermatids and spermatocytes (fertilization by these cells was 14-28 days after irradiation), about 36% were arrested during cleavage. This is about twice the proportion found in controls for the same period. Furthermore, of those that developed to the morula stage, about 27% did not form blastocyst after 72 h in culture, while only 8% of the control embryos failed to develop into a blastocyst under these conditions. The germ cells irradiated as early spermatids and spermatocytes produced about equal proportions of dominant lethals at each embryo development stage, whereas those irradiated as sperm produced dominant lethality primarily at the blastocyst stage.

At lower doses (0.9, 1.8, 2.7, and 3.6 Gy), the relatively greater sensitivity of the earlier stages of spermatogenesis becomes obvious. For these lower exposure regimes, when the germ cells used for fertilization were spermatozoa or spermatids at the time of irradiation, no difference in the efficiency of fertilization could be detected at any dose. If the germ cells used for fertilization were spermatocytes at the times of irradiation, however, the efficiency of fertilization was significantly depressed. Furthermore, embryos derived from germ cells irradiated as spermatids showed increased developmental arrest both before and after blastocyst formation; moreover, the dominant lethals manifesting before blastocyst formation were about equally distributed over all cleavage stages.

*Irradiation of females.* Female germ cells do not react to the reproduction-suppressing effects of radiation in the same way that male germ cells do, for given doses. This is hardly surprising, since oogenesis is significantly different from spermatogenesis.

X-irradiation of females. For example, two groups of CBA x CBA, and CBA x $A_{jax}$ female mice were administered 0.04 and 0.08 Gy doses of x rays. The first group consisted of one-, two-, and three-week old females. The second group consisted of two-week old female fetuses irradiated in utero with the same doses. All the females were mated to CBA or C57BL males when they were 63-64 days old. On the 18th day after the beginning of the matings, the females were killed and their intrauterine contents were examined. The experiment showed that the stage at which the embryos die (early cleavage, late morula, late blastocyst) depended on the radiation dose and the germ cell stages that were irradiated.

Gamma irradiation of females. Gamma irradiation of female mice also affects reproductive capacity. For example, female mice of the H strain were given whole-body irradiation from a $^{137}$Cs (gamma) source over a 35-day period. The total whole body dose was about 5 Gy; the mean dose rate was about $1 \times 10^{-4}$ Gy $min^{-1}$. Following exposure, the females were mated to unirradiated males of the same strain and age for a seven-week period. The females were checked from the 13th day after mating onward; those found to be pregnant were killed, and the numbers of corpora lutea, live and dead embryos were counted. The frequency of pregnant females in the irradiated group was about 60% that of those in the control group. Mean corpora lutea counts in the irradiated females were normal at week 1, but declined progressively (vs. controls) thereafter through week 7. The post-implantation mortality rate was rather low, however, ranging from 2.6 to 6.1%, with no statistically significant difference between the weeks. The pre-implantation mortality rate varied over a wide range (4 to 16%), and showed no definite trend over the weeks observed.

Alpha and chronic gamma irradiation of females. The reproductive capacity of irradiated female mice is altered by exposure to internal alpha and external chronic gamma radiation, also. In a series of experiments involving dosage regimes similar to those described for the gamma series described above, the breeding performance of two groups of irradiated hybrid (C3H/HeH female x 101/H male)$F_1$ female mice was studied. The first group was administered $^{239}$Pu at the rate of 0.19 or 0.37 MBq/kg body mass. The second group was kept in a 0.1 Gy $d^{-1}$ or 0.2 Gy $d^{-1}$ $^{60}$Co-gamma field for up to six four-week periods. The irradiated females were mated to males of the PT stock. In the plutonium series, most of the females were mated 24 hours after injection and were allowed to breed until they failed to produce any live-born litters within two months of the previous one; other females were killed after injection or at later intervals for radiochemical or autoradiographical studies. In the gamma-ray series, the mice were initially put together as trios (1 male to 2 females) in the control area and a female was moved into the radiation field when a vaginal copulation plug was found; she was removed to the control area on the 18th day of gestation and the appropriate male added to allow a mating at post-partum estrus. One day after the birth of the litter, or on the day of birth if a vaginal plug was recorded then, she was returned to the radiation field and only removed when the litter was 18 days old and ready to be weaned. She was then paired with the male again and the procedure was repeated until sterility ensued. The ovarian dose rates from the injected plutonium series were initially $8 \times 10^{-3}$ and $1.7 \times 10^{-2}$ Gy $d^{-1}$, respectively, in the two groups. Actual gamma-ray dose rates averaged around $8 \times 10^{-2}$ and $16 \times 10^{-2}$ Gy per day, respectively. The results show that both gamma-ray regimes affected the duration of fertility and the number of offspring per litter in successive four-week periods more than the plutonium regimes did. In the gamma-ray series, the percentage of fertile females dropped to zero by the fifth and eighth four-week periods (respectively, in the 0.2 Gy $d^{-1}$ and 0.1 Gy $d^{-1}$ groups). In the plutonium series, a similar drop was noted after 12 and 15 four-week periods, respectively in the 0.19 and 0.37 MBq series. The mean number of offspring per litter in the gamma-ray series dropped to one-third of that in the controls in the 0.1 Gy $d^{-1}$ group and to one-sixth in the 0.2 Gy $d^{-1}$ group. In the plutonium groups, this drop was less noticeable, reaching only two-thirds of the control level even in the 0.37 MBq group.

Effects of irradiation on other species. Exposure to radiation has adverse effects on the fertility of other species.

Rats. For example, when rats were given testicular exposures of 450 and 600 R of x rays and germ cells irradiated as spermatogonia were sampled by mating the irradiated males with unirradiated females., there was a significant reduction in litter size. The results were consistent with an dominant-lethal induction rate of between 2 and $3 \times 10^{-4}$ per gamete per Roentgen.

Guinea pigs. Irradiated (4 Gy) female guinea pigs produced fewer offspring in the first litter and the mean litter sizes were slightly reduced in the first six months; from this time onwards up to two years, there was no pronounced alteration in the overall mean number of litters per female, but the irradiated females produced a smaller mean number of offspring than controls.

Hamsters. In the female Djungarian hamster, x-irradiation (4 Gy) produced a marked sterilizing effect: both the total number of litters and litter sizes were lowered.

Summary of effects on reproductive capacity. In general, these experiments strongly suggest that ionizing radiation of all types diminishes reproductive capacity, in some cases by increasing the frequency of dominant lethal genes. There is typically an increasing dose-effect relationship, at least for the exposure regimes studied. Alpha particles and high-energy neutrons (14-15 MeV) appear to be the most effective of all the types in inducing this effect, and are from two to twenty times as damaging as gamma irradiation, for a given dose, depending on regime.

## Translocations

Reciprocal translocations are the predominant kind of chromosomal structural aberration induced by ionizing radiation. Substantial data has been collected on such effects from experiments involving the irradiation of male and female mouse germ cells.

Comparative gamma- and x-irradiation of male mice. The spermatogonia of CD1 male mice were irradiated by exposure to $^{60}Co$ (a strong gamma source) at dose rates in the range of 100 - 0.001 R $min^{-1}$ for a total dose of 100 to 800 R. A second group was irradiated with x-rays at 100 R $min^{-1}$ for the same total dose. The mice were 8-10 weeks old at the start of the study. The animals were killed at different times after irradiation. Testes were processed for cytological preparations. The data obtained from this experiment show that the yield of translocations decreases over the range of 100 - 0.003 R $min^{-1}$, with no significant difference between 0.003 and 0.001 R $min^{-1}$. The rate per Roentgen at the high exposure rate (100 R $min^{-1}$) is about sixteen times that at the lowest rate of 0.001 R $min^{-1}$ after a correction is made for the biological efficiency of gamma rays relative to x rays.

Comparative translocation studies have also been performed on male mice spermatocytes which have descended from spermatogonia irradiated by high-dose-rate x rays and low-dose-rate gamma rays. In two groups of mice, the spermatogonia were respectively given 2 Gy of high-dose-rate x rays and 2 Gy of low-dose-rate $^{60}Co$ gamma rays. The spermatocytes were scored for translocations. The frequency of translocations was about 5.2% in the x-ray group, and about 2.8% in the gamma ray group.

Unequally fractionated x-irradiation of male mice. A number of experiments have been performed on the induction of translocations in male mice germ cells by unequally fractionated x-irradiation. For example, when a 1000 R exposure was administered as 100 R followed by 900 R 24 h later, the yield of translocation (22%) was similar to that which could be obtained from linear extrapolation of the lower exposure. However, when a 900 R exposure preceded a 100 R exposure, the response was much lower (7.4%), yet still much higher than that produced by a single 1000 R exposure (4.5%).

Alpha- and neutron-irradiation of male mice. Similar experiments were performed on male mice using plutonium alpha rays and neutrons. Testes preparations were made from 8 to 60 weeks after injection of plutonium, and the spermatocytes at metaphase I were screened for translocations. The frequencies of translocation configurations observed were such that when they were plotted against total accumulated dose, no dose-effect relationship was evident. Fission neutrons, in contrast, have a quite noticeable ability to induce translocations. When male mice were also exposed to fission neutrons up to total doses of 1.2 Gy, the testes preparations show an incidence of about 7 x $10^{-4}$ translocations/gamete/$10^{-2}$ Gy.

Summary of translocation induction effects. X rays appear to be much more effective than gamma rays in inducing translocations in male mice germ cells. Higher dose rate x-irradiation (100 R $min^{-1}$) is as much as sixteen times more effective in producing translocations in mouse spermatogonia than chronic low-dose rate gamma and low-dose-rate x-irradiation; even single small (2 Gy) doses of high-dose-rate x rays produce about twice the incidence of translocations (5.2% vs 2.8%) of chronic gamma irradiation of the same dose. Spermatocytes at metaphase I appear to be relatively insensitive to the induction of translocations by alpha irradiation at the same dose as that administered in the low-dose (2 Gy) x- and gamma-ray series. Fission neutrons, however, can induce about 7 x $10^{-4}$ translocations per spermatocyte per $10^{-2}$ Gy. Unequally fractionated x-irradiation (900 R, followed 24 h later by 100 R) produces about the same yield of translocations (22%) as that which is obtained by linearly extrapolating the lower dose of the regime over the entire exposure interval. Reversing the order of the x-ray fractions gives a much lower yield, but one which is nevertheless about 60% higher than a single 1000 R exposure.

Loss or addition of chromosomes

Ionizing radiation can produce loss or addition of entire chromosomes in experimental mammals. Much of the work on this phenomenon has been on nondisjunction, which is the failure of paired homologous chromosomes to separate. The result of nondisjunction is polysomy.

The spontaneous incidence of monosomics and trisomics in most mammalian species is quite low, generally on the order of 0.5 to 1.0%. (Humans are a notable exception to this generalization, however, and corresponding care must be taken when extrapolating animal studies to man.) Thus, one can hope to obtain a particularly clean ("low noise") test of the chromosome loss- or addition-effects of ionizing radiation using non-human mammals as subjects.

Male mouse studies. X rays clearly have the power to induce nondisjunction. For example, random-bred Q strain adult male mice were irradiated with 1 Gy of x rays and mated to females five weeks after irradiation (to sample spermatocytes) or seven weeks after irradiation (to sample spermatogonia). Unirradiated controls were also run. Pregnant females from week 5 and week 7 and control matings were killed at 9-10 days of gestation and chromosome preparations made from all viable fetuses or their membranes. The overall frequency of abnormalities in the controls was 1.1% and included 2 trisomics (41,XXY), 1 triploid (60,XXY) and 3 mosaics. The frequency of trisomies alone was 0.35%. In week 5 of the irradiated series, in contrast, the frequency of abnormalities was 2.0% and included 2 monosomies (39,X), 1 trisomy (41,XY,+16), 1 triple-trisomy (43,XXY,+10,+17), 5 triploids, 2 tetraploids, 8 mosaics, and 1 with a miscellaneous aberration (40,XY,1q+). In week 7, the frequency of chromosomally abnormal embryos in the irradiated group was about double that of controls (2.8% for the irradiated) and included 2 monosomies, 7 trisomies, 8 mosaics, 1 triploid, and 2 tetraploids.

Female mouse studies. The power of x rays to induce nondisjunction in female mice, even if it exists, is difficult to detect at lower doses. For example, three groups of female mice of the same strain (Q) were irradiated with $5 \times 10^{-2}$ Gy of x rays. The first group consisted of 6-8 week old females irradiated and immediately mated, the second group consisted of 9 month old females which were immediately mated, and the third group consisted of 6-8 week old females irradiated and mated at age 9 months. When the younger females were irradiated, the frequency of chromosomally abnormal embryos was 1.5%, not significantly higher than in the controls. When the younger females were irradiated, aged, and then mated, the frequency was about 2.1%, again not detectably higher than the controls (the increase in chromosomal abnormalities in offspring with increasing female parental age is a common phenomenon among mammals, and is independent of radiation exposure). When the older females were irradiated and then mated, the frequency of chromosomal abnormalities in the embryos was about 3.6%. There is only a faint hint that the aged females were more susceptible than the controls, although the difference is not, the authors of the study report, statistically significant. Overall, then, there was no statistically significant increase in chromosomal abnormalities in the embryos of any of the groups.

Similarly, virgin female mice of an inbred CBA strain aged 6, 15, and 46 weeks were irradiated with with 2, 4, 8, or 16 R of x rays, and then mated to young males of the same strain when the irradiated mice were, respectively, 16 (Group 1) or 32-35 weeks (Group 2) old. The third group (Group 3) was mated soon after irradiation. Against controls, no statistically significant increase in nondisjunction could be detected.

These results appear to be sustained by studies on other mouse strains. For example, virgin inbred C57BL females aged eleven months were irradiated with 4, 8, or 16 R of x rays and mated with young unirradiated males of the same strain five days after irradiation. The pregnant females were killed 10 days after vaginal plugs were observed and the conceptuses were processed for chromosome analysis. The experiment provided no evidence for an increase in nondisjunctions in the irradiated group compared to the controls.

Similar experiments at exposures of 200 R show no detectable difference in the incidence of non-disjunction between irradiated and control groups.

Effects on other species. Compared to other experimental mammals, the mouse is not peculiarly sensitive to the induction of nondisjunction by exposure to ionizing radiation. Hence, it can serve as a model for this kind of radiation effect.

Chinese hamsters. Female Chinese hamsters, for example, were exposed to 20 R and 200 R doses of x rays, and the metaphase II oocytes were then examined for aneuploidy. Not a single case was found in the irradiated or control group.

In a similar experiment, 5-month old virgin female Chinese hamsters (first series) were given 50, 100, or 200 R doses of x rays, and 16-19-month old females (second series) were given a 50 R dose of x rays. The animals were irradiated at 85, 59, 35, 19, 17, 11, 9, and 7 hours before ovulation. In the first series, unfertilized eggs were collected from the animals exposed at 85, 35, and 17 hours before ovulation and processed for chromosome analysis at metaphase II; in the second series, the oocytes were collected from the 85-hour group. The data showed no significant increase in the frequency of aneuploid eggs except in the 200 R, first series group, where there was a very slight increase (3.4% in the irradiated, versus 2.1% in the controls). There was no increase in chromosome structural abnormalities except in the 35-hours eggs of the 100 R group (1.1% in the irradiated, versus 0.2% in the controls). In the second series (aged females exposed to 50 R) there was no detectable difference in the incidence of aneuploids and structural chromosomal anomalies between the irradiated group and the controls. With the exceptions noted above, there was in general no significant difference between the incidence in controls and the irradiated groups in the other regimes, either.

Northern field vole. Contradictory results have been obtained in the Northern field vole, *Microtus oeconomus*, on the radiation-induction of sex-chromosomal aneuploidy in male meiotic stages; some x-irradiation experiments provided positive evidence while the others showed no significant induction for doses up to 2 Gy. There is, however, good evidence for the induction of diploidy; the cell stages sampled on days 4 and 9 after irradiation seem to be most sensitive in this respect.

Summary of chromosome loss or addition effects. In general, acute x-irradiation at doses as low as 1 Gy produces a measurable increase in the frequency of nondisjunctions in male mouse germ cells. In contrast, the available evidence suggests that x-irradiation in the range of 200 R to female mice may produce an increase in the frequency of nondisjunction; below this dose, experimental results are ambiguous or negative for this effect. Similar results have been obtained in studies with female Chinese hamsters. There is some evidence that x rays can induce diploidy at much lower doses (as low as 2 Gy) in the female Northern field vole.

## Point mutations

Ideally, we would like to know the effects of ionizing radiation on genetic material at each individual gene locus. Systematic information of this sort is unfortunately not yet available, and is not likely to be for many years. Short of that, it would be highly useful to know the dose-normalized rate of specific-loci mutations for a few "representative" gene loci, since such a rate would provide at least a rough bound on the rate at which specific dose regimes can cause genetic damage.

Specific-locus mutations in male mice: beta irradiation. To investigate the specific-locus mutation rate induced by exposure to beta irradiation, (101 x C3H)$F_1$ wild type male mice were injected intraperitoneally with tritiated water (a strong beta source) at the rate of 1.85 or 2.78 x $10^{-2}$ MBq $kg^{-1}$ body mass, then mated to unirradiated females. Seven specific (a (nonagouti), b (brown), c (albino), p (pink-eyed dilution), d (dilute), se (short-ear), and s (piebald)) (recessive) gene loci were analyzed in the offspring for mutations. These loci represent a reasonable cross-section of the mouse's chromosomes, and are obviously responsible for distinctive phenotypical features. In some loose sense, therefore, they are a "representative" sample. Presumed mutants were bred to establish allelism of the mutations and to determine the viability of the mutations in the homozygous condition. Both post-spermatogonial and spermatogonial stages were sampled in this experiment (by the staggered breeding technique described above). The data suggest an induction rate of 4.4 x $10^{-7}$ mutations per locus per $10^{-2}$ Gy for post-spermatogonial irradiation, and 1.5 x $10^{-7}$ mutations per locus per $10^{-2}$ Gy for spermatogonial irradiation.

The distribution of mutations observed in these experiments is similar to the results of point-mutation studies done with x and gamma radiation. For spermatogonial irradiation, for example, a low frequency of mutation is observed in all regimes at the a and se loci. The mutations in the offspring from beta-irradiation of post-spermatogonial stages appear to be more evenly distributed among the loci than those arising from spermatogonial irradiation; this feature also occurs in x- and gamma-irradiation studies. In the beta-irradiated spermatogonial series, 10 out of 18 tested mutants were viable in the homozygous condition; this proportion was 3 in 11 in the post-spermatogonial group. Again, these results are not significantly different from the results obtained from x- and gamma-irradiation studies of the same loci. Mutations at the d and se loci did not occur in the same germ cell at the same time in the beta-irradiation study, although this simultaneous pair of mutations was frequently observed in post-spermatogonial x- and gamma-irradiation.

Specific-locus mutations in male mice: alpha irradiation. Point mutations are induced in male mouse germ cells by exposure to alpha radiation. For example, (101 x C3H)$F_1$ male mice were intravenously injected with 0.37 MBq $kg^{-1}$ body mass of monomeric $^{239}$Pu citrate. Thirteen weeks after injection, the males were mated with unirradiated females of the same stock and the offspring were examined for mutations at the same seven loci as in the tritium-injection study described above. At least 11 mutations have been observed in 54,679 offspring. This frequency is significantly above the control frequency, and is at least 20 times that observed for x- and gamma-irradiation of male mice, for the same dose.

Specific-locus mutations in female mice: comparative x-irradiation. X-irradiation causes specific-locus mutations in female mice. For example, maturing mouse oocytes were exposed to a total dose of 2 Gy of x-irradiation given in 20 equal fractions of 0.1 Gy each. The controls were given 2 Gy acutely. The females were then mated with unirradiated males. Genes of the offspring conceived within the first seven weeks (and later) after the last irradiation were analyzed for mutations at the seven loci described above.

The distribution of mutations in this experiment is both informative and typical. Approximately 50% of the mutations occurred at the s locus. Mutations at the a and se loci were rare. There were no simultaneous d-se mutations. The proportion of mutations which was lethal in the homozygous condition was 78%; nearly all the s and d mutations were homozygous lethal. The dominant and X-linked visible mutations included 2 viable repeats of tabby (Ta), two lethal splotch (Sp) alleles and a lethal W allele. There was a new X-linked mutation (broad-headed; Bhd); two other mutations caused a light coat color and/or spotting. Six additional mutants with irregular inheritance occurred; these had tail kinks, light coats, behavioral abnormalities, and small size, either together or in various combinations. The mutation frequency obtained in the fractionated series was lower than that obtained in the acute radiation regime, as expected.

In a similar experiment, adult female mice were given doses of 2, 4, or 6 Gy of x rays at 0.52 or 0.72 Gy $min^{-1}$ and mated immediately. Offspring conceived in the first seven days (i.e., derived from oocytes which were mature at the time of irradiation) were analyzed for specific locus mutations at the seven loci mentioned above. Mutations rates at these loci ranged from 2.65 x $10^{-7}$ mutations per locus per $10^{-2}$ Gy for the 2 Gy regime, to 6.27 x $10^{-7}$ mutations per locus per $10^{-2}$ Gy for the 6 Gy regime.

Note that these rates are similar to the rates obtained from beta irradiation of male mouse germ cells.

Specific-locus mutations in female mice: comparative x- and gamma-irradiation. Results similar to the above have been obtained from high-dose-rate x-irradiation of the oocyte stages in female mice shortly before birth. These mice were given 300 R of either x rays at the rate of 93 R $min^{-1}$ or 300 R of gamma-irradiation at 0.8 R per minute. At maturity, the irradiated animals were mated to males of the tester stock. The offspring were analyzed for mutations at the seven specific loci described above. More or less as expected, the high-dose-rate x-irradiation regime produced a higher mutation rate than the gamma regime: in the acute x-ray series 3 mutations were found among 16,194 $F_1$ progeny; in the gamma-ray series, 1 mutation was found among 37,218 $F_1$ progeny.

Induction of sex-linked recessive lethals in male mice and rats. Ionizing radiation can induce sex-linked recessive lethals in male mice and rats. For example, male mice and rats were given a fractionated x-ray doses of 5 + 5 Gy at 0.9 Gy per minute with a 24 h interval between the end of the first, and beginning of the second, fraction. As above, the irradiated males were mated with unirradiated females of the test stock; the offspring were then crossed to identify the frequency of sex-linked recessive lethals. It was found that 2/536 irradiated and 0/529 control X chromosomes carried a confirmed lethal. These frequencies correspond to a mutation induction rate of $1.9 \times 10^{-6}$ mutations per $10^{-2}$ Gy per X chromosome for single exposures (this figure contains a normalization for the enhancing effect of fractionated exposure, roughly threefold).

High dose rates enhance the induction of sex-linked recessive lethals greatly. For example, male mice and rats were given a total x-ray exposure of 450 R, given in three fractions of 100, 150, and 200 R at 10, 12, and 14 weeks of age, respectively. The measure used for estimating the rate of sex-linked recessive lethals was litter size on day 1 and on day 21. The rates obtained in the experiment are as follows. For day 1, the rate was $1.1 \times 10^{-4}$ mutations per X chromosome per R, and $2.1 \times 10^{-4}$ mutations per X chromosome per R for day 21.

Note that the dose-normalized rate of induction for the fractionated high-dose-rate study is about 100 times greater than the rate obtained in the fractionated low-dose-rate, low-dose study.

Autosomal recessive lethals in male mice: neutron irradiation. Acute and chronic neutron irradiation also induce autosomal recessive lethals in male mice. For example, adult CBA male mice were irradiated with 1.5 or 2.5 Gy acute doses, or with 2.5 Gy chronic doses at about $7 \times 10^{-3}$ Gy $h^{-1}$, 8 hours per day, for 5 days a week for 11 weeks, with 14.5 MeV neutrons. The irradiated males were then mated with unirradiated females of the same stock. Offspring from this mating were then mated to identify the frequency of recessive autosomal lethals. The study showed that there is no measurable difference in the rate of induction of autosomal recessive lethals in mouse spermatogonia between neutrons delivered acutely or chronically. The data obtained in study also suggests a rate of mutation induction from these regimes of about $2 \times 10^{-4}$ mutations per gamete per $10^{-2}$ Gy. This is about twice the rate ($1 \times 10^{-4}$ mutations per gamete per $10^{-2}$ Gy) obtained in similar experiments using fractionated, high-dose, high-dose-rate x-irradiation. It is roughly a hundred times greater than the fractionated low-dose, low-dose-rate, x-ray rate.

Autosomal recessive lethals in female mice: x-irradiation. The mature oocytes of female mice of an inbred CBA strain were administered a 250 R dose of x rays acutely. The techniques used were essentially those used in the determination of radiation-induced autosomal recessive lethals in mouse spermatogonia, described above. The irradiated females were mated to males immediately after irradiation and were allowed to produce one or two litters conceived during the first six weeks following irradiation. The data show an induction rate of about $1.0 \times 10^{-4}$ autosomal recessive lethals per gamete per Roentgen, varying slightly with method of calculation. The results are thus about the same (rate) as for acute x-irradiation of mouse or rat spermatogonia.

There is some indication, furthermore, that the germ cells of female mouse fetuses are about as sensitive as maturing oocytes to the induction of autosomal recessive lethals. For example, when female fetuses were administered 25 R $d^{-1}$ for four days from day 10 to day 13 of pregnancy plus 50 R $d^{-1}$ for a further four days from day 15 through day 18 with a total accumulated dose of 300 R, the mortality among the $F_2$ crosses was no higher than in the $F_2$ crosses which descended from the females in the mature-oocyte series described above.

Autosomal recessive lethals in female mice: gamma irradiation. CBA female mice were chronically irradiated with $^{137}$Cs in utero during one or the other of the following periods: day 10 to day 14 (Group 1), or day 14 to day 18 (Group 2) of gestation. The doses administered were 0.34 Gy/generation in the first group and 1.6 Gy/generation in the second group; the dose rates were $3 \times 10^{-3}$ Gy $h^{-1}$ in the first, and $1.7 \times 10^{-2}$ Gy $h^{-1}$ in the second group. The exposures were administered through nine generations; at the end of the experiment, the female progeny were tested for the induction of autosomal recessive lethals. The data obtained suggest an induction rate of $1.5 \times 10^{-4}$ autosomal recessives per $10^{-2}$ Gy per gamete in the group that received 0.34 Gy per generation for gestation days

10 through 14. The group that received 1.6 Gy per generation for gestation days 14 through 18 showed a rate of 0.3 x $10^{-4}$ autosomal recessives per $10^{-2}$ Gy per gamete.

Note that these induction rates are on the same order of magnitude as those for fractionated, high-dose, high-dose-rate x-irradiation of male rats and mice.

Biochemical mutations in male mice: gamma irradiation. There is some weak biochemical evidence that protein-synthesis-controlling genes are altered when male mice are exposed to acute fractionated gamma-irradiation. For example, adult male C3H males were administered 100 R, followed 24 h later by a 500 R dose of $^{137}$Cs gamma irradiation. The males were then mated to females of the 101 strain. Liver extracts of the irradiated animals were then separated electrophoretically to resolve the proteins. In 2100 $F_1$ offspring screened, three isoenzyme variants were found. It is difficult to say whether this incidence is significantly different from the control frequency.

Similar neutral to negative results were obtained from acute fractionated gamma-irradiation studies of male mice. Adult C3H males or (101 x C3H)$F_1$ were given 100 + 500 R of fractionated $^{137}$Cs gamma-irradiation. The two fractions were separated by 24 h. The irradiated males were then mated to females of the 101 strain, or T strain, respectively. Both pre- and post-meiotic stages were sampled. The blood from the $F_1$ offspring was analyzed using a thin-layer-chromatographic method. No mutants could be detected in the sample of 5786 animals.

Dominant lethals in male mice: fractionated gamma irradiation. Acute fractionated gamma irradiation has the power to induce dominant lethals in male mice. For example, adult male mice were give a fractionated gamma-ray dose of 100 R, followed 24 h later by a 500 R dose; the dose rate in each fraction was 60 R $min^{-1}$. The irradiated males were then mated with unirradiated females. The skeletons of the $F_1$ male offspring conceived during the post-sterile period were then examined after the mice were allowed to mate. Thirty seven of the 2646 $F_1$ males were judged to have dominant mutations that caused one or more rare skeletal abnormalities; 31 of these were shown to be mutants by showing that the defects could be propagated through breeding, and the remaining 6, having no progeny, were counted as mutants merely on the basis of the skeletal examination. In the breeding tests, the dominant mutations affecting the skeleton showed variable expressivity and incomplete penetrance for many or all of the effects that they caused. A number of the mutations severely affected viability. Given the experimental technique (breeding of the offspring of the irradiated animals), incomplete penetrance and decreased viability would cause an underestimation of the mutation rate. It appears from subsequent cytological studies that the skeletal abnormalities arise from reciprocal translocations.

Gamma irradiation can also induce dominant eye mutations, particularly cataracts, in mice. For example, four different groups of adult male (101 x C3H)$F_1$ mice were exposed to fractionated gamma-irradiation at 53 - 55 R per minute. The first group received a total exposure of 910 R in two equal fractions which were separated by a 24 h period. The other two groups received a 534 R exposure, followed 24 h later by a 600 R exposure. The fourth group received only single 534 R, or 600 R doses. The irradiated males were then mated to unirradiated females of the seven-locus tester stock described above. The eyes of the $F_1$ progeny were examined to detect lens opacities. All presumed mutants were bred to confirm the genetic nature of the lens opacities. The mutation rate for post-spermatogonial irradiation stages was higher than for regimes in which spermatogonia were irradiated. The experiment also showed that exposure fractionation has an *enhancing* effect on mutation rate. More specifically, the mutation rate estimates obtained from these series are as follows. Single high dose rate gamma ray exposures cause rates of 0.45 - 0.55 mutations per $10^6$ gametes per R; fractionating the gamma dose into two roughly equal fractions almost triples the mutation rate for the same total dose. (All frequencies were derived by subtracting the frequency in controls from those of the irradiated group.)

Note that this regime produces a mutation rate of the same order of magnitude as the beta regime described above.

*Histocompatibility mutations*. The histocompatibility system in mammals is controlled by a group of co-dominant genes located throughout the genome, and the action of this group (the production of cell membrane alloantigens) determines the acceptance or rejection of dermal grafts. These loci in mice are classified into two groups, distinguished operationally by graft acceptance/rejection features of the $F_1$ hybrids of the B6 and C (parent) strains that are typically used in these studies. The class-I loci, which number at least 30, have different alleles in the two parental lines and are therefore heterozygous in the (B6 x C)$F_1$ hybrid. The class-II loci have similar alleles in the two parental lines and are therefore homozygous in the hybrid; the number of class-II loci is unknown, but it is believed to be around 50. To detect new mutation in these loci, one exchanges tail skin grafts orthotopically between $F_1$ hybrid mice derived from irradiated germ cells of the parent B6 and C strains. The progeny are then classified as "gain" type mutations (i.e., grafts donated by the putative mutant are rejected) or "loss" type mutations (grafts placed on the putative mutant are rejected), or "gain-loss" type of mutations (one specificity of acceptance of the graft replaced by another). Class-I mutations are distinguished from class-II by their "loss" and "gain-and-loss" phenotypes for loci on autosomes and X chromosomes; class-II mutations could produce only "gains" unless the Y chromosome is involved.

Recent work using this technique has suggested that x rays can cause mutations at the histocompatibility loci, but frequency of such mutations is low compared to the frequencies of the seven-locus experiments described above. In particular, mice of the B6 and C strains were given 3.5, 6.5, or 3.5 + 3.0 (fractions separate by 24 h) Gy doses of x-irradiation. They were then mated with unirradiated females to produce the hybrids described above. Grafting was performed as described, and the rejection/acceptance results recorded. Using this technique/criterion, 8 histocompatibility mutations were found in the nearly 3000 (including controls) progeny tested. Two of them were clusters and therefore could not have occurred in sperm; one class-I mutation was not relevant because in occurred in the B6 genome of the unirradiated mother. Of the remaining five, two were class-I losses (1 from the 6.5 Gy group, which represented a frequency of 1/565; the other was in the 3.5 + 3.0 Gy group, which represented a frequency of 1/514). The remaining three were class-II mutations of the "gain" type; one on the control, one in the 3.5 Gy group; and 1 in the 3.5 + 3.0 Gy group. In general, the mutation rate appears to be only about 1/60 the frequency of those observed in the seven-loci group for the same x-irradiation regime. The technique does not, therefore, provide a particularly sensitive test specific-locus mutation induction.

Induction of congenital anomalies and tumors by irradiation of mouse germ cells. Congenital anomalies and tumors are induced in the offspring of irradiated mice. For example, adult female mice were administered 1.08 to 5.04 Gy of acute x irradiation and mated at intervals of 1-7, 8-14, 15-21, and 22-28 days. Their uteruses were examined at late pregnancy to detect early fetal abnormalities and late malformations. At each weekly interval, the incidence of abnormalities tended to rise with increasing dose, and at any given dose, the incidence tended to rise with time after irradiation. Changes in incidence of dominant lethals and of abnormal fetuses paralleled each other closely; in the 5.04 Gy group, the incidences in week 3 were 59 $\pm$ 5% for dominant lethals and 12.5 $\pm$ 3.1% for abnormal fetuses.

Summary of point-mutation effects. Ionizing radiation clearly has the power to induce point mutations in chromosomes. Beta-, gamma-, and x-irradiation produce similar distributions of mutations among seven specific loci for the regimes studied. Although the geography of the distribution is similar among these regimes, the mutation induction rates may differ greatly depending on nature of the radiation and the temporal distribution of the dose. The data obtained from the beta-irradiation of male mice suggests a specific-locus mutation rate of between 1.5 and 4.4 x $10^{-7}$ mutations per locus per $10^{-2}$ Gy for irradiation of spermatogonia and post-spermatogonial irradiation. Irradiation of female mice under the same regimes produces similar rates. Alpha irradiation of male mice produces a mutation rate at these same loci approximately 20 times greater than the rates observed for the x- and gamma-irradiation regimes, for the same dose.

Low-dose-rate fractionated (5 + 5 Gy) x-irradiation of males produces only about 1% the rate of sex-linked recessive lethal mutations (2 x $10^{-6}$ mutations per X-chromosome per $10^{-2}$ Gy) as high-dose-rate fractionated (100 R + 150 R + 200 R) x-irradiation (2 x $10^{-4}$ mutations per X-chromosome per $10^{-2}$ Gy).

Autosomal recessive lethals are produced at roughly the same rate in female and male mice (1 - 2 x $10^{-4}$ mutations per Roentgen per gamete) by both high energy neutrons (14.5 MeV) and fractionated high-dose-rate, high-dose x-irradiation. Neutrons, if anything, may be about twice as effective as the high-dose-rate, high-dose x-irradiation regimes in inducing autosomal recessive lethals.

A number of histocompatibility studies have suggested that histocompatibility sites are about 1/60 as sensitive to mutation from low-dose x-irradiation as the seven-loci group.

Cataracts and skeletal deformities are induced by fractionated high-dose-rate, high-dose x-irradiation at the rate of about 1 x $10^{-6}$ per gamete per Roentgen; these same disorders are induced at about one-third that rate by large single doses of x rays.

Taken together, these data show that point-mutation rates are quite sensitive to the spatial and temporal features of the dose distribution, to the identity of the chromosomes and gene loci, and to the linear energy transfer involved. None of this is particularly surprising, since the effects of irradiation of living tissue is determined by rather non-specific transfers of energy to unusually complex, dynamic biochemical systems and molecules; these energy transfers, moreover, occur at magnitudes near the binding energy of those molecules. Any extrapolations from the experimental conditions must therefore be done with explicit accounting of the extrapolating assumptions (e.g., linearity of the dose-effect relation, scaling factors, and so on) made. Within these caveats, we can say at least the following. Chronic, low-dose, low-dose-rate beta and gamma irradiation appear to be the least damaging; dose-normalized alpha and neutron irradiation appear to be more damaging by a factor of five to 100, depending on dose. Histocompatibility loci appear to be the most robust, and some X-chromosome loci the least robust, for the sites and regimes which have been studied.

### Genetic effects of internal emitters

When a radionuclide is introduced into a living animal, tissues absorb a portion of the decay transition energy(ies). The biological effects of the deposition of energy in a tissue are usually reported in terms of the activity per unit weight of animal (Bq $kg^{-1}$). However, as Chapter 3 details, the actual absorbed dose depends on the intake of activity, its transport, metabolism, and re-utilization, as well as its excretion. These factors depend on the characteristics of the material introduced, the nature of the radionuclide, its chemical and physical form, the chemical and physical form of the carrier, the method of introduction of the activity, its distribution in time, route of entry, means of introduction, the animal species, weight, sex, age, response to diet, and numerous other factors.

Since the radiosensitive structures (e.g., chromosomes) of cells are located at specific sites, the biological effects of radiation depend on the microscopic (i.e., near-molecular dimensions) distribution of energy. Radionuclides emitting alpha particles (such as uranium and radium in water supplies) cause dense ionization along their tracks, and may thus be expected to be many (generally, thought to be 5 to 20) times more effective than similar distributions of beta- or gamma-emitting radionuclides in producing tissue damage for the same absorbed dose.

## MISCELLANEOUS HUMAN DATA

There is strong evidence that individuals living or working in areas of elevated exposure to natural radioactivity have a higher incidence of chromosome damage than those who do not have such exposures.

Chromosome aberrations in peripheral blood lymphocytes in individuals living and working in areas known to have high natural radioactivity have been studied. Near Badgastein, Austria, for example, are several radon-containing hot springs. Five million liters of water from these springs are delivered daily to reservoirs; from there the water is delivered to hotels and spa houses. Almost all of the $^{222}$Rn in this water is discharged into the air. The radiation burden on the population accordingly comes from inhaled radon its daughters and from the external gamma dose created by these nuclides. Air activity concentrations in the region range from about 100 mR per year in peripheral areas to 1500 mR per year in what is known as the "thermal gallery," in a former gold mine near the town. More more than 5000 patients are "treated" per year in the gallery.

The Badgastein study is particularly revealing. Even at very low dose levels, dose/chromosome-damage relationships were observed; effects (fragments, dicentrics, and interstitial deletions) increased with exposure.

Analogous results were obtained in a study of the incidence of chromosome aberrations in peripheral blood lymphocytes in 197 nuclear ship dockyard workers over a 10-year period. These workers had been exposed to mixed neutron and gamma irradiation during the refuelling of nuclear reactors; most of the exposures were below the internationally accepted maximum permissible level of 0.05 Sv per year. The majority of the workers at the facility had not had previous significant exposures to radiation. Blood samples were taken before the individuals were classified as "radiation workers." The study showed a strong dose/chromosome-aberration correlation. Statistically significant effects of dose were evident for dicentric aberrations, acentric (chromosomes without centromeres, a pathological condition) fragments, and cells with aberrations.

Similar studies were performed on the chromosomes of 57 healthy male employees of six German nuclear power plants. These individuals were metal workers, technical engineers, or radiation protection workers involved mainly in maintenance or refuelling crews. All of them had received annual doses below the maximum permissible limit of 0.05 Sv per annum and had worked with radiation for periods ranging from one to 14 years. The exposure was mainly to external sources of gamma rays and high-energy X rays. The controls for this study were 11 healthy males with no radiation history except the "natural" background. The frequencies of dicentrics and acentrics in the radiation workers were significantly higher than in the controls.

GENETIC EFFECTS ON MAN

There are a number of ways to extrapolate the results of animal irradiation studies to man. The results of all of these methods are in surprisingly good agreement, and thus provide strong mutual corroboration. In this section, we survey those methodologies.

UNSCEAR estimates

UNSCEAR (United Nations Scientific Committee on the Effects of Atomic Radiation) uses two methods to estimate the rate of induction of genetically deleterious conditions in man by low-level, low-dose rate, low-LET ionizing radiation. The first of these is called the direct method; the second, the doubling-dose method.

The direct method. For mutational damage to human males, UNSCEAR extrapolates from the rate of induction of dominant cataract mutation in male mice following spermatogonial irradiation. The data obtained from the mouse studies strongly suggests that $12.7 \times 10^{-7}$ cataract mutations per Roentgen per gamete are produced by a fractionated gamma irradiation regime consisting of a 455 R dose , followed by a 455 R dose, with 24 h between exposures. A single dose of 534 R or 600 R, in contrast, produces $5.0 \times 10^{-7}$ cataract mutations per Roentgen per gamete (average estimate based on the data from two single gamma-ray exposure experiments involving 534 R and 600 R, respectively). These rates must be converted into those that will reflect the response under low-dose, low-dose-rate irradiation conditions in the mouse, and must subsequently be transformed into quantities that will express the rate for humans. To do this UNSCEAR assumes that the ratios between the rates of cataract mutation in high-dose-rate, high-dose, fractionated, low-LET regimes to the cataract mutation rates for single low-dose, low-dose-rate, low-LET regimes are the same in mouse and in man.

The available mouse data suggests that fractionation enhances the cataract mutation rate we would observe at low chronic doses by a factor of 1.2; and the high dose rates enhance that mutation rate by a factor of 3. Accordingly, UNSCEAR divides the fractionated high-dose-rate value ($12.7 \times 10^{-7}$ cataract mutations per gamete per $10^{-2}$ Gy) by ($1.2 \times 3$), to obtain an adjusted value of $3.5 \times 10^{-7}$ cataract mutations per gamete per $10^{-2}$ Gy of paternal exposure. Since the mouse cataract induction rate for single exposures at high rates suggests that high dose rates enhance by a factor of three, the corresponding mouse data ($5 \times 10^{-7}$ cataract mutations per gamete per $10^{-2}$

Gy) is divided by three to obtain a value $1.67 \times 10^{-7}$ cataract mutations per gamete per Gy of parental exposure for single exposures. The values for the fractionated exposures and the single exposures are then weighted by the number of mutants in the fractionation and single exposure series, to obtain a weighted average of $2.6 \times 10^{-7}$ cataract mutations per gamete per $10^{-2}$ Gy of parental exposure.

The direct method then assumes that since in man about 2.7% of all known and proven significantly deleterious dominant mutations are associated with one or another form of cataract, the reciprocal of this (36.8) can be used to estimate the overall rate of serious dominant mutations which will be expressed in the offspring of irradiated fathers. On this assumption, the production rate of serious effects in the first generation progeny for man is $2.6 \times 10^{-7} \times 36.8 = 10 \times 10^{-6}$ serious dominant mutations per gamete per $10^{-2}$ Gy of paternal exposure. That is, 10 individuals per million born in the first generation from irradiated fathers will be affected by one or another kind of clinically important serious genetic disease of induced mutational origin per $10^{-2}$ Gy of male parental exposure to low dose, low dose rate, low-LET irradiation.

Similar calculations, using an extrapolation from radiation-induced skeletal deformities in the mouse, produces a similar mutation-rate value. Thus, the two values (from cataract, and from skeletal projections) strongly corroborate one another.

There is no experimental data on the induction of either skeletal or cataract mutations in female mice. Thus, it is not possible to use the methods described above to determine induction rates of dominant deleterious mutations.

## Induction of translocations

Chromosome translocations are a second significant source of heritable deleterious disorders induced by ionizing radiation. To compute the rate of induction of balanced translocations for man, UNSCEAR used limited human data and marmoset cytogenetic data obtained in experiments involving acute x-irradiation. These data suggest an induction rate of $7 \times 10^{-4}/10^{-2}$ Gy/spermatocyte. From this, the rate of heritable translocations can be estimated as $1.75 \times 10^{-4}/10^{-2}$ Gy/gamete (assuming that the rate of recovery in the $F_1$ generation will be one-fourth of that in spermatocytes). The rates for low-dose x rays, low-dose-rate x rays, and chronic gamma-irradiation were derived by dividing the above figure of $1.75 \times 10^{-4}/10^{-2}$ Gy/spermatocyte by 4, 2, and 10, respectively. The rates thus derived are

$0.44 \times 10^{-4}$, $0.88 \times 10^{-4}$, and $0.18 \times 10^{-4}/10^{-2}$ Gy/gamete, respectively. These, in turn, were used to estimate the proportion of unbalanced zygotes that will result (assumed to be twice the above rates, based on the segregational properties of translocation heterozygotes) and the proportion of the unbalanced zygotes that will results in children with multiple congenital abnormalities (6% of the unbalanced zygotes, based on human data). These calculations suggest that per $10^{-2}$ Gy of paternal irradiation, between 2 and 10 per million children born will suffer from multiple congenital anomalies which can be attributed to the induction of reciprocal translocations in the spermatogonia of their fathers.

The direct method as described above considers only the rate of induction of dominant deleterious mutations or balanced reciprocal translocations. The rate of induction of recessive deleterious mutations and the overall impact of these mutations is unknown; some authors have urged that the effect of recessive deleterious effects may far outweigh the effects of the dominant ones. Whether they do, the consensus view is that we do not yet have enough evidence to judge the matter one way or another. For the present, then, one must regard the direct method as producing, at best, a lower bound on the rate of induction of mutations likely to produce clinically serious dominant mutations in the offspring of irradiated fathers.

If experimental animal studies are any guide, the rate of induction of minor mutations is many times that of significantly deleterious dominant mutations. Some authors have urged that collectively, these minor mutations may have a greater impact than the significantly deleterious dominant mutations. Given our present knowledge, however, there is no way to quantify, and hence, to evaluate, such a claim. Again, what we can say is that these minor mutations can only add to the overall rate of genetic detriment. Thus an induction rate limited to the effects of a clinically significant deleterious dominant mutations represents, as noted above, a lower bound.

The doubling-dose method. In order to assess the importance of the genetic effects of radiation, it is often useful to know how those effects are related to the existing incidence of genetic effects. One way of expressing this quantity is to compute the fraction by which the mutation rate would be increased by one rem of radiation. The reciprocal of this quantity is the mutation-rate doubling dose, that is, the dose required to produce as many mutations as occur under existing conditions. (The existing incidence is sometimes called the "natural" incidence, but in light of Chapter 3, it is unclear what "natural" might mean at this time, given the extent of radiocontamination of the natural environment and the fact that we cannot, in most cases, clearly identify the contribution to that background made by human activities).

The doubling dose is computed as follows. Chronic low-dose, low-dose-rate x-, beta, or gamma-irradiation of mouse spermatogonia produces a mutation rate on the order of 0.5 and 2.5 x $10^{-7}$ recessive mutations per gene locus per rem (see the description of the seven-locus beta irradiation study, above). Female germ cells appear to be much less sensitive to this kind of irradiation than male germ cells. Suppose we take the nominal average rate of mutation for both sexes to be about 0.25 x $10^{-7}$ recessive mutations per gene locus per rem.

The existing rate of human mutations appears to be on the order of 5 x $10^{-7}$ to 5 x $10^{-6}$ recessive mutations per gene locus per generation. This is, respectively, 20 to 200 times the induced rate in the mouse. This says that the fractional increase in the mutation rate per rem is between 1/200 and 1/20, or that the doubling dose (the reciprocal of this) for recessive mutations is between 20 and 200 rem. If we look at the mutation rate for dominant visible mutations, which is something on the order of 2 x $10^{-9}$ dominant mutations per rem, the doubling dose is 100 rem or more. These calculations are consistent with the data we have from the survivors of the bombings of Hiroshima and Nagasaki.

It appears, then, that the doubling dose is roughly 100 rem (or roughly, 1 Gy). The incidence of autosomal dominant and X-linked diseases in man is 1%; the increase in the frequency of these diseases as a result of radiation exposure will be directly proportional to the mutational rate. Autosomal recessive diseases occur in humans at a frequency of 0.1% or so, and the increase in their frequency will be only very indirectly related to the mutation rate. The incidence of numerical chromosomal anomalies and unbalanced structural anomalies is on the order of 0.4%, and the increase in their frequency will be directly proportional to the mutation rate. Balanced reciprocal translocations as such may not confer any risk to the carriers; the risk from the induction of such structural defects comes mainly from the unbalanced products generated by these during meiotic segregation. For diseases of complex etiology, such as irregularly-inherited dominant diseases, multifactorial diseases, and congenital malformations (collectively constituting 9%), the mutational component is of the order of 5%. Under continuous radiation exposure at roughly $10^{-2}$ Gy/generation, the population will reach a new equilibrium in the incidence of these diseases. The rate of approach to the equilibrium as well as the incidence values at equilibrium and in the first generation following exposure will be different for different classes of diseases. More specifically, the approach to equilibrium will be fastest for single-gene dominant diseases, very slow for recessive diseases, and probably intermediate for diseases of complex etiology. The first generation incidence following radiation exposure (under conditions of continuous radiation exposure) will be approximately one-fifth of that at equilibrium for single-gene dominants and X-linked diseases, nearly the same in the first generation and at equilibrium for chromosomal diseases, and about one-tenth of that at equilibrium for diseases of complex etiology. All these considerations point to a total increase in mutations in the first generation of 63 new cases of genetic diseases per million progeny and 185 cases per million at equilibrium, when the population is continuously exposed to low-dose-rate, low-dose, low-LET irradiation at a rate of $10^{-2}$ Gy per generation.

Thus, the doubling-dose method produces estimates in good agreement with the direct method. The doubling-dose method has the strength of assimilating at least some of the non-dominant, complex-etiology effects of irradiation.

Total UNSCEAR estimate. UNSCEAR's estimate for the total mutational increment expected from chronic, low-dose, low-dose-rate, low-LET exposure is the following. In the first generation following paternal exposure, there will be 20 (10 from dominant delterious, and 10 from reciprocal translocations) new cases of genetic diseases per million progeny in the first generation following paternal exposure, and 150 cases per million progeny at equilibrium when the population is continuously exposed to low-dose-rate, low-dose, low-LET external irradiation at a rate of $10^{-2}$ Gy per male parent.

For low-dose-rate, low-dose, chronic high-LET (alpha) internal irradiation (such as that from radium or uranium in water supplies) the corresponding rates should probably be multiplied by five to 20.

## ICRP estimates

The International Commission on Radiation Protection (ICRP) established a Task Group to make a quantitative assessment of the rate of induction of genetic detriment to man caused by low dose, low dose rate ionizing radiation. Because the methodology and assumptions of the Task Group were very similar to those used by UNSCEAR, the Task Group's estimates were essentially the same as UNSCEAR's.

BEIR estimates

The National Academy of Sciences regularly publishes estimates of the effect on populations of exposure to low-level ionizing radiation. These reports are commonly known as the BEIR (Biological Effects of Ionizing Radiation) reports; the last was published in 1980. The BEIR methodology is essentially the doubling-dose method of UNSCEAR, augmented by considerations of the range of doubling doses suggested by the limitations of available experimental data. Nevertheless, the BEIR results agree well with the UNSCEAR and ICRP; the results are thus mutually corroborating.

NONHERITABLE EFFECTS: CANCER INDUCTION

Heritable genetic disorders are not the only biological effects which ionizing radiation can produce in living tissues. It is well known that nongenetic structures in cells can be impaired by ionizing radiation to the extent that cell death and death of the organism are imminent. Although it is very important to understand such effects, it is extremely improbable that the dose regimes required to produce them (on the order of megaBecquerels per kilogram body mass, administered acutely) will be encountered in water supplies except in the case of a major nuclear accident or nuclear war. Those cases deserve separate and careful treatment in their own right and will by fiat taken to be outside the scope of this book.

Of all the possible nonheritable somatic effects of chronic low doses of ionizing radiation, it appears that the induction of cancer is by far (at least two orders of magnitude) the most serious. Complete evaluation of the risk of cancer requires knowledge of the dose-effect relationship. Unfortunately, very little information is available for this relation. Estimates of the effect at low doses therefore have been extrapolated from effects at much higher doses, and this method has inherent problems associated with it, as noted above.

Although the mechanisms of radiation carcinogenesis are poorly understood, the information we do have suggests that most if not all types of cancer develop as a result of the combined effects of multiple factors. These may include inherited chromosomal mutations, somatic cell mutations which are acquired after conception, and changes resulting from the action of viruses, changes in the body's regulatory systems.

The quantity of radiation which can induce cancer no doubt varies from one individual to the next. The distribution of individual dose responses for any particular population will determine the dose-effect relationship for that population. It may be that there is some threshold below which no cancer will be induced in any individual. However, it is extremely unlikely that adequate evidence will ever be collected to determine whether there is such a threshold: the size of the irradiated population which would be required for reliable statistics would run into the millions, if current rough estimates are any indication of the frequency.

As a result of such limitations, most consensus estimates of the risk of cancer induction assume that the rate from high-dose, high-dose-rate regimes can be _linearly_ extrapolated to low-dose, low-dose-rate regimes. Whether this assumption is correct, or whether it is even conservative, is beyond today's technology to determine.

_Induction of leukemia._ Induction of leukemia by exposure to ionizing radiation has been demonstrated in experimental studies on animals, and by epidemiological surveys on man. What the exact mechanism of radiation leukemogenesis is, however, is still unknown. It may involve some heritable factors, some damage to genetic material after birth, and possibly a complex of factors which depend on the irradiation regime. Current theory suggests that the blood-producing marrow of the body may be the most sensitive to leukemogenic effects.

Most of the human data we have comes from epidemiological surveys of the survivors of the atomic bombings of Hiroshima and Nagasaki, and on patients in Great Britain who were given therapeutic x-irradiation of the spine for ankylosing spondylitis. Both these studies suggest that 1-2 excess leukemias per year will be induced per rad (primarily external) exposure per million population. There are several features of this estimate which must be kept in mind. First, the data from Hiroshima _external_ exposure of mixed neutron and gamma radiation, while that from Nagasaki is external gamma irradiation almost exclusively; the ankylosing spondylitis data is for _external_ x-ray exposure. Second, the data on the actual dose received by the survivors of the bombings contains large uncertainties (up to a factor of 5). Third, both the bombing and x-ray data are for single, high-dose, high-dose-rate exposures. How one extrapolates these figures to chronic low-dose-rate, internal _alpha_ (high-LET) irradiation (which is what we would expect to find in exposures from uranium and radium in water supplies), even to within an order of magnitude, is unclear. In the absence of better information, the effect of high-LET, low-dose-rate chronic alpha irradiation is taken to be dose-linearly scalable from the high-dose, high-dose-rate, single neutron, gamma, and mixed neutron/gamma exposures. There is some, but not strong, experimental evidence from mouse irradiation studies to indicate that this assumption is correct, and there is virtually no experimental evidence to the contrary.

Even if these assumptions are accepted, a truly thorny problem still remains. Uranium, radium, and thorium are the primary contributors to the internal dose from ingestion of water contaminated by natural radionuclides (see Chapter 3). These nuclides are bone-seekers because their biochemistry is akin to that of calcium. To make matters worse, there is good evidence that at least U-238 tends to accumulate on the inner surface of the long marrow bones; thus, its local concentration is much (about 9 to 10 times) higher than it would be on the assumption that it is uniformly distributed throughout the bone mass. Furthermore, the turnover of uranium-238 in the body is very slow: its biological half-life is about 500 days. Thus, the long-lived uranium nuclides tend to remain in the skeleton, irradiating the blood-cell-producing bone marrow next to which they reside. These factors greatly enhance the local dose and hence, the rate of leukemia induction. The actual calculation of the dose may be done using Equations 3-9 through 3-33. Depending on the assumptions used, one obtains a value of roughly 5-150 excess leukemia deaths per 10 pCi per liter uranium alpha per million population per generation (30 years).

Bone cancer. Bone cancer is a relatively rare form of cancer in man, and environmental factors which contribute to the production of bone cancer are not well known. Radium dial-painters in the early part of the century had noticeably higher incidences of bone cancer than the population at large; ingestion probably came from dressing the contaminated brush tip to a point by drawing it between the lips. Other irradiated groups that have been studied include patients receiving radioactivity or radiation exposure in therapy for various diseases. Again, depending on the assumptions made, one can compute that there will be approximately 5 to 150 excess deaths per generation per million population exposed per 10 pCi per liter uranium alpha in drinking water. Thus, the rate of induction of bone cancer from natural radioactivity in water supplies is nominally on the order of the rate of leukemia from the same source.

## SUMMARY OF BIOLOGICAL EFFECTS

The biological effects of ionizing radiation depend on whether the dose is internal or external, on the nature of the species, on the nature of the nuclide and its intrinsic decay regime, on metabolic factors, on individual variation, on the spatial and temporal distribution of the dose, and on the linear energy transfer of the energy involved. Currently, there is no well controlled experimental evidence on the effects of chronic low-dose-rate exposure regimes for man; all the good experimental evidence we have comes from relatively high-dose, high-dose-rate, short-term exposures of small experimental mammals, primarily mice. Extrapolation from one irradiation regime to another must therefore explicitly identify the extrapolating assumptions (e.g., linearity of the dose-effect relation, similarity of effects across species, and so on) involved.

The consensus scientific opinion holds that the most serious long-term effect of external exposure to ionizing radiation will probably be damage to the genetic material of fathers; that estimate says 10 to 150 offspring per million population per $10^{-2}$ Gy of chronic paternal exposure to low-dose, low-dose-rate, low-LET (gamma-, x-, and beta-) external radiation will suffer from seriously debilitating or lethal genetic damage. For internal exposures arising from the ingestion of natural radioactivity in drinking water, the primary health detriment will come from radiation-induced leukemia and bone cancer. Best estimates suggest that 10 and 150 leukemia and bone-cancer deaths will occur per generation per million population per 10 picoCuries uranium or radium alpha per liter drinking water.

*None of this is to say, however, that any of these estimates are correct, and the fact that no well controlled experimental evidence is available for exposure to chronic low-dose-rate ionizing radiation must temper any inference or policy decision.*

The first four chapters of this book have been devoted to surveying what is known about the physical and biological effects of exposure to ionizing radiation. There is clearly substantial uncertainty in our current knowledge of those effects. Yet even if there were no uncertainty about these matters, it is not obvious what radiation protection policy should be set, because such decisions intrinsically involve questions about the importance of human health relative to other things in which we have an interest, such as the cost of electrical energy. The next chapter is an introduction to that important and complex topic.

NOTES

1. The data used in this chapter come from UNSCEAR (United Nations Scientific Committee on the Effects of Atomic Radiation, *Ionizing Radiation: Sources and*

*Biological Effects*, UN, 1982). The interpretation of that data is largely my own, but nothing said here substantively disagrees with UNSCEAR estimates.

# 5
# The Role of Values in Radiation Protection Policy

## INTRODUCTION

The last four chapters surveyed what we know about the physical and biological effects of exposure to natural radioactivity in water supplies. And as the last chapter intimated, one of the most deleterious features of typical radiocontamination of water supplies, low-dose-rate uranium-produced alpha radiation, is not currently regulated. The question of whether it should be, and if so, in what way, clearly involves issues which go beyond the mere scientific account of these effects. To the extent that radiation protection policy is not capricious, it tries to capture what is in the interest of the public's health. Precisely *what* is in the interest of the public's health is a matter of considerable dispute even when we can concur about the relevant scientific facts, however, because it necessarily involves an attempt to identify *why* something is in the public interest, and this effort raises difficult questions about *values*.

Sociologists and philosophers have generally recognized a fundamental distinction which must be made at the outset of any account of value-related behavior. On the one hand, there is a part of value language which prescribes or urges action, or assigns value to those actions. For example, the remarks "Protect human health!" prescribes a complex of activities; "It is wrong that children should be allowed to die of environmentally induced leukemia!" assigns value to a perceived or possible state of affairs. In contrast, there is a part of value language which tries to *explain* or *justify* these prescriptions or value-assignments. For example, the phrase "A radiation protection policy is right if and only if it produces the greatest possible good" does not prescribe a particular policy nor does it directly assign value to a particular kind of action. Rather, such a statement provides a general criterion or justification for taking or prescribing particular actions; the phrase is therefore part of an explanatory or justificatory scheme for identifying and supporting particular actions or prescriptions. We may therefore regard it as part of a theory of value-related behavior. An examination of theories of value behavior is even a level more abstract than the theories themselves. Yet it is precisely at this level that many disputes over value issues in health policy exist.

This chapter, then, surveys the role which values play in radiation protection policy. The aim is not to show that there is a privileged general basis in value theory for radiation protection policy, but rather to outline issues that have been and will be disputed, and which will strongly affect the evolution of radiation protection policy, even when we can agree about the physical and biological evidence.[1]

## AN IMPORTANT CRITERION

There is a widely held view that the radiation protection standards we ought to set can hardly be mysterious or even difficult to obtain, because what is in the interest of the public's health may be ascertained by examining (through polls and surveys) what people generally believe is in their interest. Indeed, almost all contemporary sociological investigations of such issues are based on this methodological assumption. We shall call this view the public opinion theory of radiation protection policy.

One very appealing argument can be brought favor of the public opinion theory. In particular, it seems reasonable to assume that the opinion of the public has been subject to and has withstood the test of criticism over time. Given that deliberation about radiation protection policy and its implementation are not fundamentally hindered by economic or political constraints, and that the public has a forceful role in and commitment to the development of wise radiation protection policy, it seems reasonable to believe that the robustness of that policy ought to improve with time. If a radiation protection policy proposal were wildly at odds with public opinion, the argument concludes, very likely it would be so because the proposal fails to take into account various important considerations which public opinion, by virtue of surviving repeated test and challenge, already considers.

Whether this account as stated is as clear or defensible as we might desire it to be, it makes a very important point: any account of radiation protection policy which fails to somehow take into account what people actually believe about such matters can hardly be considered adequate.

NATURALISM

One of the most common defenses of the public opinion theory holds that that theory is correct simply because all we mean by a phrase like "Limiting the concentration of uranium in drinking water is in the public's interest" is "Limiting the concentration of uranium in drinking water is desired by the public." It would follow on logic alone that to determine what is in the interest of public health, all we would have to do is to determine what the public desires on this matter, say, through polls or surveys.

Let us call this defense of the public opinion theory the "meaning" defense. The meaning defense is based on a general type of value theory called naturalism. Among other things, naturalism holds that all radiation protection policy dicta such as "Clean water is in the public interest," "Keeping water clean is for the public good," "Clean water is a personal right," and so on, are translatable into terms or phrases that behave in our language like the phrases "radium is radioactive," "uranium is toxic," and "the ozone layer strongly attenuates incident ultraviolet radiation" do: all of these, naturalism claims, can be determined to be true or false (to our practical satisfaction) by the methods of ordinary scientific observation and experiment (e.g., polling and surveys, Geiger counter measurements, toxicity studies, optical measurements, and so on).

Westermarck's theory.[2] One of the most influential proponents of naturalism in this century was Edward Westermarck, an American anthropologist of uncommon erudition. Westermarck's research spanned a wide range of cultures. Primarily from this research he concluded that not only are there no intercultural views about value of health (or anything else, he argued), but there is no truly interpersonal value assigned to health within the same culture, either. Nonetheless, Westermarck noted, in every culture most individuals assign some (although not necessarily the same) value to health. To Westermarck, there appeared only one natural way to account for all these observations: when an individual says that a policy or action (say, a radiation protection policy) is desirable he merely means that the policy or action is such that that individual has a tendency to feel approval toward any agent (person) who acts in accordance with it. On this view, then, to determine whether a radiation protection policy is desirable, I merely ask whether I feel approval toward anyone who acts in accordance with that policy. For example, for me to determine whether a proposed standard for uranium in drinking water is desirable, I merely determine whether I would tend to feel approval toward anyone who tried to help ensure that this standard were met. This account, Westermarck often claimed, reflects the way people actually use that language.

Whatever may trouble us about this view, it implicitly posits at least three criteria which can hardly be ignored by an adequate account of radiation protection policy. First, even within the same culture, there is apparently a wide divergence of opinion about matters concerning the value of health. Recent debates about whether human life should be preserved by heroic measures, whether the life of a fetus ranks equal with a mother's, and how we should handle the tradeoffs between the cost of electricity and the carcinogenic effects of nuclear fuels show that this divergence of opinion is not likely to be resolved soon. Any theory of radiation protection policy which does not take this diversity of opinion into account will no doubt be considered unsatisfactory by most people. Second, Westermarck's theory is relatively clear: even if it is just mistaken, there is little doubt about what it says. Any critic of Westermarck's theory must accordingly achieve at least this kind of explicitness to be truly forceful. Third, Westermarck's theory tells us at least one way we can evaluate it: it is true only if it conforms to the way people actually talk about the value of health.

Despite these strengths, two general challenges have been raised to Westermarck's view. The first of these questions whether Westermarck's theory adequately handles the role doubt plays in reasoning about value of health; the second questions whether reasoning about value of health depends so heavily on the judgment of individuals as Westermarck suggests. We consider these objections in turn.

The "simultaneity" problem. Does Westermarck's theory adequately handle the role which doubt plays in our reasoning about the value of health? On his theory, if I now have a tendency to feel disapproval of a given radiation protection policy, and if I know that I have this tendency, then I cannot have any doubt that that radiation protection policy is wrong, because the theory says that "having the tendency to disapprove of the policy" and "holding the policy to be wrong" mean exactly the same thing; similarly, if I have a tendency to feel approval toward a given radiation protection policy, and know that I have this tendency, then I cannot doubt that the policy is right. This accounting seems rather questionable, it could be argued, because I can both have a tendency to approve of a radiation protection policy and question whether I am correct in asserting that this policy is right. Suppose, for example, I truly know that I have a tendency to feel approval of the limited contamination of drinking water by surface mining runoff because it seems necessary to the economic well-being of a certain community. I may still believe that such a policy is wrong or at least entertain the possibility that I may be mistaken if I learn that the contamination is clearly causing birth defects in the communities who use the runoff as a source of drinking water. If such an example merely makes sense, it would appear, Westermarck's view cannot be correct.

Against this kind of objection, which we shall call the "simultaneity" objection, proponents of Westermarck's view typically offer one of two arguments.

The "feeling-state" defense. The more radical of these holds that it is in principle impossible for us to both doubt a radiation protection policy and think it right because of what we mean by "doubting," "feeling approval," and so on. What we mean when we say that we have a certain feeling about something, this argument goes, is that we are in a certain "feeling state." Feeling-states, like classical physical states (e.g., the position, mass, and velocity of a particle) are unique, exhaustive, and mutually exclusive, because they are (mathematical) functions of time. Just as a (classical) particle cannot have two physical states at the same time, a person cannot simultaneously have two feeling-states. Hence, the argument concludes, we cannot both feel doubt about, and approval of, a given radiation protection policy proposal.

Let us call this counterargument the "feeling-state" defense. Even if we were to grant that it is formally correct, it is not as forceful as it might at first seem. In particular, it is a fact that we sometimes doubt, and (perhaps not simultaneously) feel approval of a radiation protection policy, and this fact an adequate theory of reasoning about the value of health must take into account. The feeling-state defense, however, gives us no insight into the relation between doubt about that policy and our feelings of approval or disapproval about that policy. If the feeling-state defense is to be satisfactory, that is to say, it must at least explain why our intuitions about the possibility of the simultaneity of doubt and approval can persist, that is, it must show that the intuition rests on a mistaken, perhaps implicit, belief or inference, and this the feeling-state defense does not do. Let us call this line of argumentation the "explicitness" objection.

The defense from psychology. There is a less radical argument which proponents of Westermarck's view offer against the simultaneity objection. On this argument, it is just a psychological fact that it is impossible for us to approve of a radiation protection policy or action and at the same time doubt that that action or policy is correct. Unlike the feeling-state defense, on this view there is no semantic necessity in the "fact" that we do not have doubt and approval of one and the same radiation protection policy: we just happen to be psychologically constituted so that this state of affairs never occurs.

Let us call this counterargument to the simultaneity objection the "defense from psychology." In contrast to the feeling-state defense, the defense from psychology gives us a little insight into the role which doubt plays in reasoning about the value of health; specifically, it asserts that doubt and belief about a matter cannot occur, simply because of the way humans are built. Such a claim is at least in principle testable because it talks about a matter of fact. It is thus to a degree protected from the explicitness objection. In spite of this virtue, the defense from psychology suffers from a distinctive limitation of its own. In particular, the heart of the defense, and hence its cogency, rests on a claim about human psychology. Since it is precisely this point which we want to be explained, the defense from psychology as it stands comes down to an unsubstantiated (but nevertheless testable) denial that we can have both doubt about, and approval of, a radiation protection policy. To be forceful, the argument must appeal to some generally established fact or evidence about doubt and approval, and this it does not do. Although there is no reason in principle why such evidence could not be produced, it has not been forthcoming from proponents of Westermarck's view.

<u>The "multiple-grounds" problem</u>. Let us suppose, in any case, that Westermarck's theory can be defended against the charge that it is incompatible with our apparent ability to both doubt and approve of the same action or radiation protection policy. The problems which the theory has with the role of doubt in reasoning about the value of health do not thereby vanish. It has been claimed by some critics, in particular, that on Westermarck's view there is <u>only one</u> ground for doubt about a radiation protection policy issue, namely, that I doubt that I have a tendency to approve (or disapprove) of the policy. But clearly, this criticism continues, there are other grounds for doubt: I may have some doubt that a policy is correct because I may have reason to believe that my information about it is based on an incorrect understanding of the relevant facts. For example, I may approve of a uranium standard for drinking water of not more than 50 pCi/l, but wonder whether my approval is entirely correct because I know that statistical features of the evidence supporting this standard make it disputable. Because this

objection alleges that there can be multiple grounds for doubt, we shall call it the "multiple-grounds" objection.

The general thrust of the simultaneity, explicitness, and multiple-grounds objections is to challenge proponents of Westermarck's view to provide a more natural, convincing, and illuminating accounting of the role of doubt in reasoning about the value of health. This account unfortunately has not been forthcoming.

The "anti-egoism" objection. Even if Westermarck's theory could somehow be defended against the charge that it does not adequately account for doubt's role in reasoning about matters of the value of health, the theory faces another set of problems. In particular, the theory holds that a radiation protection policy is desirable for me only if I have a tendency to feel approval toward it. Strictly speaking, then, if someone else tends to feel disapproval toward that same policy, the policy is wrong only for him. Thus, on Westermarck's account, when someone says that something is wrong, strictly speaking, he never can be in a position of contradicting someone else. This result casts Westermarck's claim that he is merely defining utterances concerning the value of health in an embarassing light: we do in fact hold that we are contradicting someone who says that a certain radiation protection policy is right if we say that that same policy is wrong. If I say, for example, that a ceiling of more than 50 pCi/l of uranium alpha in drinking water is right, and my colleague says that a higher ceiling is acceptable, neither of us supposes that we are not directly contravening the declaration of the other.

Perry's theory.[3] This concern with Westermarck's view, which we shall call anti-egoism (no perjorative intended) objection, has troubled many naturalists who otherwise find substantial agreement with him. They have sought to modify Westermarck's formula by removing the dependence of that view on strictly individual approval (or disapproval). The writings of Ralph Barton Perry provide a good example of this kind of refinement. Perry's theory claims that when we say that a policy is desirable we simply mean that the policy is the object of a favorable attitude, or a favorable interest, on the part of somebody, all things considered. By "all things considered," Perry means "by all parties affected." Furthermore, Perry requires, a policy is desirable only if there can be no other policy which bears on the topic of consideration which could be more conducive to the happiness of all affected.

Perry's theory clearly dodges the anti-egoism objection, because it does not require the meaning of utterances concerning the value of health to depend only on individual tendencies or approval. It thus represents something of an improvement over Westermarck's formula.

Perry himself offers two additional arguments for his theory.

Arguments for Perry's theory. First, he notes that there is substantial agreement among educated and reflective people about the things which are desirable. If people were to talk about the value of health in accordance with Perry's account of that language, Perry continues, the way they would express themselves would be essentially the same as it now is. Let us call this argument the "consensus" defense.

Second, he argues, what we accept as evidence for a policy being desirable agrees well with what his definitions prescribe that evidence should be. Let us call this argument the "prediction" defense.

The consensus and prediction defenses are quite good if they are sound. But in fact, critics have pointed out, they are not nearly so impressive as they first appear.

Incompleteness of Perry's theory. The consensus argument in particular seems to be contravened by the way in which people actually talk about the value of health. For example, one of the most widespread rules regarding the value of health involves the obligation to preserve the of the health of children, and that rule binds even distant relatives. Perry's account does not predict the application of the "health-preservation" rule to distant relatives, because relatives have little to no stake in ("are not affected by") the health of children distantly related to them. Thus, Perry's critics contend, the consensus defense is not sound.

Let us call this line of reasoning the "incompleteness" objection. Proponents of Perry's view could, it appears, rejoin this criticism by arguing that not only do distant relatives have a stake in the preservation of the health of children, everyone does. The survival of children is essential to community survival and well-being, this rejoinder would go, and hence it is in the interest of everyone, distant relatives included, to preserve the health of children.

The "community-survival" defense. This rejoinder to the incompleteness objection, which we shall call the "community-survival" defense, is on the surface highly plausible. But on closer analysis, it can be seen to trade on a claim for which further argument must be presented: if distant relatives and/or total strangers were indifferent to the health of given children, the communities containing those individuals would suffer significantly. Although this premise may hold true for small, subsistence-type living groups, it is hardly clear that it is necessary for large, heterogeneous populations such as that of the United States. In particular, it appears that a child's distant relatives or total strangers can pay no attention to a child's health, but the population manages to survive reasonably well in spite of this state of affairs. Thus, the community survival defense is not particularly forceful, and is at best of limited applicability.

Inadequacy of the prediction defense. The second of Perry's arguments, the prediction defense, is similarly open to question, his critics claim. The prediction defense claims that people take into account just those things which would be the object of a favorable attitude on the part of all affected by them more than any other course of action. But as a matter of fact, this criticism begins, people do take evidence into account other than that which bears on whether a proposed radiation protection policy is conducive to the interests of all parties affected, when trying to decide whether a radiation protection policy is correct. One's intuition, for example, is a widely adduced source of evidence for many people. Thus, the prediction defense is not nearly as forceful as it would at first appear. Let us call this line of reasoning the "insufficient-evidence" objection.

The insufficient-evidence and incompleteness objections strongly question whether the consensus and prediction arguments are good defenses of Perry's theory. Even if these objections could be overcome, critics have questioned whether the very claims made by the theory (apart from the defenses of that theory) can possibly be correct. We turn to the most forceful of those objections.

A problem with Perry's utilitarianism. On Perry's view, in particular, a health policy is desirable only if it will probably be more conducive to the harmonious happiness of mankind than any other policy. But as a matter of fact, some critics have claimed, people are not committed to this view, whatever they may at first think. Some thoughtful, conscientious, informed people consult their "intuitions" when making decisions concerning the value of health. Consider in particular, the following. Suppose that a physician in a small rural town confines a pregnant patient to the town's only hospital, which in fact is owned by the physician. The physician believes that the mother and fetus will certainly die if the pregnancy is allowed to continue full term. He accordingly decides that an abortion is the only defensible medical approach to the problem. Unfortunately, a radical right-to-life group learns of his intentions and threatens to bomb the hospital to prevent the abortion, if necessary. Let us suppose that this group has already been associated with clinic bombings; the physician therefore has good reason to believe that its threats will be carried out. The physician is thus faced with a serious dilemma. Clearly the greatest good will be done if he sacrifices the life of the pregnant patient: he will at least save the hospital, and with it, all the good that it can do. The physician contemplates sacrificing the patient, but his intuition says that such a move would be wrong. In feeling this, he is appealing to something other than just those considerations which will produce the greatest good. If this kind of situation is merely possible, Perry's theory is not adequate. We shall call this line of reasoning the "anti-utilitarian" objection, because it is intimately related to a theory of value (specifically, a theory of obligation) known as "utilitarianism."

The anti-utilitarian objection questions whether Perry's theory describes the way people actually behave. There is a rejoinder to this kind of criticism which is not entirely free of problems itself, but at least allows something like Perry's theory to answer some of the less forceful concerns about it. It may be, in particular, that Perry's account should be viewed only as a formal model for predicting the actions of individuals in situations in which they must judge the value of health; accordingly, the psychological idiom in which his theory is cast should not be taken as a literal account of the psychology involved in those actions. If this is the case, then Perry's theory should be read as implying something like "a radiation protection policy is desirable only if all parties affected by it would behave as if they believed that the policy promoted the greatest possible happiness for all affected." (If this is what Perry meant, a critic could of course object, he should have used such a formulation.) In any case, such a formulation would blunt the force of any criticism which questioned whether Perry was literally describing the way in which people talk about the value of health.

However we may decide this point, there is a refinement of Perry's theory which can overcome some of the above objections. According to this refinement, when we say that "radiation protection policy A is better than radiation protection policy B" we simply mean "Any individual fully informed of the consequences of adopting A and B, in a calm frame of mind, under no threat or duress, and in all other respects normal, would prefer A to B." On this view, for example, if we needed to decide whether a limit of 50 pCi/l were preferable to a limit of 100 pCi/l for uranium in drinking water, we would simply determine what an individual fully informed of the consequences of each of these standards would prefer.

This version of naturalism, which we will call the "rational observer" formulation, nominally avoids some of the problems which plague Perry's view because it implicitly asserts that whatever preferences "calm, completely informed, normal" people have defines what utterances concerning the value of health mean: unlike Perry's account, the theory does not try to prescribe in advance what those preferences are (e.g., a preference to do the greatest good). Thus, it cannot run into trouble with incompleteness, insufficient-evidence, or anti-utilitarian objections. In this sense, the rational observer formulation of naturalism constitutes an advance over Perry's theory. Nevertheless, it has drawn unanswered fire on at least two grounds.

_A problem with the consensus requirement_. First, the theory presumes that all "rational" observers (in the sense of the requirements imposed by the theory) will in fact have the same preferences among candidate policies. But as a matter of fact, there is substantial disagreement among rational, informed, and conscientious people about matters of radiation protection policy. The current debate over the concentration of uranium which should be in drinking water shows that the agreement presumed by the rationalist formula is not likely to be reached soon. Let us call this objection to the rational observer formulation of naturalism the "anti-consensus" objection. Proponents of the rational observer theory have had nothing to say about this problem.

_A problem with preferences in Perry's theory_. Second, it is hardly clear, critics of the rational observer theory have argued, that mere _preference_ by rational observers, the anti-utilitarianism objection notwithstanding, constitutes a complete basis for determining whether a radiation protection policy is desirable: apparently some rational observers believe that they also need to take into consideration whether the radiation protection policy promotes the greatest possible happiness. Let us call this the "insufficiency-of-preference" objection. Proponents of the rational observer theory, unfortunately, have had little to say about this objection, either.

_Summary of unanswered questions in naturalism_. At the least, then, naturalism has at least two serious problems (the anti-consensus and the insufficiency-of-preference objections) to solve. Whether it can overcome these difficulties remains to be seen.

## NONNATURALISM

Driven by problems of the kind outlined above, some writers, whom we shall call nonnaturalists, have proposed a theory of what utterances concerning the value of health mean which begins by presuming that such utterances behave in a way quite distinct from the way phrases like "radium is radioactive" and "uranium is toxic" behave. In support of this claim, they point to what they perceive to be failure of naturalism to produce an adequate account of the way in which value utterances behave, adducing arguments of the sort we have been examining. Indeed, nonnaturalists take the failure of the naturalist's approach to demand an alternative theory. Nonnaturalism, at least as it applies to utterances concerning the value of health, can be summarized in three theses. (1) If a utterance concerning the value of health can be defined at all, its defining phrase must contain at least one value term itself; indeed, many nonnaturalists have gone farther than this: they assert that there are some value terms which are simply undefinable, because those terms refer to (value) properties which are simple and hence unanalyzable. (2) Even utterances concerning values are not testable in the same way that sensible properties such as being radioactive or being toxic are. (3) There is at least one thing which has an unanalyzable value property, and we know that that thing has that property. (Concerning precisely what this thing and property are has been a source of some disagreement among nonnaturalists.)

Nonnaturalism is immune to the charge that it is just inconsistent with the nature of human desires, preferences, doubts about the value of health, because nonnaturalism does not allow a role for these features in reasoning about the value of health. But even if it avoids these particular problems, nonnaturalism must answer at least two thorny questions.

Is nonnaturalism even intelligible? First, the very feature that renders nonnaturalism immune to the problems which naturalism faces can be turned into a criticism of nonnaturalism itself. Naturalism, because it appeals to something we can observe, is at least intelligible. In contrast, the nonnaturalist account asserts that value properties cannot even be observed: no experience of the world around us, the nonnaturalist maintains, can make it more or less likely that we should accept any utterance containing a value term as true. One may accordingly wonder how it is that we ever get the concept of these properties. Ostensibly, the way we get all of our concepts involves being shown examples of those concepts. We know what "being radioactive" is, for example, because among other reasons, we have been shown examples of radioactive things. We may have an indirect acquaintance with certain concepts, of course, through other concepts. For example, we could know that uranium oxide is radioactive because we know that uranium is radioactive and that uranium's radioactivity is not diminished when it combines with oxygen. But even in this case, a strong argument could be made that our knowledge of whether uranium oxide is radioactive depends our having seen at least some examples of radioactive compounds. On the nonnaturalist account, however, at least some value terms are neither definable nor observable, and hence involve concepts that could never have arisen from experience of at least some particular examples of those terms. Is such an account, critics charge, even intelligible? Since this objection challenges the nonnaturalist's account of the genesis of our at least some concepts about matters of value, we will call it the "genesis" objection.

The second question difficult question which nonnaturalism must answer is: what empirical reason can the nonnaturalist ever give for a claim such as "Reducing the concentration of radioactive species in drinking water is a desirable policy?" Since "being desirable" is not observable according to the nonnaturalist, there is no observation which we can make to ascertain whether such a claim is even more or less likely to be correct. Because this objection concerns whether there is an observation to which we can can subject the nonnaturalist view to determine its verisimilitude, we shall call it the "non-observability" objection.

There are two general approaches to the genesis and the nonobservability objections which nonnaturalists take, and there is by no means agreement among them about which to take. Some nonnaturalists, furthermore, do not sharply distinguish answers to the two objections, and some do not appear to perceive a need to provide an answer to the objections at all.

The first type of answer which some nonnaturalists give to the genesis and nonobservability objections actually provides little insight into how we come to have insight into matters of value, given that they are non-observable. The answer simply says that we do have such concepts and are aware that we do, because the human mind has the capacity to have them, and in fact does have them. As to the second question, this account, which we will call the "a-priori-knowledge" ("a priori" = "independent of experience") theory, maintains that correct utterances involving the value of health are so because they are necessary. Such propositions, that is to say, cannot possibly be false, in the same sense that the laws of logic or mathematics cannot be false. The human mind, the "a-priori-knowledge" theory further maintains, can identify correct utterances concerning the value of health because it the capacity to know that these propositions are necessary and does in fact know that they are so merely if they are understood. Thus, for example, on the a-priori-knowledge account if I hold that clean water is a personal right, I know that it is, simply as a result of understanding what that right means.

The intuitionist theory. The second type of answer to the genesis and nonobservability objections which nonnaturalists sometimes give might be called the "intuitionist" theory. Proponents of the "intuitionist" version of nonnaturalism tend to regard the "a-priori-knowledge" theory as contrived, and just possibly false. Like the a-priori-knowledge theory, the intuitionist account posits that in fact we do have concepts about the value of health and use principles concerning the value of health. Unlike the a-priori-knowledge theory, the intuitionist claims that often enough we are not exactly sure why we have these concepts and principles, and often are not conscious of what these principles and concepts are. We have this knowledge through some kind of "nonperceptual intuition" or "emotional intuition," intuitionists sometimes maintain.

Problems in the intuitionist formulation. There is something inherently disconcerting about both the "a-priori-knowledge" and "intuitionist" accounts of how we know the value of health, and the history of the knowledge theory (epistemology) demonstrates that the issue is not likely to be resolved soon. Although the literature on this dispute is extensive, we need to at least survey its outlines. We consider the intuitionist account first.

A problem with beliefs. The intuitionist account of our knowledge about the value of health claims that we have some sort of direct, albeit possibly unconscious, knowledge of the value of health and "facts." There appears to be substantial evidence that such knowledge is at best indirect, however. Consider, first of all, our knowledge of the decisions, acts, and states of mind involving the value of health of other persons. It seems quite clear that we do not have direct knowledge of those decisions, acts, and states. Instead, whatever knowledge we may have of the acts or deeds of others is mediated by our beliefs about what those states or deeds are. When I know, for example, that my colleague believes that a concentration of 10 pCi/l or lower uranium in drinking water is desirable, I have, among other things, a belief about my colleague's beliefs; I do not directly have his beliefs as such.

Analogously, it appears that we do not have direct knowledge of even the related features of _our own_ minds and acts: our knowledge of the the value of health as it relates to our own acts and mental states appears to be mediated by our _beliefs about_ those acts and states. Someone who wishes to defend the intuitionist view is therefore obligated to clearly show that our knowledge of the value of health as it concerns at least our own acts and mental states is _not_ mediated by our beliefs about those events or acts. Let us call this line of reasoning the "mediation" objection. Proponents of the intuitionist version of nonnaturalism have had, unfortunately, nothing to say about this objection.

_Problems with ambiguous situations_. There is a second problem which the intuitionist account faces that can be best considered with an example. Suppose that in the course of his research a physician comes upon evidence hinting that there is no level of exposure to radioactivity in drinking water which does not produce lethal effects in at least some (perhaps a very small proportion of) individuals. The hint is somewhat ambiguous. The research funds available to the physician are limited and would almost surely be exhausted if diverted to the resolution of this ambiguity; there is certainly no guarantee of success in any case. There are many investigators in the area in which the physician is working. The physician decides that since there is the possibility that other investigators will devote attention and resources to this problem, and that it is plausible that they will do so, he will not pursue the resolution of this problem himself. Time passes, and he continues to read the professional journals to find some mention of the problem he had noted. Let us suppose that there is none. On the one hand, he may feel vindicated. "After all," he may reason, "if the problem had been real, others would have seen it and would have mentioned it somewhere." But at the same time, he may wonder whether the problem was real but has been ignored simply because he is the only one who has seen it. The dilemma persists in his mind.

Let us ask where in this physician's experience the direct intuition of the value of health existed. The more we look at the the example, the more difficult it is to see where this direct recognition occurred; in fact, by the design of the example there is no such place. Provided it merely describes a possible state of affairs, such an example calls into direct question whether the intuitionist account is really all that forceful. Let us call this argument the "ambiguous situation" objection. Proponents of intuitionism have had nothing to say about this problem, either.

Further troubles for the intuitionist. There are two additional problems with the intuitionist account. Equally informed and conscientious people sometimes have conflicting views about what is desirable in radiation protection policy. If there is such a difference, the intuitionist must answer two serious questions. First, how is it possible to have such conflicts, given that responsible, intelligent people have direct (i.e., nearly certain) knowledge of what is desirable? Second, given that we do sometimes, but not always, have direct knowledge of what is desirable, how can we distinguish our mistaken intuitions from our correct ones?

To answer the first question, intuitionists often try to draw an analogy between intuitions about the value of health and other faculties, particularly vision. There are disputes possible among otherwise normal, informed adults, intuitionists argue, because the faculty of apprehending concepts and principles involved in assessing the value of health can be faulty, much as vision can be faulty. We allow that although there are differences in what people think they may see in certain situations, there is nevertheless a direct sense involved when they see correctly.

Although it is difficult to know just how to take this argument, suppose we were to grant it. This still leaves the second question, of how we are to identify correct intuitions about the value of health, distinguishing them from the ones which arise from the malfunction of our intuitions about the value of health. Let us call this the "discrimination" objection. Unfortunately, no intuitionist to date has given this matter serious attention.

Summary of problems with the intuitionist account. On the whole, then, proponents of the intuitionist formulation have yet to meet two serious objections--the ambiguous situation and discrimination objections. Whether they can do this remains to be seen.

Problems for the a-priori-knowledge theory. The "a-priori-knowledge" account of how we know the concepts and principles involved in assessing the value of health does not fare much better than the intuitionist one, although it suffers from its own distinctive limitations. We examine the more important of these.

On the a-priori-knowledge view, recall, we have direct, conscious, conceptual insight into the necessity of correct judgments about the value of health. This kind of insight is alleged to be analogous to the kind of insight we have into the necessity of mathematical propositions.

To fairly evaluate the a-priori-knowledge proposal, we need to remind ourselves what this analogy amounts to. Consider, in particular, what is involved in saying that "the sum of two and two is four." In this case, we may say without doubt or hesitation that whatever is the sum of two and two is four. Our lack of doubt and hesitation about this matter has at least two features. First, being the sum of two and two requires being four. In general, when we say that something's having a property A requires that it also have the property B, we can say "anything that is A is also B;" furthermore, if a single thing that is A is not B, evidently having A does not require having property B.

Second, a single property, "being four," is involved in "being the sum of two and two." The a-priori-knowledge account claims that judgments concerning the value of health are like arithmetical judgments. If an action has one or more properties of some (possibly special) sort, that is to say, then having these properties requires having a value property, and it is immediate insight into this necessity that the "a-priori-knowledge" thesis claims we have.

Difficulties in the analogy with arithmetic. There is something particularly disturbing about the a-priori-knowledge analogy. First, considerations involving the value of health are typically not dependent on just one or two (more generally, a definite number of) features, but instead depend on an indefinite number--whatever is relevant to health. The property of being, on the whole, desirable, is just such a property. We do not, for example, say that the policy of allowing limited contamination of drinking water by radioactive species is, on the whole, desirable, without first considering the cost and consequences of an indefinite number of alternative arrangements including the possibility of alternate forms of government involvement in the administration of such a policy, purification technologies both present and proposed, and comparable public health concerns. We can eventually conclude that a given health policy is on the whole desirable not because there is any single property or even definite collection of properties on the basis of which we can reach this conclusion; we can reach the conclusion simply because we can sometimes discover that no issue beyond those we have already considered is involved. This feature has no analog in our knowledge of the necessity of mathematical propositions (if we indeed even have such knowledge). Thus, the claim that judgments concerning the value of health are like mathematical judgments is at best problematic. Let us call this the "indefiniteness" objection.

Good-making properties. Some features of a situation which have to do with the value of health, of course, are not of the sort just described. Some authors who hold the a-priori-knowledge view have accordingly claimed that in such cases a single distinctive aspect of the situation under consideration makes it desirable. These individual, value-related, distinctive features are called "good-making aspects" of the situation. A physician's knowingly unremunerated effort to save a child from cancer, for example, is good simply because it involves helping another without regard to personal benefit. Into the "good-making" aspects of such situations, a-priori-knowledge proponents claim, we may plausibly claim to have a priori insight based solely on a few, definite features of those situations. Let us call this the "good-making" defense.

For those situations in which a single (more generally, a small definite number of) "good-making" features determines all the important features of the situation, the good-making defense rejoins the objection that an indefinite number of features of the situation must typically be considered in evaluating that situation. Even if the good-making defense is granted, however, there are still profound differences between the way we judge whether a public health policy is desirable and the way we judge the necessity of "two plus two is four." In the former, in particular, we begin with the unrefined assumption that anything which promotes health is desirable. We then begin to consider the differential cost of maintaining or providing for health. Furthermore, public health policy is typically related to a complex body of law, and the relation of the policy to that law must be considered. In the case of arithmetic judgments, we merely consult the laws of arithmetic. In short, a judgment concerning the value of health is rarely so simple as arithmetical judgment. Thus, the analogy which proponents of the a-priori-knowledge theory try to draw between our judgments in arithmetic and our judgments about the value of health is at best oversimplified. Let us call this the "oversimplification" objection.

*How can people disagree if the a-priori-knowledge view is correct?* There is a second general difficulty with the a-priori-knowledge version of nonnaturalism which is worth examining before we go on. This account asserts that intelligent people who think carefully and patiently will often see, and reach agreement on, the necessity of utterances concerning the value of health. However, it is not clear that this state of affairs is even achievable. Even the people who have devoted their lives to the study of medical ethics disagree with one another about the most fundamental features of reasoning concerning the value of health. For example, some say that health is an "intrinsic" (in and for its own sake, without regard to anything else) good; others claim it is not.

A problem with false judgments. Given that people merely disagree about the value of health, furthermore, the a-priori-knowledge account faces a third problem. Even if there are a-priori insights concerning the value of health, there are apparently false insights, too. If so, a problem which plagues the intuitionist account now rears its head. If we have false and veridical insights concerning the value of health, how does the a-priori-knowledge account propose we distinguish between the two? To date, no convincing account has been produced by those who hold the a-priori-knowledge view. This is not to say that the a-priori-knowledge account is just wrong, but only to point out that a crucial part of that program remains to be developed.

Summary of problems with the a-priori knowledge view. On the whole, then, the a-priori-knowledge version of nonnaturalism has at least the indefiniteness and the oversimplification objections to answer. Proponents of the formulation have to date had nothing to say about these questions. Whether they can, remains to be seen.

## EMOTIVISM

The difficulties faced by the naturalist and nonnaturalist programs have driven some writers to question whether there is not some very deep problem which those programs share. Is it just possible, these critics rhetorically ask, that the assumption that utterances concerning the value of health talk about properties of persons, that is, make testable claims about preferences, states of affairs, desires, and so on, is just mistaken?

Mere denial of the correctness of the correctness of naturalism and nonnaturalism does not by itself constitute a refutation, of course. A positive theory (explanation) of decision-making concerning the value of health is required. In this century, a major alternative to the naturalist and nonnaturalist programs was developed C. L. Stevenson.[4] We will call this theory the emotivist theory for reasons which will become apparent. The theory is typically articulated (and, in part defended) as follows.

First, the theory notes, a distinctive feature of thinking about values is that it involves attitudes. Roughly speaking, we say that someone has an attitude toward something when he is for or against that thing, or that he seeks or avoids it, or that he is emotionally concerned about it. There is a distinctive subclass of attitudes, the theory continues, which is concerned with forms of behavior and with persons on account of their behavior. These include a disposition to be indignant toward others if they behave in a certain way, or an inclination on one's own part to behave in a certain way, and to feel guilty if one does not. If I say that a concentration of 10 pCi/l uranium in drinking water is undesirable, for example, according to the emotivist theory I am inclined to be indignant toward anyone who behaves in such a way as to promote or implement a higher concentration, or I am inclined to try to promote this standard and am inclined to feel guilty if I do not. We may call this subclass of attitudes "value-specific attitudes."

## The "magnetic attraction" thesis

The emotivist notes that there are obviously settings in which the sharp utterance of phrases such as "Get up!" or "Fire!" will cause a quick and immediate reaction of hearers of those utterances, assuming that they are more or less normally constituted. Similarly, utterances such as "We must spare children from leukemia!" elicit a favorable attitude toward the action to which the utterances refer. In the jargon of the emotivist theory, these value-related utterances "magnetically attract" favorable or unfavorable attitudes. This evidence strongly suggests to the emotivist that one of the principal roles of speech acts about the value of health is to "magnetically attract" specific attitudes.

Emotivists have offered several arguments to support this thesis. The first of these we consider turns on an argument developed by the English philosopher G. E. Moore at the beginning of this century.[5] At the heart of Moore's argument is a criterion which he believed could be used to determine whether two terms, A and B, mean the same thing. To determine whether A means the same thing as B, Moore urged, pose the following question: "Is everything which is A also B?" Then ask yourself whether the resulting question is "intelligible" or "significant," or what it means to doubt that the answer is affirmative. If the question is intelligible or significant, Moore claimed, then A and B <u>do not</u> mean the same thing, for if A and B did mean the same thing, the resulting question would be no more significant than "Is everything A also A?," nor can we understand what it would mean to doubt an affirmative answer to the question. For example, to determine whether "1 nanoCurie" means the same as "$10^{-3}$ microCurie, we simply try to determine whether the question "Is 1 nanoCurie $10^{-3}$ microCurie?" is intelligible or significant. If that question is neither intelligible nor significant, Moore's criterion holds, then "1 nanoCurie" means exactly the same as "$10^{-3}$ microCurie." Moore believed that no naturalistic definitions of value terms could pass this criterion. For example, if we were to define the term "desirable" as "is desired by somebody," the question "Is everything that is desired by somebody also desirable?" is both significant and intelligible. On the basis of Moore's criterion, emotivists have argued, utterances concerning the value of health must have some function other than stating facts, and they propose that that function is to "magnetically attract" favorable or unfavorable attitudes.

Let us call this the "synonomy" defense. There are some fairly serious problems with it. The first of these has to do with precisely _what_ Moore's criterion says. For example, it is unclear what Moore means by "intelligible" or "significant." There is some ordinary sense, at least, in which the question "Is every P also a P?" (e.g., "Is 1 nanoCurie also 1 nanoCurie?") is intelligible, namely, we _understand_ what the question means. Presumably, then, Moore must be taking "intelligible" to mean "can be raised with some point." This raises the question of precisely what we mean when we say that a question can be raised with some point. _If we knew_ that "1 nanoCurie" did not mean the same as "$10^{-3}$ microCurie," of course, we would know that the question "Is 1 nanoCurie also $10^{-3}$ microCurie?" can be raised with some point. But this answer is truly irrelevant to the interests we have in this context, because we are trying to formulate a criterion which will allow us to _discover whether_ "1 nanoCurie" means the same as "$10^{-3}$ microCurie." Perhaps what Moore means is this: "1 nanoCurie" and "$10^{-3}$ microCurie" do not mean the same for a person if, when he asks himself "Is everything which is 1 nanoCurie also $10^{-3}$ microCurie?" he is _doubtful_ of the answer, or the answer to the question does not obviously seem to be to him affirmative.

If the latter is what Moore meant, he was obviously mistaken about at least one type of situation in which we frequently find ourselves. Consider the above example. Suppose it were the case that we did not know that 1 nanoCurie and $10^{-3}$ microCurie are identical. Then we could be doubtful of the answer to the question "Is 1 nanoCurie also $10^{-3}$ microCurie?." Yet in fact the expression "1 nanoCurie" means the same thing as "$10^{-3}$ microCurie."

Moore's test is not a good one, therefore, for cases in which we are in doubt about whether two terms mean the same, when in fact, unknown to us, they do mean the same. The test may be a reasonable one, however, for the case in which two terms mean the same, and obviously so, to us.

Let us suppose that Moore's criterion is able to establish that there must be some function of utterances concerning the value of health other than that of merely stating facts. Given this, the correctness of the criterion still does not drive us to the view that utterances concerning the value of health "magnetically attract" favorable or unfavorable attitudes. At best, the criterion merely allows us to say that utterances concerning the value of health state some facts and might attract favorable or unfavorable attitudes.

_The poetic-language analogy._ Emotivists have often wanted Moore's criterion to justify the stronger thesis that utterances concerning the value of health in fact attract attitudes. On this count, it appears that the emotivist theory needs further help. To buttress the synonomy defense, therefore, emotivists have occasionally appealed to an analogy. Value-utterances, the analogy goes, are like poetic language. "Lily of the valley," for example, is frequently suitable for poetic purposes (e.g., it has alliteration and internal rhyme) whereas "Convallaria" (the genus of the lily of the valley) often is not. Some words clearly have a special emotive force, the argument goes, and utterances concerning the value of health are among them.

There are serious difficulties with this argument, which we shall call the "poetic analogy" defense, too. On close inspection, it does not seem that the emotional effects of a word like "lily of the valley" sufficiently resemble the effects of utterances concerning the value of health. In particular, it is difficult to believe that a term like "lily of the valley" attracts anything like the attitude which "the suffering of children" does. The argument by analogy to poetic terms, then, does not seem to provide much support for the emotivist's claim that one of the functions of utterances concerning the value of health is to magnetically attract the corresponding favorable or unfavorable attitudes.

## _The "expressivist" thesis_

In addition to holding that a principal function of utterances concerning the value of health is to evoke corresponding attitudes, emotivists believe that value language has a further important feature: to have a conviction concerning the value of health (or to express that conviction), they urge, is to _have_ (or _express_) the corresponding attitude. We will call this the "expressivist" thesis. According to this thesis, for example, if I say that I strongly support a ceiling of 10 pCi/l for uranium in drinking water, then I am, among other things, expressing my personal attitude that I support a ceiling of 10 pCi/l. In defense of such a claim, emotivists have offered several arguments. Although these arguments serve to valuably sharpen the emotivist's account of the role which attitudes play in behavior concerning the value of health, it will be argued that none of them are as conclusive as their proponents desire.

The "partitioning" argument. The first of these which we consider posits that all human cognitive acts, entities, and agencies can be exhaustively and exclusively categorized as either beliefs or attitudes. Either convictions concerning (the) value (of health) are beliefs that something has a certain property (for example, in the fashion described by the naturalists) the argument maintains, or they are primarily attitudes. If they are beliefs, the account continues, they must be beliefs something like those which are held in science. Obviously, there is much disagreement about the (relative) value of health, however, and no identifiable methodology exists for resolving these disagreements. Such disagreement could not persist, the argument claims, if convictions concerning the value of health were like scientific beliefs, because scientific disagreements are in principle resolvable, or, at least, it is always possible for the disputants in such disagreements to agree on the conditions under which they could resolve their dispute. Therefore, since convictions concerning the value of health are not like the beliefs we hold in science, they must be attitudes.

For the sake of brevity, let us call this argument for the expressivist thesis the "partitioning" argument. The argument has at least one virtue: it calls our attention to the fact that there is substantial disagreement among even thoughtful people about the value of health. It is not as forceful as it might at first seem, however: various types of naturalism can explain diversity in opinion about the value of health with at least as much force as the partitioning argument. For example, a naturalist theory which held that "high health costs are more desirable than widespread bone cancer" means "a calm, completely informed person would on the whole prefer high health costs to widespread bone cancer" requires us, among other things, to consider our attitudes (preferences). Since it is difficult to know all the relevant facts in such a situation, we would expect naturalism to anticipate considerable disagreement in the preferences of even serious-minded people about it and other situations concerning the value of health. Thus a crucial feature of partitioning argument--that the only source of disagreement possible among serious-minded people about the value of health must be differences in attitudes alone--is highly questionable. And as a result, the partitioning argument is a weak one.

The argument from conviction. A second argument is frequently offered by emotivists in support of the claim that to have a conviction concerning the value of health is to take the corresponding attitude concerning a policy or action. According to this argument, whenever a person is honestly convinced that a policy or action is right, he is almost always more inclined to adopt that policy or take that action. No theory of motivation other than one which asserts that to have an conviction concerning the value of health is to have the corresponding attitude, can explain this fact, this argument continues. Suppose, for example, we know that limiting the concentration of uranium in drinking water to 10 pCi/l will contribute more to the overall happiness and health than any other ceiling. A reasonably strong case could be made that we would adopt the ceiling only if we had the attitude of being interested in overall happiness and health. Thus, the argument concludes, having an conviction about the value of health is just taking the corresponding attitude.

Let us call this argument for the expressivist thesis the "argument from conviction." To evaluate it, we must keep in mind what the emotivist's interest in it must be. It may be true, as the argument from conviction claims, that attitudes play an important role in typical behavior concerning the value of health. But the argument from conviction (and any other argument which attempts to defend the expressivist thesis) must go farther than this: the argument must establish that at most, there are only two principal functions of utterances about the value of health: to attract attitudes about the value of health and to express those attitudes, and this the argument from conviction does not do.

The argument from motivation. A third argument which emotivists sometimes offer in defense of the expressivist thesis is closely related to the argument from conviction. Consider, the argument begins, what seems to be involved when we think a person "really believes" what he says about the value of health. For example, consider what we think to be involved when a person asserts that we should do all we can to prevent environmentally induced leukemia. We judge that he truly believes this if he actively and frequently attempts to persuade others his view is correct, perhaps contributing money and personal time to attempting to reduce the incidence of the disease, or perhaps clearly foregoing personal gain to promote its reduction. In general, then, we judge that he has this conviction if he actually seems to live by this conviction, that is, is motivated by the corresponding attitude. This is just what we would expect if the expressivist thesis were true.

Let us call this argument for the expressivist thesis the "argument from motivation." It cannot be ignored, for something about it is surely correct. It does seem, in particular, that in many cases, we are inclined to believe that someone truly holds (in all senses) a conviction just in case he lives by it. But the argument from motivation is not conclusive, for reasons much like those considered above. Defending the expressivist thesis, in particular, requires showing that we may identify having an conviction concerning the value of health with taking a specific attitude concerning it. What the "motivationist" argument just considered shows, however, is only that there is some close psychological connection between the two, not that the two are identical. A critic of the expressivist thesis can in fact show this to be the case by arguing that it is easy to imagine how people could behave as though they had a conviction concerning the value of health without having the corresponding attitude: people might simply have been trained to automatically do whatever they think promotes health, and on such an account, no identification of behaving as though one had a conviction concerning the value of health and taking the corresponding attitude is necessary.

The "intent-of-persuasion" argument. A fourth argument is frequently offered by emotivists to support the expressivist thesis. Consider, the argument begins, what our aim is in arguing a point about the value of health with someone. We are trying to affect is what that person feels about the value of health, that is, the attitudes which he has. If we don't achieve this aim, we imagine that our argument has failed. We believe that we have won the point when our opponent agrees with us in attitude. For example, suppose that we were trying to persuade someone that we ought to do whatever we can to prevent lung cancer. Suppose our opponent is a hardened smoker who grudgingly admits that our arguments have great force, but nevertheless says that he is still unconvinced, and that he will probably not act in a way that helps prevent the disease. In all likelihood, we will consider our efforts to have failed. Thus, the argument concludes, having an conviction concerning the value of health is just taking the corresponding attitude.

For the sake of brevity, let us call this argument the "intent-of-persuasion" argument. The argument is worthy of careful inspection because it does seem that at least in some instances, we do intend to alter the attitudes of people whose behavior we are trying to change in deliberation concerning the value of health. The intent-of-persuasion defense fails, unfortunately, to respect a very important but simple distinction. In particular, we must distinguish between what a person wants to do and what he ought to do. It is certainly possible for a person to do or to want to do what he ought not to do, and it is possible for a person to not want to do what he ought to do. Given these two possibilities, there are at least two possible situations in which we could find ourselves when we are trying to persuade a person that he ought to do something. The first type of situation concerning the value of health, of course, is the one urged by the intent-of-persuasion defense; in this case, we really succeed in changing the driving attitudes of our hearer. In the second type of situation, we are trying to persuade a person that he ought to do something, in spite of the fact that he does not want (that is, does not possess the attitude which drives him) to do it. He may eventually agree with us that he ought to do a certain thing but nevertheless not do that thing. We could win such an argument, that is to say, but fail to change a person's driving attitudes about the matter. Thus, it is not obvious that winning an argument about the value of health changes attitudes in the sense that the intent-of-persuasion argument desires.

The argument from evidence. A fifth argument is sometimes offered by emotivists in support of the expressivist thesis. Consider the nature of the facts which are considered "relevant" in any discussion about the value of health the argument begins. For example, suppose that I am trying to persuade a friend that we must keep drinking water free of uranium. As part of my effort to persuade, I invoke a number of "facts" which I consider relevant to the situation: ingesting uranium causes bone cancer and leukemia, and these cancers are typically quite painful. These facts are just the ones that are thought likely to influence our opponent's attitudes concerning uranium in drinking water. Thus if the example is any guide, arguments concerning the value of health are expected to influence attitudes. This is exactly what we would expect if having a conviction concerning the value of health is just taking the corresponding attitude.

Let us call this argument for the expressivist thesis the "argument from evidence." The argument from evidence does make at least one point which any adequate theory about utterances concerning the value of health should address: we often try argumentation strategies which we think likely to influence our opponent's attitudes. Despite this virtue, the argument just seems to be mistaken about the facts of persuasion in disputation about the value of health. In particular, we do not consider an argument about the value of health successful just in case it changes an attitude. Consider for example, the following. A wants to influence B, who is a legislator. Pointing a gun at B's head, A threatens to kill B unless B supports a more restrictive standard for uranium in drinking water. B may vote for the more restrictive standard, but no one would seriously consider A's argument a relevant one.

The "harmonious attitude" argument. A sixth argument is sometimes offered by emotivists in support of the expressivist thesis. When we are faced with an decision-making situation involving the value of health, this argument begins, we think over all its aspects, noting our reaction to them. We then sum up how we feel about the situation as a whole. If it isn't clear how we feel, or if we have conflicting feelings, we re-examine the facts, trying to see relevant relationships which were overlooked before. Our job is finished only when we have managed to get our conflicting feelings to "speak with one voice." To come to a conclusion, then, is to harmonize our attitudes, that is, obtain a coherent determination of what we "want" to do. This is just what is to be expected if the emotivist theory is correct.

For the sake of brevity, let us call this argument the "harmonious attitude" defense of the expressivist thesis. On the surface, it appears to be reasonable. And in at least one sense, it overcomes some of the limitations of the defenses of the expressivist thesis previously examined, because the role which attitudes are alleged to play according to the harmonious attitude defense is somewhat less direct and more sophisticated than that proposed by any other defense of the thesis considered so far. Unfortunately, the harmonious attitude formulation still oversimplifies some of the relevant facts of decision-making concerning the value of health. For example, when we are trying to decide whether we ought to adopt a 10 pCi/l standard for uranium in drinking water, we pretty clearly are not trying to decide merely what we want to do. We may want, for example, to adopt a higher ceiling for uranium because we know that the monetary cost of meeting the lower ceiling will be high and have a depressing effect on the economy; nevertheless, we may (reluctantly) agree that protection of the public health is more important than the high cost. Thus, we may in fact decide that what we want to do must be set aside in favor of what we (regrettably) "must" do.

One might try to overcome this objection by refining the expressivist thesis in the following way. When we deliberate about the value of health, this refinement might run, we are trying to make our impersonal attitudes about the value of health, "speak with one voice."

This refinement of the expressivist thesis does not imply, in contrast to the the formulations of that thesis we have looked at so for, that what we ought to do and what we want to do are the same. It thus represents a real advance over the limitations of those earlier formulations. Unfortunately, it is reasonably clear that such a proposal would not be accepted by modern emotivists (although it is a notion espoused by some traditional philosophers, notably Francis Hutcheson and David Hume). We shall have more to say about this idea later on, for some part of it clearly seems to be correct.

Apart from the difficulties we have examined in specific arguments for the expressivist thesis, there are further problems with the notion that having an conviction about the value of health is just taking the corresponding attitude. These problems arise when we try to get more precise about just what kind of attitudes are expressed by utterances concerning the value of health. Such problems are not conclusive arguments against the thesis. But even if they are not, they at least show that the theory of attitudes has not yet been well enough developed to ensure its clarity. We consider the two most important of these.

An early form of the emotivist theory argued that the attitudes referred to in the expressivist thesis are actually emotions. This proposal was abandoned early on even by emotivists because it became clear that many occasions, an utterance concerning the value of health is not intended to express an emotion, nor does the hearer of that term attribute an emotion to the utterer of the term. The formula was accordingly refined as follows: sentences which report attitudes about health merely express the speaker's overall (driving) motivations, pro or con, concerning the value of health. On this view, for example, if I utter "An absolute ceiling of 10 pCi/l uranium in drinking water is not desirable," I simply mean "On the whole, I am not inclined to behave in a way that promotes an absolute ceiling of 10 pCi/l uranium for drinking water."

This form of the expressivist thesis, which we shall call the "motivationist" formulation, does not imply that some emotion attends all utterings concerning the value of health. Thus it represents an improvement over the formulation of expressivist thesis which holds that attitudes are just emotions. Unfortunately, the motivationist formula is just too simplistic. In particular, it may sometimes be the case that I know that 10 pCi/l for ceiling for uranium in drinking water is not desirable, yet nevertheless act in such a way as to bring it about: I may have friends in the water-treatment business and succumb to the temptation to promote their economic interests, quite apart from what I may believe is a desirable standard. Thus my driving attitudes (that is, the attitudes which ultimately determine and dominate my behavior) may not correspond to what I think is on the whole desirable. If it is even possible to act in a way that is at odds with what I believe to be correct, having a conviction concerning the value of health cannot be the same as having the overall (driving) motivation to act in accordance with that conviction.

Problems with emotivism as such. We have considered a number of difficulties which various versions of, and particular arguments for, the "magnetic attraction" and "expressivist" theses respectively face. Perhaps most, if not all, of those considered are superable. In any case, the emotivist's arguments serve to sharpen our understanding of what he has in mind. But even if the problems considered thus far can be solved, there are three general considerations, which, taken as a whole, appear to conclusively reject the emotivist's claims that the only functions of utterances concerning the value of health are to have a "magnetic effect" on attitudes or to express those attitudes.

The first of these questions not only the emotivist, but any theory which asserts that no important function of utterances concerning the value of health is bound up with the task of stating facts. The conviction has long persisted, the argument goes, that utterances concerning the value of health do, among other things, state facts. No other non-fact-stating feature of language has enjoyed this privileged status. Grammarians have not, for example, persisted in the belief that statements like "Would that cancer did not exist!" attempt to state facts. If the task of utterances concerning the value of health were obviously not to state facts, it is very difficult to understand how the belief that such utterances state facts could persist. At the very least, the argument concludes, the emotivist theory owes us an explanation of the error of our convictions, not merely a simple denial of them. Let us call this the "privileged status" objection. To date, no emotivist has answered this objection.

Second, people in fact think that statements concerning the value of health are true or false (i.e., state testable claims concerning policy, actions, preferences, desires, and so forth). Hearers of utterances concerning the value of health do not think that when a statement concerning the value of health is being made, the speaker is merely voicing his personal attitude. The emotivist account, however, just baldly ignores this feature of behavior. At the least, the burden lies with the emotivist to explain this feature, even apart from the question of whether the belief that utterances concerning the value of health state testable claims is correct. Let us call this the "testability" objection. To date, no emotivist has answered this concern, either.

Third, if the emotivist account of the role of utterances concerning the value of health is correct, certain types of behavior related to the value of health become just unintelligible. For example, suppose I exclaim to a friend who already shares my view, "The attitude of the government toward public health is unacceptable!" If the role of utterances concerning the value of health is to have a magnetic influence on attitudes, my exclamation is completely irrational. Similarly, consider the same remark made to myself. What rational thing am I doing if the emotivist account of utterances concerning the value of health is correct? Let us call this argument the "rationality" objection. To date, not emotivist has attempted to answer this objection, either.

On the whole, then, the emotivist has yet to answer at least three serious objections. Whether it will succeed remains to be seen.

## PRESCRIPTIVISM

Driven by problems like those just discussed, some writers have urged a theory of utterances concerning the value of health which tries to preserve some of emotivism's insights while softening the claim that the only function of utterances concerning the value of health is to attract or express the corresponding attitudes. Perhaps the most articulate proponent of this revised view is R. M. Hare.[6]

Hare's theory runs as follows. First, value statements should not be regarded as descriptions; for example, they are not like the utterance "putting bolt A into hole A will help produce strong shelves." Instead, they should be regarded as a type of "prescribing," something that is akin to instructions for building user-assembled, prepackaged shelves, such as "insert bolt A into hole A." Prescribing, the theory continues, is quite different from having an emotional influence on a hearer's attitudes: one has successfully (and minimally) prescribed merely if one has told someone to do something and that person has understood. The theory openly accepts the fact (in contrast to emotivism) that the hearer's attitude may remain unchanged by this understanding.

Second, general statements concerning the value of health (for example, "we ought to promote human health") are a distinctive kind of prescription: they are universal. Value-related prescriptions, that is to say, are intended to hold for all time, and for all persons and relevant similar circumstances. Thus uttering a general prescription concerning the value of health amounts to implicitly prescribing one's own behavior.

Let us call this view the "prescriptivist" theory of value-utterances. This theory clearly overcomes some of the limitations of the emotivist view. For first, it is not susceptible to the criticisms of the "magnetic attraction" thesis discussed above, because it simply does not assert that one of the two principal functions of utterances concerning the value of health is to have a magnetic effect on attitudes. It would seem, in addition, that the theory is immune to the problems faced by the claim that the only other principal function of utterances concerning the value of health is express attitudes which are specific to the value of health, because the theory does not make this assertion, either. Despite this appearance, it is not clear that the theory is sufficiently unlike the emotivist's thesis that having a conviction concerning the value of health is just taking the corresponding attitude to avoid the difficulties that befall the expressivist thesis. According to Hare, for example, to express agreement to a prescription, or to accept it, is to express one's resolve or decision to carry out that prescription. Whether the difference between "resolve to carry out what a prescription demands" and "having the corresponding attitude toward the prescribed action" can be drawn sharply enough to allow the prescriptivist theory to escape all of the criticisms leveled against the expressivist thesis is difficult to assess, and proponents of prescriptivism have had little to say about the issue.

## MULTIFUNCTIONALISM

The criticisms of the emotivist theory and its derivatives (such as prescriptivism) which we have considered collectively suggest that the view that the only role of utterances concerning the value of health is to have a magnetic influence on attitudes, to express an attitude, or to prescribe behavior, is just mistaken. These criticisms do not deny, of course, that at least in some instances attitudes or prescriptions are involved in behavior related to the value of health. Given this, one might try to sympathetically refine the emotivist or prescriptivist account in the following way. Even individual utterances concerning the value of health, this refinement might begin, may not always do the same job or have the same meaning; and they may do more than one job in a single occurrence.

We will call this proposal multifunctionalism. Multifunctionalism squarely faces the criticisms leveled against the emotivist view. It openly admits that at least in some instances, language which sounds like it is talking about the value of health may not be playing the role that true utterances concerning the value of health play. (Even the emotivist can hardly deny the truth of this suggestion.) But one of the very features which makes this formulation of multifunctionalism attractive also detracts from its utility: the theory as formulated so far is palatable primarily because it says almost nothing about utterances which concern the value of health. A thoughtful emotivist, for example, could respond to it by saying "What you say about the function of certain terms in our language is correct, but I still maintain that when used in their value-related sense, the only function of utterances concerning the value of health is to have a magnetic influence on, or to express, the corresponding attitudes."

If multifunctionalism is to provide a forceful reply to this kind of objection, its account of how values are expressed must be sharpened. One way to do this is to be somewhat more specific about the kinds of jobs which terms could have. As a start, one could propose that there is difference between the immediate end or use of terms on the one hand, and the derivative use or end of terms on the other. We will call these jobs, respectively, the direct and the derivative jobs of terms. The difference between the direct and the derivative jobs of an utterance is best be explained by example. Suppose that I notice that a friend is disappointed about the outcome of a referendum on nuclear waste treatment after he has campaigned arduously. I may wish to cheer him up. One way to do this is to compliment him. I might say "Your arguments in favor of Proposition 0 were some of the most rational and convincing I have heard in several years." By making such a remark, I have done several things. I have: (1) made a statement about the proponent's efforts; (2) complimented him on his work; and (3) possibly cheered him up. My overall aim in this was, of course, to cheer him up. The way in which I chose to cheer him up, however was somewhat indirect: other goals were achieved along the way to the cheering-up. In fact, the means to the cheering up may have been more successful than the cheering-up. In any case, we will say that making the statement and having it understood was the direct job of the speech act, whereas the cheering-up was derivative. In general, we will say that the effect of a speech act is direct if (a) for the type of context in which it occurs, the effect is accomplished by the speech-act alone, without benefit of favorable circumstances other than the hearer's familiarity with the language, (b) it was intended, and

(c) its accomplishment is not causally dependent on other effects of the same speech acts with properties (a) and (b). Other intended effects of a speech-acts which lack either property (a) or (c) will be called _derivative_. Given this distinction, we can sharpen the multifunctionalist thesis as follows. Value terms, the multifunctionalist proposal would go, have different _direct_ jobs, some of them in addition to making statements (naturalism) or expressing attitudes (emotivism).

To give this formulation some plausibility, it is helpful to look at an example and try to determine whether in this case multiple direct jobs are being performed by utterances concerning the value of health. Consider in particular the following. Suppose that during a heated debate with a colleague over the relation between uranium ingestion and bone cancer, I lose my civility and make some disparaging remarks about my colleague's values and attitudes. A mutual friend who has been listening to the debate intervenes and chides me: "You ought to apologize to your colleague for those remarks."

What has the friend done? What are (were) the direct job (s) of his remark? The naturalist will try to convince us that the friend was stating a fact and trying to ensure that this fact was understood. The emotivist will argue that the friend is trying to have a magnetic influence on my attitude or is expressing an attitude taken himself.

Let us suppose for the sake of argument that the friend knows how to use English precisely, and has not stated his intentions, desires, and so forth, in some obscure, non-standard way. Given this, let us also concede that in some sense, the remark of the friend is intended to extract an apology.

If the multifunctionalist thesis has anything distinctive to say about such an example, it must claim that there are multiple directs jobs of the friend's remark. Stating precisely what these direct jobs are is by no means easy. Is _extracting an apology_ the direct job of the remark? This seems highly unlikely, for there are a number of ways in English of extracting apologies, and none of these seem to have been employed by the friend. For example, one could _order or command_ that an apology be given, but then an expression such as "Apologize now!" would be more nearly appropriate. Similarly, the friend does not appear to be trying to extract an apology by _threatening_, for it would be a simple matter for him to hold a gun to my head and command "Apologize now!" The friend is not _requesting_, for then the appropriate formulation of the friend's interests would be something like "Could you, for my sake, apologize to your colleague?" Nor is the friend _expressing a wish_, for then something like "I wish you would, for my convenience, apologize to your colleague" be more to the point. The friend is not entreating, because then something like "Horner, I beg you, for my sake if nothing else, to apologize!." One can say, in any case, that the friend has urged, expressed a preference, advised, and exhorted me to make an apology. Could we say, then, that urging or advising, and so on, were the direct jobs of the utterance? There are problems in claiming so. If the direct aim had been to advise, the friend (given our assumption about his competence in English) would have said "I advise you to apologize." A similar problem exists for the proposal that the direct aim of the friend's remark was to urge, for "I urge you to apologize" would have more nearly appropriate. Analogously, with the proposal that the direct aim of the utterance was to express a preference.

If we are forced to determine what the direct aim of the friend's remark is, we would probably say that it was to make a statement to the effect that disapproval of failing to make an apology can be objectively justified. This proposal, of course, raises the question of what exactly we mean by the phrase "objectively justified." Presumably, something is objectively justified just in case there is some fact or feature of the situation which the friend observes which permits him to infer (possibly in light of some principle related to the value of health) that apologizing is objectively justified. But such a proposal, because of its reliance on some factual feature of the situation under evaluation, would come precariously close to the naturalist's account, and thus be susceptible to the very problems (in naturalism) which the multifunctionalist is trying to avoid. Furthermore, if one adopted the above proposal, the urging, advising, and other jobs of the friend's statement would be derivative, and the multifunctionalist analysis would accordingly fail to say anything forceful against the emotivist (or naturalist) account of the example. Thus, there do not appear to be multiple direct jobs of utterances involving the value of health in such situations; because of this, the multifunctionalist thesis, at least as formulated, remains suspect.

Driven by such problems, some authors have offered a further refinement of multifunctionalism. It has been suggested, in particular, that the direct job of "You ought to apologize" is identical with a special kind of urging, advising, preference-stating, and so on.

Let us call this the "refined multifunctionalist" thesis. This refined formulation points more sharply at what direct jobs we should be looking for than the formulation of multifunctionalism considered above does. It is thus potentially something of an improvement over the previous formulation. Such an account, if it is to be cogent, requires some articulation of a criterion for distinguishing the special use of terms relating to the value of health from the non-value-specific use of the same terms. Various proposals have been offered to this end. For example, it has been suggested that use of utterances concerning the value of health implies that competent persons would agree with one's advising or urging. It has also been suggested that use of utterances concerning the value of health (in a non-misleading or unclear way) implies that the speaker would give the same advice in all similar situations. And, it has been suggested that the proper use of "ought" in a sense specific to the value of health implies that the speaker's advice is based on reasons, that it is in conformity with the relevant recognized rules related to determining the value of health or with principles concerning the value of health which the speaker holds. Some writers have said that the special force of utterances concerning the value of health is to make some sort of claim. More specifically, it has been suggested that a statement concerning the value of health claims that the attitude expressed demands priority in the direction of behavior, and that this demand can be justified in an appropriate way. Other writers have urged that the attitude declared will commend itself to anyone who considers the facts, allows them to register on his sensibility, and the fact that the attitude has survived and will survive the impact of criticism. It has also been suggested that the attitude declared is unbiased, that it is based on adequate knowledge of the nature and effects of the kind of thing toward which it is directed, and it will be shared by others who are also unbiased and knowledgeable.

If we take these suggestions sympathetically and as a whole, we can distill the following: at the least, the multifunctionalist approach must hold that utterances concerning the value of health sometimes have one direct job such as advising, expressing a preference, urging, and so on, and sometimes another; there is not just one direct job that utterances concerning the value of health always have, but several different ones; and the direct jobs of sentences concerning the value of health are special jobs, because of the special types of claims made and the inferences authorized.

Is this formulation adequate? It appears, unfortunately, that more work is needed. First, we really do need to know exactly what descriptions should be given of the direct aims or jobs of this or that statement concerning the value of health, and this the multifunctionalist theory, even as refined, does not do. Second, we need to know exactly what distinctive types of implications and claims, if any, statements concerning the value of health make. There may be such distinctive features, but proponents of multifunctionalism have had little say on the issue. We must accordingly regard multifunctionalism at present as a plan, not a mature theory.

Relativism

Some writers have taken the difficulties faced by naturalism, nonnaturalism, emotivism, prescriptivism, and multifunctionalism to be fatal. These difficulties, they further assert, show that value-related decision-making is of a highly subjective, personal sort. We would accordingly be mistaken, they infer, to seek an account of value-related decision-making in which is the same for all normal reflective people. Any theory of value which hopes to survive, they conclude, must embrace the "fact" that at least in some instances, there are "equally valid" but contradictory opinions about matters of value. In such cases, it is urged, decision-making considerations concerning the value of health and the conclusions they lead to must be regarded as relative to individual beliefs.

Let us call this view value-relativism. Roughly speaking, then, value-relativism is the view that conflicting opinions about values, and about the value of health in particular, are sometimes equally valid. There are several features of this view which must be clarified if we are to learn anything from it.

First, the relativist formula is, however it might first appear, a statement about statements concerning values, but is not an statement containing value terms itself. It is not, for example, like the statement "Harming human health is wrong!;" it is more akin to a claim like "All statements containing value terms should be expressed in complete sentences."

Second, the relativist thesis is in at least one way modest. It does not, for example, imply that no opinions about say, the value of health, are valid for everybody. It merely claims that some opinions about the value of health are not more valid than some other opinions that conflict with them. For example, it might be possible on the relativist's view that we could come to two opposite conclusions about the level of uranium we should allow in drinking water, and there would be no way to judge which was correct; indeed, the relativist would maintain, they are both correct.

Third, the relativist's view is not merely the claim that different people sometimes have conflicting opinions about the value of health. At least this much is claimed by the formula, but the formula goes further than this. It holds that at least in some instances conflicting views about the value of health are equally valid. By the phrase "conflicting opinions" the relativist means the following. Two opinions can be "conflicting" only if they are about the same subject. For example, suppose that Mr. X holds that any level of uranium in drinking water is acceptable, because he believes that this life is full of suffering, and that life hereafter is to be reached as soon as possible; allowing high uranium levels in drinking water will help, X believes, hasten us to that other life. Mr. Y holds that their is no life hereafter, or if there is, we know too little to make consideration of it part of mortal deliberation. Y accordingly holds that we should help prevent human suffering by limiting the amount of uranium in drinking water. Can we say that X and Y have "conflicting opinions"? In a sense, of course, they do, because they disagree about the nature or existence of life after death. But that disagreement is not really a disagreement about the same subject, given the same premises. If Y believed that this life is full of suffering, and the one hereafter is not, he might in fact agree with X that we should do everything possible to hasten our own deaths. The disagreement between X and Y, that is to say, is possibly not over the peculiar inference concerning the value of health we are allowed to draw, given the premises, but rather, over whether the non-value-related premises of the argument are probable. In any case, we will restrict the interpretation of the phrase "conflicting opinion" to the case in which the disputants agree on all the premises or relevant facts of the deliberation yet hold contradictory views about the conclusion.

In using the phrase "equally valid," furthermore, relativists are not making the rather pedestrian claim that two claims about the value of health are equally plausible in light of the facts known at present; the theory is more radical by far. It asserts, in particular, that given the best possible methodology for assessing the value of health, there will still be cases in which two conflicting views equally merit our commitment. On the relativist's view, for example, we might come to two opposite conclusions about whether we ought to use public funds to help discover a cure for leukemia; no matter how much money we had, no matter how inexpensive it might eventually be to discover a cure, no matter what we might learn about the cure itself, the relativist might maintain, there may be no resolution of the dilemma. Both positions could in principle equally and forever command us.

There are, as one might expect, more and less radical formulations of the relativist view possible. The most radical asserts that there are conflicting opinions about the value of health and that there is no unique rational method in decision-making concerning the value of health; let us call this formulation <u>value-skepticism</u>. The less radical formulation asserts that there is a unique rational method in decision-making concerning the value of health, but still holds that there are cases in which conflicting opinions about the value of health are equally valid; we will call this kind of relativism value-indeterminism. We consider these in turn.

The distinguishing feature of the two types of relativism obviously concerns whether there is a unique methodology for determining the value of health. Skeptics about the value of health have claimed that there is no single correct method. Whether there is, of course, must be examined. One way to begin to address this issue is to look at the practice of normal, reflective people and see whether there is something approaching a single method in their behavior concerning the value of health. In principle, we would want to look at a large number of examples of such behavior to determine whether there is a common methodology. For the present purpose, we will direct our attention to a single example, hoping that it is representative. Suppose that a certain company, Radioactive Waste Management (RWM), proposes to use a ditch on Mr. A's land to convey what is at the time perceived by the experts to be harmless waste water. In return for the use of the ditch, RWM agrees to pay Mr. A what he considers to be a very generous one-time fee. The proposal is effected. Two years later, Mr. A's young daughter drinks water from the ditch and goes blind. Subsequent tests about the cause of the daughter's blindness are inconclusive. Mr. A nevertheless demands that RWM pay for the daughter's medical expenses, compensate her for her loss of sight, and restore the ditch to its original condition. In return, Mr. A proposes to return the fee paid to him by RWM. RWM refuses every part of Mr. A's demand.

How do we go about thinking through the issues related to the value of health in this situation? On the surface, at least, one is inclined to believe that since there is no clearly demonstrable connection between the contents of the ditch and the daughter's blindness, RWM is not obligated to restore A's ditch or compensate his daughter. Whatever we believe, in any case, we do form an initial attitude about the merits of the case. Forming attitudes, it seems, is a part of decision-making methodology for determining the value of health. (In this sense, we should note, the naturalist and emotivist accounts of this type of decision-making are not wholly mistaken.)

Second, in making a decision about the above case, it appears that we appeal to some general principle in the case: the facts of the case alone would point to no opinion unless they were conjoined with some principle concerning the value of health. In the above example, one of these principles might have been something like "X has no claim against Y unless it can be shown that Y jeopardized the health of someone for whom X has responsibility."

In light of the fact that we do seem, perhaps invariably, to use principles in decision-making concerning the value of health, one might reasonably wonder whether forming an initial attitude about the situation is really essential to decision-making methodology in assessing matters involving the value of health. Wouldn't a straightforward appeal to principles concerning the value of health, without forming some attitude about the situation, that is to say, be sufficient? There are at least two reasons, such reasoning might go, that a method which consisted strictly of an appeal to elementary principles might be desirable: it would be simpler, and it would less directly susceptible to the criticisms that trouble emotivism, nonnaturalism, and naturalism.

There is much to be said in favor of this suggestion. Perhaps most everyday situations concerning the value of health involve no more than appealing to a simple principle. But it is highly unlikely that this can be a complete account of decision-making methodology in matters involving the value of health (if one exists) for several reasons. Let us consider, for example, the following situation. Suppose I make a promise to my colleagues that I will devote fifty percent of my research time to investigating the effects of alpha radiation on human bone tissue. To do this, let us suppose, I must spend not an insubstantial part of my personal income on the research. Shortly after I make this commitment, one of my children falls gravely ill, requiring expensive medical care which commands all my financial resources. What am I to do? On the one hand, it seems that there is a principle which says that I must do what is in my power to promote health generally; on the other, a principle which claims I must attend to the health of my family. There is not any overriding principle which tells me which of these two principles to follow. To make a decision, it seems, I need to augment principles with something else.

There is a second reason why we should believe that decision-making methodology in matters concerning the value of health consists of more than an appeal to elementary value principles. We hardly ever have completely precise rules about the value of health in mind when we appeal to them. It is unlikely, for example, that anyone has formulated, in a fashion that includes all the necessary exceptions and qualifications, the principle that one should protect the health of others. Yet there clearly are exceptions to the rule of health-keeping: refraining from self-defense cannot considered binding by anyone. We do not ordinarily invoke these qualifiers, nor are we usually aware of them, unless we are dealing with uncommon circumstances. The qualifiers, it seems, are brought to our attention as a consequence of feeling uneasy about what simple principles alone direct us to do when applied to the case at hand. At the very least, then, these qualifiers are of a rather different sort, or play a different role in our thinking, than everyday principles concerning the value of health do. More than a simple appeal to everyday principles, it thus appears, is at least sometimes involved in actual decision-making by normal, reflective people.

There is yet a third reason why we should believe that reasoning about the value of health involves more than an appeal to simple principles. We occasionally find that we have to give up a principle, at least as it happens to have been formulated up to this point. For example, we may have believed up to a certain point that we should remove uranium from drinking water at all costs. Should new evidence show that the tradeoffs between uranium-free drinking water and other health effects are on the whole unfavorable, we will likely modify or abandon our principle. In any case, a simple appeal to principles at hand does not appear to be the way that reflective people always make decisions about the value of health.

On the whole, then, we do something more than appeal to simple principles already in hand when we make at least some decisions about the value of health. What can we say about this additional element? Among other things, it appears to involve _accepting the promptings of attitudes_ we find in ourselves. An example will help show this point. Imagine that I work for a large nuclear fuel manufacturer located near a small town. Suppose further I discover that in ignorance, employees of my firm have been flushing low-level radioactive waste into the town's wastewater system. After treatment, the wastewater discharges into a reservoir which is the principal source of the town's drinking water. As a result of these practices, radiotoxic concentrations of the waste materials are detected during a proprietary assay of the drinking water. I am approached by a member of the press who asks me to confirm or deny a "rumor" he has heard about radioactivity being found in the town's water supply. On first inclination, I am inclined to invoke the principle "Tell the truth!" But suppose I am aware that the local public health authorities are political appointees who have a distinguished record for ignoring controversial public health problems; if history is any guide, furthermore, I know they will actively try to downplay or simply ignore the problem, and have a ponderable chance of succeeding in their efforts quite apart from any merit, scientific or otherwise, their efforts may have. I feel that public awareness of this problem must be carefully managed if public health, in the long run and considering all relevant matters, is to be protected. My initial inclination is therefore moderated by my feeling that greater harm could be done by telling the truth to the reporter than by carefully orchestrating the way in which responsive officials are made aware of the problem.

What can we say about this example? Perhaps most importantly, my appeal to a principle ("Tell the truth") is augmented by accepting the promptings of feelings, attitudes, and so forth. This method involves an appeal to principles augmented by attitudes, preferences, a "sense" of obligation, and so forth.

Something more needs to be said about the attitudes which we use to moderate our decisions. Not just any attitude, apparently, is given the privileged status of moderator. We do in fact discount attitudes for several reasons. What can we say about these reasons?

First, we discount an attitude, inclination, or preference if we think it is impartial. In the case sketched above, for example, I would feel obliged to ignore my inclination to withhold information from the press if I thought it merely served the narrow interests of my employer. A view, attitude, or preference is impartial just in case the behavior of the parties affected would not be changed if their roles were interchanged. A reasonably good test of the impartiality of a view, preference, or inclination is whether the utterer of such a view, preference, or inclination is willing to advocate a general principle corresponding to this attitude; if he is not, we question his impartiality.

The requirement of generality (that is, whether the utterer is willing to advocate a general principle) is not an easy one to formulate. We may imagine, in particular, a clever but essentially self-serving person who could formulate a principle which met the technical requirement of generality but which de facto gave some people an advantage over others. For example, I might advocate a principle which read something like "Any person who is over six feet, five inches tall, who was born in Western Kansas on the 23rd of June, 1947, has black hair and a prominent scar at the base of his left thumb," and so on, until de facto I have described myself to the exclusion of anyone else, "shall have the privilege of dumping any amount of radium chloride into the drinking water supply of New York City." Surprisingly, attempts to formulate a general criterion of adequacy which can eliminate this kind of nonsense have run into rather serious difficulties, not the least of which involves coming to agreement about what kinds of facts might make relevant contributions to whether a principle had a general or particular formulation.

Second, we discount an attitude if we think it is uninformed. An attitude is uninformed, if, given that we are impartial, we would not have had the attitude if we been disabused of false beliefs, or if our true beliefs had been more vivid. An attitude is informed, if would stand up in the face of a vivid awareness of relevant facts, given that we were impartial. For example, I would regard my strong inclination to support a ceiling of 10 pCi/l for uranium in drinking water as uninformed if I were to discover that there are no deleterious effects even at 1000 pCi/l.

Third, an attitude is discounted if it is a consequence of an abnormal state of mind. The notion of "normal" is of course somewhat vague. In general, a healthy state of mind is one in which we are alert, responsive to fact, free of repressive burdens. We reckon our minds are less healthy when we are suffering from illness, insanity, fatigue, anger, grief, or depression, among others.

Fourth, an attitude is discounted if accepting its prompting would lead to having a system of principles concerning the value of health which is excessively complex. At the least, that is to say, we do demand that guiding principles are simple enough to be understood by persons of reasonable intelligence, and simple enough to serve, at least in the most commonly occurring cases, as a guide for conduct.

To summarize, the everyday decision-making methodology for assessing matters concerning the value of health which we have been describing consists of the following. We decide everyday, common problems involving the value of health by appeal to principles and by accepting the promptings of qualified preferences, feelings of obligation, and so forth. We occasionally correct our principles if they are incompatible with our undiscounted (critically considered) attitudes, and we rely on such attitudes to fill out and moderate the application of the principles we use. Whatever we decide, our judgments must be consistent, and the particular ones we use must be generalizable in a way that does not simply result in the ad hoc promotion of our own personal interests. Let us henceforth call this the "Standard Method."[7]

If this account does with high probability represent the way in which people actually go about making decisions about the value of health, then the claim of value-skepticism that there is no unique method for deciding matters involving the value of health is just mistaken, and with that value-skepticism falls. Whether it does, in any case, is a matter open to empirical research. At present, the "jury is still out" on the issue: we must wait the evidence to come in on value-skepticism's distinctive claim that there is no unique method.

What can be said about the more restrained form of relativism (value-indeterminism), which holds that although there is a unique method for making decisions about such things as the value of health, there are nevertheless equally valid but conflicting opinions about it which can arise from the employment of this method?

To make headway on this question, we must first tighten our understanding of the Standard Method a little. In one respect, our formulation of the Method is incomplete: as formulated it tells each of us how to decide whether a given judgment about the value of health is valid (that is, the judgment must be consonant with our principles and the undiscounted (critically considered) attitudes that may arise in us during our deliberation about that judgment), but it does not make clear whether a person's judgment about the value of health is valid if it satisfies the tests of the Method made by him but does not satisfy the same kinds of test applied by others. This presents something of a problem for us, because it seems that although we personally may come to a decision about a judgment about the value of health using the Standard Method, someone else may come to the opposite conclusion using just the same method. There are thus at least two possible formulations of the Standard Method which we must consider. The first, which we shall call the _relativized Standard Method_, holds that a judgment about the value of health is correct just in case it is consistent with generally accepted principles concerning the value of health and is harmonious with _my_ critically considered attitudes. The second, which we shall call the _universal Standard Method_, holds that an judgment about the value of health is correct if and only if the judgment is consistent with generally accepted principles concerning the value of health and is harmonious with the critically considered attitudes of _everyone_.

_What if the Standard Method is correct?_ Which of the two formulations we decide to be correct will influence our evaluation of value-indeterminism. In particular, value-indeterminism appears to be consonant with the relativized Method, because both allow us to expect differences in the decisions concerning the value of health of equally informed and conscientious people. It is reasonably clear, in contrast, that value-indeterminism is logically _in_compatible with the universal Standard Method. Consider, in particular, the following. Suppose that Mr. A uses the universal Standard Method and concludes that "Keeping the concentration of uranium alpha activity at or below 10 pCi/l in drinking water is desirable." By using the universal Standard Method, he is also asserting that desiring that the uranium alpha activity concentration in drinking water to be kept at or below 10 pCi/l, _on the part of everybody_, must meet the conditions that are set. Value-indeterminism holds that it is possible for someone else, say, Mr. B., to adopt the attitude that keeping the uranium alpha concentration at or below 10 pCi/l is not desirable because the attitude does not meet the (same) appropriate conditions. We would, if value-indeterminism were true, just hold Mr. A's claim to be false. The universal Standard Method therefore prohibits precisely the state of affairs which the value-indeterminist claims can occur; value-indeterminism and the universal Standard Method are therefore inconsistent with one another.

Some "common-sense" arguments for value-indeterminism. The above arguments address only the question of the logical compatibility of the Standard Methods with value-indeterminism. This leaves open the issue of which of the Methods is more likely to be correct. At least two arguments suggest that the relativized Standard Method is the one more likely to be correct or desirable. First, common opinion holds that that there can be great variations in beliefs about the value of health and hence it is likely that everybody's judgments, however critically considered, will not necessarily agree on all health-related issues. Second, if we decide that the universal Standard Method is the correct one, then we will require universal consensus on a policy before we can implement it. Now it is highly doubtful whether any policy proposal can gain the support of everyone. Imagine, for example, trying to obtain at present concurrence even in the medical community about the relation between uranium ingestion and bone cancer, given that much research remains before we truly understand that relation; we would expect, if not demand, substantial disagreement among the experts. If we decide, in contrast, that the relativized Standard Method is the correct one, then we will be able, at least in principle, to implement policy without requiring universal consensus.

Some methdological problems. These arguments do not conclusively show that value-indeterminism is correct. To show this, we would have to show that there are two people whose critically examined attitudes actually conflict. Even if two individuals can be produced who allegedly have conflicting attitudes about exactly the same issue, there are some serious philosophical/methodological problems which we will have to solve before we dare be convinced. For example, how are we to be sure that the relevant considerations are present to the minds of both, with requisite vividness? How can we be sure that all the same facts are believed by both, that both have the same attitudes, and that neither harbors false beliefs which bear on these attitudes?

There may be a less direct way of examining this question. If there were good agreement among modern psychological theories to the effect that a person's attitudes are not strictly a function of his information and his state of personal needs or wishes and his normality, then these theories would give us room to believe that critically examined attitudes about issues concerning the value of health could vary. Thus, although actual examples of conflicting attitudes under exactly the same conditions might never be produced, the theories would still support the value-indeterminist's contention.

Unfortunately, modern psychological theories do not provide us with uniform guidance on this point. Gestalt theory, for example, holds that attitudes toward a given health-related situation will be identical, and personal needs and interests do not play a differentiating role. Psychoanalytic and stimulus-response learning theory, in contrast, claim that attitudes can be conflicting, depending on personal development. Some current research in psychology, moreover, suggests that fundamental orientations may be adopted from parents early in life, and that these may have a permanent influence on attitudes. This research also suggests that identifications or emotional relations with important figures in one's life and feelings of security play a role in the formation of one's attitudes. If these latter suggestions are correct, it does seem that we can specify some features in the life of an individual that would have the effect that his attitudes now could be different from what they were in his earlier life. It thus seems plausible <u>on this evidence</u> that the attitudes of "equally informed and critical" people could differ, a view nominally consonant with the relativized Standard Method. But on the whole, modern psychology does not provide unambiguous guidance on the issue of whether the value-indeterminist's view is likely to be true.

This situation notwithstanding, the matter is by no means closed. Anthropological evidence, for example, suggests that in a wide spectrum of cultures, personal conflicts and maladjustments, the attitudes of one's relatives, success or failure to achieve social status, and so on, influence the beliefs of individuals about the value of health.

Value-indeterminism thus finds some support in ordinary thinking about the value of health and in some, but not all theories of personal development and in some, but not all anthropological research. At best, then, the jury is still out on whether the relativized Standard Method describes the way in which people actually make decisions about the value of health.

## HOW DO WE DETERMINE WHAT IS DESIRABLE IN HEALTH POLICY?

The considerations which we have examined so far concern what the general rules for reasoning about the value of health are, that is, are concerned with the question of _how we determine whether given examples of reasoning about health policy are valid_. Such a question speaks directly to concerns about the correct _form_ of reasoning about public health policy. An adequate theory of public health policy must also try to circumscribe the _content_ of health policy insofar as it concerns the value of health. Traditionally this has been cast as the job of determining _what is desirable_ in health policy. We accordingly turn to surveying answers which have been given to this question.

It will be useful to introduce some terminology at this point to avoid extreme repetitiveness later on. We will call something _intrinsically desirable_ ("desirable for its own sake") just in case that thing is desirable, taken just for itself, viewed abstractly, and viewed without respect to any consequences its existence will or may produce. A thing will be called _instrumentally desirable_ just in case it is desired for the sake of something else. For example, heart surgery is not intrinsically desirable (we would not desire it, no matter what our health), but it is desirable for the sake of producing better health in many circumstances. In contrast, it may be that life itself is intrinsically desirable because it is presumed by any other thing that is desirable.

The dispute over what is intrinsically desirable has a very long history, extending at least to the "golden age" of Greece (circa 500 B.C.). A number of writers have held that only _pleasure_ is intrinsically desirable. Others have claimed that there are other things which are intrinsically desirable, including health, human life, sacrifice for one's community, or a distribution of happiness among people in proportion to their merits or needs. We will examine only those proposals which could be or have been construed as having consequences for theory of public health.

### _Hedonism_

Hedonism is the view that something is intrinsically desirable if and only if it produces pleasure. However hedonists use the term "pleasure," they must give some account of what they mean by the term, if simply to ensure that their view is reasonably well defined. Let us briefly examine this problem.

We do know how to correctly apply the terms "pleasant" and "unpleasant" in a number of cases. Most people find eating ice cream pleasant, but undiluted vinegar unpleasant. In spite of knowing how to use these terms correctly in some cases, it is not an easy matter to be sure what we are really saying about our experience when we say it is pleasant or unpleasant. Attempts by psychological theorists, unfortunately, are diverse on this point. There is unanimity at least today that pleasantness is not a stand-alone element of experience like a color patch or a sound. (To convince yourself of this, just try to examine pleasantness by itself, in the way you can inspect a red patch by itself. It cannot be done. Pleasantness always seems to be pleasantness _of_ something, of an activity, or some other element of experience.)

There are two classes of psychological theories about pleasantness at present which have some hope of providing an adequate account of that aspect of experience.

The first of these attempts to draw an analogy between the pleasantness and the "dimension" of a sensory experience. In this context a "dimension" of a sensory experience is something like the pitch or loudness of a sound, or the shade, brightness, or saturation of a color. The dimensions of a sound, like pleasure, are not distinct experiential elements; they are somehow discriminable aspects _of_ experience which allow us to order experiences in a series. (Pleasantness is not entirely like dimension in this sense, of course, because if the dimension of a sensation is reduced to zero, that experience simply ceases to exist, whereas we can imagine an experience which is neither pleasant nor unpleasant.)

Let us call members of this class of theories "dimension" theories. Psychologists who have held this view of pleasure, unfortunately, have not agreed which features of experience can have or affect the dimension of pleasantness. Some, for example, have said that only feelings (joy, fear, anger, etc.) qualify; others have suggested that only "bright" feelings in the chest, "dull" feelings in the lower back, and so on, can play this role; still others have urged that there are complexes of sensation which determine the dimension of pleasantness.

Although dimension theories are in some respects informative, they are in other respects rather unsatisfactory. The principal problem with them is that the alleged dimension-like quality is quite elusive, if not entirely so, when we try to inspect it. It seems very much unlike other dimensions in this respect. Concerns of this sort have led some authors to a second type of theory.

According to this second class of theories, the phrase "x is pleasant" simply means "x is a part of my experience that I wish to continue on its own account;" similarly, "x was pleasant" means "x was a part of my experience that I wished at the time to continue on its own account." The _intensity_ of the pleasure, the account continues, is just intensity of this wishing to continue. A proposal similar to this one holds that saying something "is pleasant" is to say that it elicits effortless heed or attention, absorbtion in what we are doing or experiencing.

Let us call members of this second class of theories "intensity" theories. Intensity theories talk about a much less elusive feature of experience than dimension theories do, and in this sense they represent some improvement over dimension theories. But intensity theories unfortunately seem to ignore an important fact about pleasantness. In particular, it seems that always when we like an experience for itself or are absorbed with it, there is always some subjective experience present. If, for example, hearing news about a breakthrough in the treatment of cancer is pleasant, there is always some emotion present, some swelling of pride, some joy, and so on. Or if hearing about a setback in cancer research is unpleasant, there is associated with this experience some disappointment, grief, and so on. This correlation seems to occur whenever an experience has the feature of being pleasant or unpleasant.

To explicitly accommodate this feature of pleasantness, the hedonist thesis can be cast as follows: something is intrinsically desirable (undesirable) if and only if and to the degree that it is an experience with a subjective element that the person at the time wants to prolong (terminate or avoid) for itself.

Such a formulation sets no restrictions on the kind of thing that pleasure (enjoyment) may be. It _may_ include wine, women, and song, but just as easily may include preserving one's health, helping to discover a cure for leukemia, promoting the economic well-being of others, and so on.

At the least, the hedonist thesis has plausibility because in general, we do think it is is a good thing for people to have enjoyable experiences, and not to have disagreeable ones. Enjoyable experiences, whatever else we would argue, are at least highly important components of whatever it is that makes living worthwhile.

_Psychological hedonism_. We must thus ask whether there are good reasons for believing the hedonist thesis. A number of writers have obviously thought we should, and have urged that certain theories of motivation can provide support for the view. Roughly speaking, these motivationist defenses hold that hedonism is correct simply because humans are motivated to do something if and only if doing that thing produces pleasure. This way of defending (and possibly articulating) hedonism is called psychological hedonism, and it has a long history. Epicurus, for example, held that hedonism is correct because "living things, so soon as they are born, are well content with pleasure and are at enmity with pain, by the prompting of nature and apart from reason. Left to our own feelings, then, we shun pain and seek pleasure." Similar reasoning was used by other hedonists, including Bentham and Mill. In general, writers who have taken this view have held that volition or desire is always and solely determined by pleasures or pains prospective, present, or past.

There are clearly three possible theories of motivation, which could be tied up in the general idea of psychological hedonism, one each corresponding to whether the pleasures or pains are, respectively, prospective, present, or past. Our task is to determine to what extent any of these theories of motivation could be used in defense of the view that a thing is intrinsically desirable if and only if, and to the extent that it produces pleasure (the hedonist thesis). That task, of course, involves two distinct questions: on the one hand, there is the question of whether any of the proposed theories of motivation is true; and on the other, the question of whether the hedonist thesis is logically entailed by any of these theories of motivation. No currently popular theory of motivation, it will be argued, provides a foundation for hedonism.

Teleological hedonism. The first theory of motivation we need to consider, then, holds that a person wants, desires, or is motivated to attain a state of affairs if and only if he believes that that state of affairs will be enjoyable to him. Since this theory concerns prospective pleasure or pain, we will call it the teleological version of psychological hedonism.

Problems in teleological hedonism. Are there are good reasons to believe that teleological hedonism is true? The question is not without difficulties. Consider, in particular, the evidence which is adduced in favor of the view. If we ask people what they think is worthwhile in life, they are apt to mention things like health, the respect of one's friends, having money, having skills and talents, and so on. Such things are not obviously identical with "anything which is pleasant or enjoyable." At the least, then, teleological hedonism must explain how in fact the two are identical.

Whatever arguments may be given in favor of teleological hedonism, the very claim the theory makes about human motivation is problematic. It seems, in particular, that there are cases in which we seek to do something without regard to the pleasure or displeasure we think it might produce. For example, consider our thoughts about the verdict which posterity may render about our handling of the problem of radioactive waste. Are we truly indifferent to the matter, as teleological hedonism would suggest, because it concerns a time after which the issue could not possibly produce pleasure or pain for us? Although the verdict of posterity may not always be on our minds, neither is it obvious that we are indifferent to the matter, either. Or consider evidence which strongly suggests that people have made and carried out decisions which they regarded as likely to bring personal disaster. Crisis situations provide the best examples of such behavior: newspapers frequently report cases in which people enter burning buildings, knowing that their own pain and possibly death is likely and immediate, to save the lives of other creatures in spite of the fact they are not even acquainted with the creatures they save. At the very least, teleological hedonism owes us a convincing explanation of how such events are not in direct contradiction to its claims.

To these obvious objections, teleological hedonists frequently argue that people who make such sacrifices do so because they think that acting as they do will produce happiness and will help avoid unhappiness for themselves.

There is something troubling about such a rejoinder, however, particularly if it is claimed to be complete. For if it is a claim about why people actually act as they do, then the teleological hedonist is obligated to show what evidence distinguishes his interpretation of the events from one which ascribes non-hedonic motives to the individuals involved. Only if there is such evidence, or if it is possible that there is such evidence, can teleological hedonism be regarded as a testable claim about human behavior. In general, such an articulation has not been forthcoming from the proponents of teleological hedonism. Until it is, we must at least suspend judgment about whether this defense of hedonism can be evaluated by ordinary scientific methods.

*Motivation by current thoughts.* The second type of psychological hedonism which we consider holds that a person is motivated to bring about a given state of affairs if and only if the *thought* of that state of affairs is *currently* pleasant or attractive to him; and a person will prefer one course of action to another if and only if the thought of what it will bring about is more attractive or pleasant than the thought of what other possible acts will bring about. Since this theory of motivation concerns current thoughts, we will call it *motivation by current thoughts*. Motivation by current thoughts, even if it is true, has no bearing on hedonism. It does not tend to show that people take an interest in <u>events</u> only insofar as the <u>events</u> are believed to be pleasant. For even if we grant that people take an interest in producing an event only if the <u>thought</u> of it is pleasant, the fact that the <u>thought</u> of an event is pleasant does not require that the thought is a thought of a pleasant <u>experience</u>. The thought of finding a cure for leukemia, or the thought of the verdict of posterity on our handling of radioactive waste, for example, may both be pleasant thoughts, but the experiences of which these are thoughts need not be pleasant (or painful). We may accordingly conclude that whether motivation by pleasant thoughts is true is irrelevant to whether hedonism is likely to be true.

<u>Hedonism-by-conditioning</u>. The third general theory of motivation which is sometimes invoked to support hedonism holds that the strength of one's present interest in a particular kind of occurrence is a function of past enjoyments. Let us call this type of theory <u>hedonism-by-conditioning</u>. We can get some idea of the attractiveness of hedonism-by-conditioning by considering a simple example. Suppose I am faced with a choice of research projects in radiobiology, some of which are like projects which have brought me recognition in the past, whereas others are like ones which have brought controversy. Motivation-by-conditioning claims that I will tend to choose those projects which are like the ones which have brought recognition (assuming that I enjoy recognition). For such an example, motivation-by-conditioning seems on the surface highly plausible.

To evaluate this theory, we need to carefully keep in mind what the hedonist's interests in it are. In particular, the hedonist must show that a perfectly informed person would prefer one thing, for itself, to another thing, if and <u>only if</u> the pleasantness of the former were greater than that of the latter. A distinctive feature of hedonism-by-conditioning, then, is its claim that we will tend to do <u>only</u> those things which produced pleasantness in the past. At the least, if this theory is plausible, it ought to find some support in modern psychological theories of motivation. To determine whether it does, it would suffice to review whether any theory of motivation which has gained the confidence of modern practitioners can support hedonism-by-conditioning. There are just three general theories of motivation which enjoy popularity today: stimulus-response (SR) theories, Freudian theory, and Gestalt theory. We discuss these in turn.

<u>Stimulus-response theories</u>. There are a number of stimulus-response theories. In general, such theories consist of a set of laws which propose to connect

certain aspects of an organism's environment ("stimulus fields"), certain needs ("drives") which the organism has, and possibly other factors. These laws allow us, given an organism's history, a description of the organism's drives, and a description of the stimuli acting on the organism, to predict the organism's behavior, including (where relevant) its value-related behavior. On the simplest SR theories, a response R will tend to occur in connection with a stimulus S if and only if a response similar to R has occurred in temporal proximity to a stimulus like S on occasions when a drive reduction or satiation has occurred nearby in time. For example, if I have been given food immediately after publicly defending a certain standard for uranium in drinking water, I will, _ceteris paribus_, now tend to defend that standard when a similar occasion arises.

The example may seem odd, but it is in fact representative of what SR theories have to say about human behavior. To a large degree, this oddity of SR explanations derives from the fact that early SR the experimental work consisted of studies of rat and pigeon behavior; these animals are presumed to have to deal with little more than the satiation of their most fundamental biological needs in most SR settings.

We should not discount SR theories for this oddity: for all we know, they may ultimately be correct. Nevertheless, any proponent of such a theory is obligated to answer to the charge that the theory's explanation strongly appears to be an oversimplification of certain types of apparently complex behavior, including, for example, why someone might repeatedly engage in certain activities (e.g., defending a position on uranium in drinking water) which do not appear to directly help satiate his most fundamental biological drives.

In some variants of SR theory, this issue is consciously handled with some sophistication. A version of SR theory defended by Clark Hull,[8] for example, distinguishes two kinds of drives. The first of these, called _primary_ drives, are the fundamental organic needs, such as the need for water. Hull's theory is not committed to any particular list of primary needs; the discovery the list is left to experiment. In contrast, the theory also recognizes the existence of _secondary_ drives, which are not identical with, and need not immediately derive from, organic needs, but are learned. In the historical genesis of any secondary drive for any individual, the theory nevertheless maintains, there was at least one primary drive involved.

The nature of secondary drives (in Hull's sense) is likely to be of more interest to psychological hedonism than his account of primary drives, because it is at least implausible (though possible) that value-related behavior stems directly from fundamental biological needs. We can in any case imagine a creature much like ourselves who could survive easily, and thus could be thought of as satisfying to a non-trivial degree its biological needs, but nevertheless was devoid of any behavior which anyone would want to call value-related.

What does Hull's theory of secondary drives say about such apparently complex behavior as my defense of a uranium standard? Although Hull's own writings on such issues are often puzzling, we may extract at least the following from them. In general, if a stimulus has been associated by the individual with the evocation and reduction of drive-stimuli, subsequent occurrence of that stimulus will tend to produce these same drive-stimuli on its own account. Thus, for example, if my defending a certain uranium standard has been associated with the evocation and reduction of drive-stimuli (being fed by the local Sierra Club chapter after I have defended a uranium standard to them, for example) I will be inclined to subsequently repeat my defense because that very activity will tend to evoke and reduce these same drive stimuli. On Hull's view, furthermore, the degree to which I am inclined to defend the theory now can, in fact, depend on such features of my behavioral history as how strongly praised I was in the past, how close in time the praise or repudiation came to the time at which I made the defense, and possibly other factors.

In many ways SR theory of the Hullian sort could roughly explain facts about value-related behavior we wish to accommodate. The theory (ies) can accord praise, blame, reward, and so on, something of the role we are commonly inclined to give them. The theory is also consonant with the highly plausible suggestion that one's values are affected by one's personal interests, because it allows that the frustration of them will be punishing, whereas the satisfaction of them will be rewarding. In general, then, Hull's theory provides a conceptual framework which is at least roughly consistent with a number of features of value-related behavior we wish to explain.

On the whole it might thus seem that Hull's theory or something close to it could provide support for hedonism-by-conditioning. But this appearance is deceiving. Hull's theory asserts that we will tend to do those things which have been associated in our experience with the evocation and reduction of drive-stimuli. It does not, thereby, obviously assert the thesis of hedonism-by-conditioning: that we will be motivated to do that and only that which produced enjoyment in the past. For Hull's theory to support hedonism-by-conditioning, however, an additional premise must be added to the effect that the evocation and subsequent reduction of drive-stimuli is identical with enjoyment. Determining whether this is the case will at least depend on getting a clearer idea of what enjoyment is than we now have.

Even if we are unable to get a clearer idea of what enjoyment is, there are features of it which make the claim that enjoyment and evocation and subsequent reduction of drive-stimuli are the same highly implausible to just false. Consider thirst. We can well imagine thirst as a stand-alone element of experience. We cannot similarly imagine enjoyment by itself, because enjoyment must always be thought of as enjoyment _of_ something. This objection very strongly suggests that no identification of evocation and reduction of drive-stimuli (e.g., thirst) can be the same as enjoyment.

On the whole, then, SR theories do not promise to provide the kind of support which hedonism-by-conditioning requires.

_Psychoanalytic theory_. A second theory of motivation which commands a following today is Freudian psychoanalytic theory.[9] According to this theory, humans are born with a collection of instinctive drives or dispositions (hunger, thirst, sex, and so on) called the _id_. In the course of normal personality development, the theory holds, the satisfaction of the id comes to be mediated by a second psychological agency called the _ego_. The role of the ego is to make it possible for the organism to recognize that it is distinct from the rest of the world and to implement interaction with the outside world in a manner that can at least partially satisfy the id. Early (nominally, at age 3-5 years) in normal personality development, the Freudian theory continues, the ego realizes that it is dealing with a world in which direct and immediate satisfaction of the id's needs is simply not possible. Freud's own version of this idea takes a highly particular form: the child strongly desires sexual access to the parent of the opposite sex, but also realizes that it is not powerful enough to physically overcome the parent of the same sex in an attempt to gain this access. The ego accordingly adopts a strategy aimed at "destroying" the interfering parent: it (metaphorically) "assimilates" the parent of the same sex by psychologically incorporating that parent. The effect of this assimilation is to deny the _otherness_ of the interfering parent, making that parent a "part" of oneself. In this process, the theory claims, a new psychological agency called the _superego_ is formed in the ego. The superego contains, among other things, the parental values and has the power to direct the value-related behavior, including any beliefs which the child has about the value of health.

To people who are not familiar with the kind of evidence and idiom that its practitioners and theorists entertain, the Freudian theory of behavior appears bizarre, if not without motivation. Our interest here is not to evaluate whether the theory is likely to be true. For even it is is, Freudian theory cannot support hedonism-by-conditioning. There is no guarantee, given the way that the Freudian theory describes the inculcation of values in the individual, that the dictates of the superego are likely to produce pleasure. Quite the opposite is true: even assuming that the direct satisfaction of the id's and ego's desires produces pleasure (and Freud's theory, strictly speaking, does not claim this) the very genesis of the superego ensures that it stands in the way of allowing certain kinds of pleasure to occur. If anything, then, the Freudian theory of behavior tends to augur against hedonism-by-conditioning.

Gestalt theory. A third theory of motivation which enjoys some popularity today is the Gestalt theory.[10] Gestalt theory arose from a certain kind of attempt to understand the nature of perception and has since been extended to cover value-related and other types of behavior. As a result, Gestaltists are inclined to invoke certain kinds of evidence about perceptual behavior as a paradigm of behavior in general. For example, when an untrained, caged chimpanzee is given a long ferruled pole in several pieces and a piece of food is placed at some distance from the animal's cage, the animal can often assemble the pole and use it to reach the food. Being able to do this, Gestaltists claim, shows that the chimpanzee has a conception of the entire pole quite distinct from any biologically primary or any learned response. The laws of experience are often like this, the theory continues: our experience and behavior typically contain highly complex structures which do not arise out of an individual's simple biological needs or learned behavior. There is just an unanalyzable "wholeness" to much of behavior which psychology must accept, Gestaltists urge, and any attempt to derive laws about its components leads to artifice and hence misunderstanding of the nature of behavior.

According to the Gestaltists, value-related behavior in particular is no different in this respect than any other kind of behavior. We have, they claim, certain kinds of complex value-related behavior already built in to us. For example, some Gestaltists have argued that when a queue for food has formed, there is a feature of behavior telling us that we ought to go to the end of the line. Similarly, some Gestaltists claim that we are simply built (at least in normal situations) so that we inclined not to take food from a hungry child or to do anything which would harm the health of the child.

Even if these particular features of human behavior exist as claimed, and more generally, even if Gestalt theory is likely to be true, all of the experimental evidence claimed to be in favor of, and certainly the Gestalt program itself, go directly against against hedonism-by-conditioning. Hedonism-by-conditioning assumes that value-related behavior is built out of elements which have been experienced or have already occurred in the individual's history, whereas Gestalt theory claims that we "know" what is appropriate without any prior exposure. On the whole, therefore, we may conclude that Gestalt theory does not lend support to hedonism-by-conditioning.

We have looked, then, at the claim that there are theories of motivation which can provide some support of hedonism, and have found that claim wanting. Moreover, no version of psychological hedonism has hedonism as a consequence. Thus, the program of arguing for hedonism by invoking psychological hedonism is at best problematic.

The "semantic" defense. There is a defense of hedonism which does not try to appeal to psychological hedonism. Briefly put, this line of reasoning claims that hedonism is true by definition. Let us call this the "semantic" defense of hedonism. Effectively, the the semantic defense says that "intrinsically desirable" means the same as "pleasant." If this claim about the meaning of "intrinsically desirable" is correct, of course, the thesis of hedonism--that a thing is intrinsically desirable if and only if and to the degree that it is pleasant--necessarily follows.

The semantic defense is clearly not susceptible to the problems which attend psychological hedonism, because those problems are limited to the empirical question of how humans are motivated. In this sense, the semantic defense may be considered some advance over the psychological defenses of the hedonist thesis. Nevertheless, the distinctly naturalistic flavor of of the semantic defense presents distinctive liabilities. As such, the semantic defense possesses both the strengths and weaknesses which naturalist definitions have, and these issues have been already discussed.

Problems with hedonism as such. Quite apart from problems with any specific arguments for the hedonist thesis, there are considerations which cast considerable doubt on whether the very claims made by hedonism are likely to be correct. The thrust of these objections is to show that it is possible, even probable, that some things are intrinsically more desirable than others, even when their pleasantness is not greater. These arguments further try to show that that are some things that are pleasant which are intrinsically undesirable.

To cast these considerations precisely, we need to look at the thesis of hedonism a little more closely. In particular, we should note that it implies that everything that is pleasant is intrinsically desirable. Let us call this the PID ("pleasantness implies desirability") thesis. Historically, many writers have questioned PID, and hence have argued that hedonism, as it has typically been formulated, must at least require some refinement. Their objection has been that reflective people will not only be indifferent to certain kinds of pleasures, but will in fact positively object to them. Since there is no other evidence which can count as strongly for or against PID as the beliefs of impartial, informed, calm people, this type of objection continues, PID is not likely to be true, and hence hedonism as typically formulated needs some refinement. Consider, for example, the following case. Suppose we discover (as we are in fact likely to) that high concentrations of uranium in drinking water causes a high incidence of bone cancer and leukemia in children, but poses relatively little difficulty for adults. Bone cancer and leukemia are frequently fatal and invariably very painful. We can imagine without too much effort someone who received great enjoyment in causing children pain and hence condoned practices which led to high concentrations of uranium in the drinking water supply. Such pleasure, a critic of hedonism could maintain, derives from the suffering of innocents, and must thus surely be undesirable, even repugnant, to informed, sensitive, reasoning people. Thus, the objection concludes, PID is just untenable.

How can the hedonist respond to such criticism? He cannot ignore it, but he can say this much. He can agree that it would be better if taking pleasure in the suffering of others did not occur. But not taking pleasure in the suffering of others would be better not because such pleasure is intrinsically undesirable; rather it is a symptom of an unhealthy state of mind, which in the long run will almost surely cause unpleasantness in any sentient being. The pleasure itself, the hedonist might say, is still desirable.

Let us suppose that this counter is forceful. The claims made by hedonism face a second problem. In particular, some of hedonism's critics urge, pleasure cannot have anything like the privileged status which hedonism attributes to it. Pleasures are transitory, they note, and what is transitory can hardly be of intrinsic worth. For example, suppose an economically depressed town invites a surface uranium mine to operate near the source of the town's water supply. The mine opens, unemployment vanishes, and, within a few months, the health and wealth of the town are the envy of many. Two years after it opens, however, the market for uranium collapses. The mine's operator declares bankruptcy and abandons the mine. Pumps which once kept the mine free of standing water cease to operate. Water fills the mine, then leaks slowly into the town's water supply. The economy of the town collapses. Health problems appear with ominous frequency. How, in light of such consequences, a critic of hedonism may ask, can the brief pleasure of short-term economic health be considered intrinsically desirable?

The hedonist can respond to this, at least in part, by at several observations.

First, he could point out, the same kind of argument would show that pain is not a bad thing, because it too is transitory, but no one (save maybe some very unusual novelists or people suffering from mental disorders) would seriously entertain such an argument.

Second, the hedonist might urge, the argument that pleasure is transitory may be correct, but without force. In some sense, all things in life are transitory, for life itself ends.

Third, the hedonist can argue that at least some enjoyments can be more than transitory. Health, knowledge, friends, and personal commitments are enjoyments. If these are attained, more than transitory pleasure will ensue.

Of these counterarguments, the third is surely the strongest and most interesting, because it hints at what hedonists think would happen if everyone were left to choose pleasure according to his interests. Many hedonists do believe that at least thoughtful people would choose precisely those pleasures or activities which would be recommended by many non-hedonist theories, just because they believe that as a matter of psychological fact, the promotion of health, the pursuit of knowledge, the nurturing of friendships, interesting jobs, and so on, are more productive of pleasure than, say, transient economic well-being. Some hedonists have gone so far as to try to defend hedonism on the grounds that people would in fact seek the more sophisticated pleasures if left to choose. To the extent that hedonism rests on such appeals, however, it is unfaithful to its principle, because the "higher" pleasures can be so only if they happen to be more productive of pleasure than the the "lower" pleasures. Thus to defend hedonism by an appeal to the level of pleasures which people would naturally select if given a choice is to invoke the hedonist principle to support itself. At best, such a defense is circular. More cautious hedonists have asserted that there may be pragmatic, non-hedonic reasons for urging the higher pleasures over the lower.

We need, therefore, to pursue the problem of how or in what sense the hedonist can legally assert that some pleasures are better, worse, or the same as others.

Typically, hedonists have had something rather complex in mind when they have asserted that one state of affairs is better than or preferable to another state of affairs if and only if it is more pleasant. Most hedonists have been inclined, and usefully so, to try to answer such questions as whether one system, say, of health-care distribution, is better (because more pleasant) than another, whether heroic measures should be used to prolong the life of the terminally ill, whether the long-term economic benefits of nuclear generation of electricity compensate for the health risks associated with such an industry, and so on.

Hedonism faces a number of special problems in attempting to deal with questions of this complexity.

First, it is difficult to estimate how enjoyable future experiences (such as cheap electricity) will be for other people or even for ourselves, because people have different likes and dislikes, and even the likes of a single individual can change.

But more importantly, it may be that the very notion of "being on the whole more pleasant" is just meaningless. If that is so, the claim of hedonism to be able to formulate and address major social issues is ill-founded. We must examine the distinctive problems associated with this matter.

Let us look at the phrase "is pleasant." One way of defining the phrase "x is pleasant" is "x is an experience with a feeling element that the person wants to prolong at the time." Similarly, we could define "x is more intensely pleasant than y" as "x and y are both experiences with feeling elements, and the wish to the subject of x to prolong x at the time is more intense than the wish of the subject of y to prolong y at the time." There is something of a problem about how to decide that the wish of one person to prolong his experience is stronger than the wish of another person to prolong his experience.

Let us suppose, however, that we know what it means to say that one person's experience at one moment is more intensely pleasant than that person's experience at the same or another time. Knowing this, however, does not tell us what it might meant to say "x's day of relief from economic want was more pleasant than x's day the risk of cancer from uranium contamination of water." We need to examine whether there is any meaning in such comparative judgments about the total pleasantness of stretches of experience.

We can assign meaning to this type of judgment in at least two types of situation.

First, we can say without controversy (but not without convolution) that x's enjoyment of a day of relief from economic want was as a whole more pleasant than his enjoyment of a day of risk of cancer from uranium in drinking water if at no instant during the day (more generally, during a given time interval) x's enjoyment from relief of economic want was less intense than his enjoyment of risk of cancer (more generally, "...from activity B..."), and if at at least one moment during the day (more generally, during the given time interval), x's enjoyment of relief from economic want ("...from a..."), was more intense than his enjoyment of risk of cancer ("...from B..."). Let us call this "Criterion 1."

There is a second situation in which we can attach a definite and plausible meaning to claims about comparative pleasures when the pleasures do not involve the same activity. We would probably be willing to say that x's day of relief from economic want was more enjoyable than his day of risk of cancer, provided that, by rearranging segments of the day (time interval), we can have the situation which satisfies Criterion 1. Let us call this "Criterion 2." Criterion 2 is a little less intuitive than Criterion 1 (at the least, it depends on understanding Criterion 1) but once understood, it is probably not too controversial.

Can every situation be handled in a fashion which satisfies the second criterion? Unfortunately, we cannot hope for as much. Figure 5-1 illustrates a case case in which the repositioning of curve segments will not satisfy Criterion 2.

Is there any criterion by which we can show that one of these activities is on the whole more pleasant than the other? If we had a defensible, clear way of stating that the "net amount of enjoyment" which the day of relief provided was greater than the "net amount of enjoyment" provided by the day of risk, then we would have grounds for saying that the enjoyment of the day of

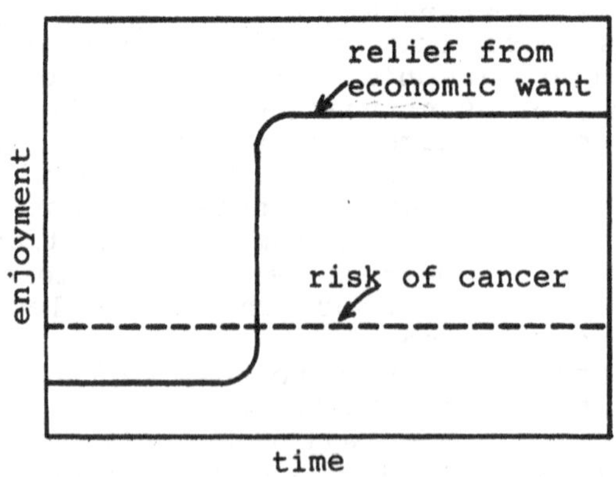


Figure 5-1. A case in which curve-segment repositioning will not satisfy Criterion Two.

relief was greater than the enjoyment of the day of risk. But to do this, we must first assign a meaning to the phrase "net amount of enjoyment." One plausible approach to this problem would be to say that activity the day of relief is more pleasant than the day of risk just in case the area under the curve for the day of relief (more precisely, the integral of relief curve for the day) is greater than the area under the curve for the day of risk (more precisely, the integral of the risk curve over the day). Let us call this criterion the "area" criterion.

Unfortunately, the area criterion goes beyond what we are justified in asserting about the experiences which give rise to the curves for relief and risk. In particular, our evidence, which is just a collection of comparative, non-metricized judgments of pleasantness, justifies us only in placing one point of a curve relatively higher or lower than some other point, but not in placing it higher or lower than the other curve to any definite extent. To use the "area" criterion, unfortunately, we would need to be able to place one curve higher than the other to a definite extent at each instant in time, that is, we would need to be able to express enjoyment as a real (number)-valued function of activity or experience, because the area criterion presumes that enjoyment is the integral of a real-valued function. Comparative judgments of enjoyment do not allow us to infer even the existence of such a metric or function. Thus, it appears that there are comparisons which the hedonist would like to make, but cannot.

Is this kind of consideration truly an objection to the hedonist thesis? Not obviously so. It may be that we cannot even know in all cases, regardless of whether we are hedonists, that one activity is preferable to another. If it were the case that we happened to know that one thing were preferable to some other thing but could not determine on the hedonist view whether the first was more pleasant than the second, of course, the hedonist thesis in particular would be in some trouble. But it is not clear that hedonists are in such an unenviable position.

The task hedonism sets for public health policy, then, is not an easy one. Decisions about what uranium standard for drinking water should be set, for example, must on the hedonist account involve attempts to resolve, even in the absence of adequate scientific information, the tradeoffs among risks to health, the cost of electricity, the cost of water purification, employment levels, a large body of standing law, and the overall advantages and disadvantages of government involvement which would almost surely attend any decision we would make on this matter. This should not necessarily be construed as an objection to hedonism, however, and may very well be viewed as one of its strengths: as a matter of fact, hedonism forces us to explicitly take into account almost all of what we intuitively believe is relevant to deliberation about matters of value in health issues.

On the whole, hedonism survives rather better than any of the other bases for public health policy we have examined so far. It is not without problems, of course, but any theory of public health policy which entails a complete denial of all the insights which hedonism provides will, if there is any justice in our account of these alternatives, face quite serious difficulties.

## HEALTH POLICY AND OBLIGATION

We have devoted considerable attention to surveying current views about two general questions of broad import for public health policy: (1) What form does deliberation about public health policy, insofar as it concerns value-related matters, take?; and, (2) How do we identify what is desirable in public health policy? The questions of what form policy deliberation takes and how we identify what is desirable in health policy strike many people as hardly distinct or important questions at all. In contrast, the question of (3) What is right or wrong in public health policy proposals, and what our duty or obligation to our own health and to the health of others is, have always been regarded as fundamental. Yet the three questions must be kept separate: although they may be closely coupled in some cases, they are not identical. We would be hard-pressed to do ourselves a greater disservice than to allow further confusion into matters which are by nature murky and difficult.

*The language of obligation*. On first consideration, it might seem that the great variety of "duty" and "obligation" terms used or implied in health policy dicta (e.g., "We *ought* to prevent leukemia," "It is one's *duty* not to harm the health of others," "It is *wrong* to cause others physical pain") would involve correspondingly many distinct issues which an adequate theory of obligation would have to address. The task is fortunately not so perverse as this. In particular, *most* questions about duty or obligation in health policy matters can be reduced to questions strictly about obligation. For example, the sentence "You ought to support a uranium standard of 10 pCi/l" means approximately the same as "You are obligated to support a uranium standard...," which is also approximately the same as "It is right for you to support a uranium standard...," and so on.

What kind of idea is expressed by obligation-terms in these contexts? At the outset, most writers have held that there is some distinction between terms like "obligatory," "duty," and "wrong," on the one hand, and terms like "is good," "is intrinsically desirable," "is preferable," and so on, on the other. Let us call terms like those in the first group "mandatorial" terms, because they invariably are associated with questions of mandate; we will call terms like those of the second group "adjectival," because they merely ascribe a value-feature to some act or policy, but are not, as such, necessarily bound up with mandates. These are but names, however: our first task in this section is to clarify the difference between these two types.

There are at least two considerations which make it clear that identifying mandatorial terms with adjectival terms simply isn't possible. These considerations may be best seen through an example. Consider the phrase "It is obligatory for you to do everything you can to fight leukemia in children" versus the phrase "Fighting leukemia in children is intrinsically desirable." First, the phrase containing the term "obligatory" clearly applies only to persons, whereas the phrase containing the expression "is intrinsically desirable" may be sensibly applied to certain chemotherapeutic agents. Second, if it is obligatory for us to fight leukemia, it must be possible for us to do so; in contrast, even if fighting leukemia is intrinsically desirable, it may not be possible (we hope that it is possible, of course) for us to do anything about it.

Moore's definition. It thus appears that no reduction of mandatorial terms to adjectival ones is possible. Some authors have proposed that although there is no identity between mandatorial and adjectival terms, mandatorial terms can be defined in terms of adjectival notions along with some non-value-related concepts. We will call such definitions "hybrid." G. E. Moore, for example, urged that "right" means nothing other than "cause of a good result;" similarly, he claimed, "I am obligated to perform a certain action" simply means "This action will produce the greatest possible amount of good in the Universe;" one's "duty," he concluded, "can only be defined as that action which will cause more good to exist in the Universe than any possible alternative." Thus, for example, if I say that I am obligated to support a standard of 10 pCi/l uranium in drinking water, I simply mean, on Moore's account, that supporting such a standard will produce the greatest possible amount of good in the Universe.

Problems in Moore's definition. Hybrid definitions have the virtue of capturing at least one feature of our beliefs about obligation: we obviously think that the notion has a value-related feature. It is even plausible that the definition of mandatorial terms must contain some non-value related features. But even given this, it is not obvious what the definition of obligation should be. Any attempt to define terms of obligation in Moore's fashion, in particular, will run into trouble quickly, because the definition does not accord with how people actually use words like "right" and "obligation." Consider, for example, the following. A wealthy benefactor of a college dies, leaving a large sum to the institution. It is generally known to the officers of the college that the benefactor wanted this money to be used by the college to pay for postgraduate leukemia research, to be done by individuals who come from impoverished families. Despite what is known about the benefactor's desires, the bequest legally contains no restrictions on how it is to be spent. As time passes, fewer and fewer individuals who meet the conditions of the benefactor's desires can be found. General educational expenses plague the college. Officers of the institution begin to eye the leukemia-research bequest with an interest not entirely consonant with the benefactor's. Obviously, they reason, the bequest would do greater good if used to meet the general educational costs.

Is it clear that the officers should use the bequest to meet general educational expenses, as Moore's definition requires? Not obviously so. If this example is merely plausible as stated, then Moore's criterion does not describe the way people actually use terms like "right" and "obligation."

Moore's definition of "obligation" is therefore wanting. It is wanting precisely because it tries to identify obligation with doing that which will produce the greatest good, and as a matter of fact, people just do not invariably use "obligation" in that way. Can we improve on Moore's definition? One plausible approach to this problem is to note that whatever we mean by "obligation," it seems to be tied up with the notion of being "bound" to do something. But "bound" in what sense? In particular, we surely do not mean "bound" in the sense of "causally necessitated:" a person may be bound to protect the health of children and simply fail to do it. "Bound" does merely mean "legally bound," either, for we can easily imagine cases in which the laws are so bad (e.g., laws requiring that we cause the physical suffering of innocent people) that we are obligated not to obey them. Indeed, it does not seem that "bound" even involves being obligated by <u>someone</u>, given the way most people use the term "obligated:" the sense in which we may be bound to help alleviate the suffering of children appears to be imposed on us by no one in particular.

_The "culpability" definition._ It has been suggested by some authors that we can capture the notion of "being bound" in the mandatorial sense as follows. Phrases like "It is your duty to protect the health of others," they urge, mean "If you don't protect the health of others, you will be culpable." A subtler formulation has sometimes been proposed: "It is your duty to protect the health of others" means "Protecting the health of others is the one thing you can do with the result that you will escape all culpability except possibly on grounds of your motivation."

Will this definition, which we shall call the "culpability" definition, work? At least it provides a clear criterion for determining when we are obligated to do something. And, it seems to accord with at least one feature of most people's beliefs about obligation: if we don't do what we are obligated to do, we are culpable, except possibly in the case in which our motives are beyond question.

Nevertheless, there are serious problems with the culpability definition, whatever else we may conclude about it. Consider in particular the following. Suppose a physician determines that a child has developed an untreatable cancer. Until a matter of a few weeks before death, the physician knows, no ill effects will be noticed; death will come swiftly and nearly painlessly. In a very small proportion of cases, this cancer has been known to spontaneously disappear. Suppose we think it is the physician's duty to inform the child and its parents. The physician believes that knowledge of the child's condition can only cause grief to the child and its parents. He accordingly decides not to inform them of what he considers inevitable; the pain of knowing, he reasons, is in any case greater than the consequences of ignorance.

The culpability definition says that the physician is exempt from culpability; we believe, by hypothesis, that he is not. If this example is merely _possible_, then the culpability definition is inadequate.

Both Moore's definition and the culpability definition of obligation are attempts to define mandatorial terms in terms of some value and some non-value terms. Neither of these approaches is without serious difficulties. Considerations like this have led some authors to believe that _any_ hybrid definition of mandatorial terms is ill-fated. It may be, they urge, impossible to develop an adequate hybrid definition. And in any case, they claim, there is an alternate way of looking at obligation-terms which does not take the tack of the above two attempts, and hence does not suffer their problems: we can, the story goes, find in the Standard Method a clear and unproblematic account of mandatorial terms.

<u>The Standard-Method definition</u>. What does the Standard Method say about our judgments about mandatorial matters? It appears that we test candidate judgments about our duty by comparing them to our "feelings of obligation." These feelings appear to be of two kinds. First, I have impulses demanding <u>me</u> to act that are independent of my "subjective wants;" they tend to be succeeded by feeling of guilt or remorse if I fail to obey them. For example, if I believe that I must do what I can to help prevent environmentally induced leukemia in children, I feel obligated to promote this prevention; furthermore, it may be patently implausible that this has anything to do with my subjective wants (since I may not currently know anyone who has leukemia, and I may not have, nor desire to have, children). Second, it appears that I test my judgments about my duty against my impulses to demand that someone else act in accordance with those judgments. If I believe, again, that we are obligated to prevent environmentally induced leukemia, I do not feel that this obligation has force only for me; I also demand that others act in accordance with the obligation. This impulse to demand is, again, independent of my subjective wants and involves a readiness to compel the action in question by some sort of sanction if the object of the demand is not met.

Whether fact appeal to the kinds of impulses alleged in forming our judgments of obligation is open to empirical investigation, and a thorough defense of the claims just made about how we form those judgments would thus have to appeal to the results of such research. Pending the outcome of this investigation, we may hazard a guess about how the definition of "obligation" would go in the idiom of the Standard Method. When we say that "X is obligated over all to do A," we simply mean that "It is objectively justified for X to feel obligated to do A, and for other persons to demand that X do A," where "X is objectively justified in feeling obligated to do A" means "X's attitude corresponding to feeling obligated to do A satisfies all the conditions that would be set, as a general policy, for the endorsement of attitudes governing or appraising choices or actions, by anyone who was intelligent and factually informed and had thought through the problems of the possible different general policies for the endorsement of such attitudes." Thus, for example, when we say that we are objectively justified in supporting a particular standard for uranium in drinking water, we simply mean that our attitude in support this standard satisfies all the conditions (e.g., thoroughness of research on the issue, fairness, universalizability, etc.) that would be set, as a general policy, for the endorsement of attitudes governing or appraising choices or actions, by anyone who was intelligent and fully informed and who had thought through the consequences of adopting the standard and alternatives to it.

This definition, which we shall call the "Standard Method" definition, has at least two strengths. First, it is testable; if the definition is wrong, there is in principle an unproblematic way of determining that it is. And second, the Standard Method definition forces us always to consider our impartial feelings of obligation, a feature of value-related deliberation which Moore's definition and the culpability definition dismiss by mere fiat. Thus far, then, the definition seems to fare relatively well.

Problems with the Standard Method definition. The Standard Method definition is not entirely free of problems, however. Consider, for example, the following. Suppose that a physician has examined a patient suffering from a rare form of liver cancer which is difficult to diagnose and about which which almost nothing is known. The physician advises and prescribes a series of operations at substantial cost and suffering for the patient. The surgery, unfortunately, produces no constructive results. Was the physician's behavior consistent with his obligations? We are inclined to say that it was. But according to the Standard Method definition of obligation, it seems that he did not discharge his obligations, since if he had been fully informed about the disease, he would have known the futility of surgery. Thus, the Standard Method definition and our intuitions appear to collide in examples of this sort.

Consider a second example. Suppose that in the course of otherwise relatively routine operation, a surgeon finds himself at a point at which he must administer additional doses of an anesthetic. The anesthetic is known to severely depress respiratory and circulatory functions in some patients, but it is in general not possible to identify these individuals in advance of the operation. Furthermore, he knows that there is always some risk in protracting anesthesia. The surgeon could take precautions to protect the patient's vital functions (e.g., administer some adrenalin) against the effects of the anesthetic, but doing so would protract anesthesia. Now it so happens, unknown to the physician, that the patient is not at all sensitive to the anesthetic. Ought the physician to take the time to protect against the possible ill effect of the anesthetic? We are inclined to say that he should, simply because he does not know what effect it will have. But according to the Standard Method definition, the physician is not obligated to take precautions, because if he were fully informed, the proper procedure would be to press on without the unnecessary, and potentially risk-inducing precaution.

The above two examples appear to be serious problems for the Standard Method definition, in the sense that they describe intuitive judgments which are at odds with what the Standard Method definition appears to predict. More specifically, the Standard Method appears to require us to know everything relevant to the situation at hand if we are to escape culpability, and this kind of knowledge we frequently do not have.

The problems may not be as bad as they seem, however, but solving them takes some effort. To make some headway on them, we need to distinguish the contexts in which we make statements about a person's duty or obligation.

Contexts of mandatorial terms. There are a number of contexts in which we make mandatorial statements. We may, in certain cases, make them before an action has taken place, stating our beliefs or giving advice about what one ought to do. For example, I may urge a friend in the legislature that she ought to vote for a highly restrictive standard for radionuclides in drinking water. In such a case, we are looking to the future, concerning ourselves (and others, possibly) about what ought to be done. For contexts of this sort, the theory of obligation which we have been considering (deriving from the Standard Method) works well. It is plausible in the course of deliberating such a matter, that is to say, that I ask whether my feeling obligated is "objectively justified" in the sense of "meeting all the conditions that would be set." It also is plausible that the statements I make in which I express my decision after I have reached it are explained by this account: I am inclined to say, upon reaching a decision, "After reflection, I have decided it is my duty to urge my friend, the legislator, to vote for a more restrictive standard." Furthermore, in anticipative contexts, the Standard Method requires us to do and to know what normal thoughtful people could know about such situations. We will say that a person is using mandatorial terms in the anticipative sense if he is using those terms in such contexts.

In contrast, we may make statements about duty or obligation after the fact of some particular action. In such cases, we are criticizing something that has already been done. We will call such contexts and evaluations "retrospective." Retrospective evaluations can often take into account more of what we think is relevant in assessing a situation, and these factors can influence our decisions about whether someone is culpable in case they did not do what we thought was their duty. In retrospective contexts, we often distinguish the reasons why people fail to do their duty, and on the basis of these distinctions, are able to assess culpability in ways we could not in anticipative contexts.

The Standard Method tells us, more specifically, two principal reasons why people fail to do their duty. One is that they did not know what it is. Sometimes failure to have known one's duty is considered a reflection on character; sometimes it is not. If knowing one's duty would have been relatively easy, we may fault someone who did not bother to know it. We would, for example, fault the physician who knowingly neglected the safety of his patient. If, however, knowing the objective facts about a situation would have been very difficult, we sometimes excuse a person for not having known his duty. We would be hard pressed, for example, to find culpable the physician who provided compassionate, attentive, state-of-the-art treatment to a patient who nevertheless died, even when we believe it is the duty of the physician to save the lives of his patients.

The second reason why people fail to have done their duty is motivational. They knew what their duty was, but did not have a desire to do it, or were strongly motivated to do something that was at odds with their duty. Usually a person who failed to do his duty for this reason is considered culpable. For example, a driver who drank too much would be held culpable by most people. But this is not always the case: we would probably judge leniently, if we condemned him at all, the case of the physician who chose, because of his desire to avoid unmitigatable suffering, not to inform the parents of the child dying from liver cancer.

In short, the Standard Method works well in retrospective contexts, too. It recognizes that in such contexts, one may know more than one does in anticipative contexts. But the Method does not require to have known what we could not have known; it only requires us to have known and to seek to have known, what normal thoughtful people could have known in the situation. This is precisely the heart of the confusion in the putative counterexamples to the Method described above: the counterexamples collapse the distinction between anticipative and retrospective contexts and further suppose that the Method talks only about retrospective knowledge.

There is a third context in which we use mandatorial terms in which we are neither strictly advising or urging someone about future value-related behavior nor issuing retrospective judgments. Consider the case of the physician faced with ambiguous evidence about the role of uranium ingestion in bone cancer, detailed above. Suppose he concludes that the most reasonable posture on the matter is not to pursue the resolution of the ambiguity, although he continues to be perplexed. He seeks our advice about what he should do.

Suppose that we think his conclusion is mistaken; we think that it is his duty to pursue the resolution of the ambiguity. Yet we do not want him to act against his conscience. We accordingly advise him: "Since you are honestly convinced that not pursuing the resolution is your obligation, you should act accordingly." In saying this, we are not informing him of what we think his duty in the situation actually is. But we are in a sense advising him. The above example falls outside the contexts captured by the notion of "anticipative" and "retrospective." It involves a third, advisory, context in which we are using terms like "obligation" and "duty." We shall henceforth call such uses of mandatorial terms "advisory." Advisory contexts are not, then, contexts in which we are stating what a person's duty is.

The distinction we have made among the anticipative, retrospective, and advisory senses of mandatorial terms would seem to exhaust the contexts in which we use those terms. Nevertheless, some writers have urged that there is a fourth context, which we shall call the putative one, in which we use mandatorial terms. This context can perhaps be best seen by considering an example. Suppose that a physician sincerely thinks that it is his duty not to inform a terminally ill patient of his situation. Let us suppose that we believe he is deeply mistaken about his obligation, in spite of his beliefs about it. It is not, therefore, the physician's anticipative or retrospective or advisory duty not to inform the patient. In some sense, however, we think that the physician is really doing his duty if he does not inform the patient, because he is doing that which he sincerely thinks is right. There thus appears to be a sense of "duty" other than the anticipative, retrospective, or advisory. This context or sense is often called the "putative" sense of obligation.

What can be said about this argument? First, let us note that it depends on a premise which is open to serious question: that it is a person's duty to do that which he thinks is his duty. Counterexamples are easy to find. Consider for example, the following. Suppose that an obviously confused person tells us he has concluded that he has a duty to commit suicide. Does this belief make it his duty to commit the act? Obviously not. The argument, therefore, is far from telling.

A second argument is sometimes offered to defend the view that there is a significant sense of "duty" other than the anticipative, advisory, or retrospective. It has been argued by some writers that the putative sense of "duty" is other, and indeed more important than the anticipative one, because it appears when we say that, all things considered, a person ought to do something, we mean that he ought to do that thing in the putative sense, not in the anticipative or retrospective sense. We do believe, for example, that the physician administering anesthesia (see example above) ought to take precautions against the possibility of depressing the patient's vital functions, and this is not the physician's retrospective duty. Furthermore, the argument continues, don't we think it is the duty of the physician whose conscience tells him not to inform the terminally ill of their condition to follow his conscience, and isn't this neither his retrospective nor anticipative duty?

Such an argument is confused on several points. First, it ignores the fact that for the agent trying to decide what his duty is, the question whether he should do his duty in one sense or some other sense cannot arise (by the very description of the example, he is ignorant of the conditions or facts which would allow him to make the distinction). Such a distinction is available only in hindsight, and to people who are more fully informed of the facts. Second, the general principles of duty are a trivial extension of the notion of anticipative duty. Specifically, it is a person's overall duty to do A if and only if doing A would have been his anticipative duty if the particular facts of the situation had been what he believed them to be, or would have believed them to be if he had explored the situation as a man of character would do. Thus, if we know what a person responsibly thinks about the facts of his particular situation, and if we know the correct principles of anticipative duty, we can infer his retrospective or putative duty. Therefore, since the true principles of duty are decisive for what is a person's retrospective or putative duty, we may confine our attention to the former to determine the latter. Therefore, even when putative duty exists, it is not something truly distinct from anticipative duty.

Thus far, we have been talking about mandatorial terms in the sense of "overall duty or obligation," as in the sentence, "Everything considered, it is your duty to help prevent leukemia in children." Certainly, the Standard Method definition assumes that this is the sense in which we use words like "obligation" and "duty." The notion of "duty" is not always used this way, however. For example, we might say, "I agree that I have _an_ obligation to help prevent leukemia, but my daughter has become seriously ill and I am the only one who can take care of her. So I have a _greater obligation_ to take care of her." Some writers have urged that obligations used in this sense, which they call _prima facie_ obligations, are thus of a distinct kind. If that assertion is correct, we must allow yet a fourth (or fifth, depending on how forceful we feel the arguments for senses beyond the anticipative, retrospective, and advisory are) sense of "duty."

Is the notion of a prima facie obligation really distinct from the senses we have been discussing? We can make a reasonably strong case that it is not. Suppose, as above, that I do have an overall obligation to help prevent environmentally induced leukemia in children, but would not be obligated to do this if my daughter were ill. There is, that is to say, a mitigating factor to my overall obligation. Does this imply that there is some distinct type of duty in such a situation? Not obviously so. For I can easily express this situation in the idiom of an overall duty. My overall duty, I can say, is simply this: "I am obligated overall to help prevent environmentally induced leukemia in children, except when those relatives for which I am immediately responsible are ill and I am the only one who can effectively care for them; in the latter case, my overall duty is to care for my charge." The problems which prima facie obligations appear to pose for the definition of mandatorial terms given above can all be solved in a manner analogous to this example. Thus, prima facie obligations are simply one kind of overall obligation, not a distinct creature in its own right.

On the whole, then, we may make the following observations about obligation. The Standard Method definition fares as well or better than any other. On this definition, someone of good character will try to find out what his obligation is, and then do it. Having done this, he can escape censure, because no failure on his part could be attributed to a fault of character.

## Egoism

We have taken a long excursion into distinctions among the general meanings of mandatorial terms, hoping to clear some of the murkiness which attends that part of our language. We now turn to the first of several theories of obligation which we will examine; with any luck, our efforts with the former will repay in the latter.

The first of these which we consider is called _egoism_. Roughly speaking, egoism is the view that a person is obligated over all to perform action A if and only if A is, among all the actions he can perform, the one that will produce states of himself of maximum intrinsic worth. The practical implications of such a principle can differ widely, depending on what one assumes has intrinsic worth. For example, if I have the view that health is of no intrinsic worth and hold the egoist principle, I am not likely to look after my own health (or anyone else's); if I think that health has intrinsic worth, maintaining my health will be, at least on the egoist theory, one of the obligations which I must take quite seriously.

Although the egoist principle initially has the ring of unmitigated selfishness about it, as a matter of fact an egoist _could_ be someone whose overt behavior appeared little different from the altruist's. For example, if bringing about the health of others promoted states of intrinsic worth for an egoist more than any other activity, then that egoist would, on the egoist principle, be obligated to promote the health of others, possibly to the exclusion of his own health. Such a state of affairs would be unexpected, to say the least, but it is consonant with the egoist principle.

A number of arguments have been brought by various authors in support of egoism. Most, if not all, of these are much like the kinds of arguments which have been offered in support of hedonism. We consider the most important of these now.

_The "semantic" defense_. It has been argued that egoism is true by definition. The phrase "A is obligated to do X," this argument goes, simply means the same as "A is, among all the actions X can perform, the one that will produce states of X of maximum intrinsic worth." On this view, for example, when I say that I am obligated to protect my own health, I simply mean that among all the actions I can perform, protecting my own health is the one action or kind of action that will produce state of myself of maximum intrinsic worth." Thus the argument concludes, egoism is true by definition. Let us call this the "semantic" defense of egoism.

_Problems in the semantic defense_. Is the semantic defense sound? There are at least two strong reasons for believing that it is not.

First, if there is any sense in our discussion of the definitions of mandatorial terms above, then it is at the very least not obvious that the semantic defense is reasonable. The very difficulties which people have in defining such terms, that is to say, show that the definition is not self-evident. Without a convincing argument for the claim that obligation simply means what the egoist thesis says, the semantic defense lacks any force.

More importantly, if the Standard Method definition of mandatorial terms is defensible, the semantic defense of egoism is just mistaken. The force of the semantic defense, that is to say, depends on just how thoroughly it is able to reveal the alleged error of the Standard Method definition. As stated, the semantic defense provides no such analysis, and to date no such analysis has been provided by its proponents. As it stands then, the semantic defense is not convincing. Proponents of this defense have had nothing further to say about it. Until they do, we must regard it as unsubstantiated.

*Psychological defenses*. Egoism is more frequently defended on psychological than on semantic grounds. There are two psychological defenses of egoism. The first of these begins by assuming that a person is motivated to produce a certain situation if and only if he believes it will be, or will produce, a desirable state of himself; and he will be motivated to produce one situation in preference to another if and only if he believes it will be, or produce, on the whole, a more desirable state of himself. Since a person is motivated only to produce desirable states of himself, the argument continues, he will hardly feel obligated to act, or think in a way that will conflict with what he thinks is his own welfare. The egoist principle follows.

There is something particularly disconcerting about this argument, which we will call the "teleological" defense of egoism. This defense, as stated, is an argument for the claim that a person will demand of others that they seek *his* welfare. The egoist thesis has no such consequence. Thus, the teleological argument is at the least misdirected.

Moreover, the teleological defense is an argument for a theory of obligation which contains, or which implies a claim, that is particular to an individual person, and thus it is not a *general* ("universalizable") rule. Since our interest here is in analyzing only general principles of obligation, the teleological defense is irrelevant.

The teleological defense is questionable on yet another ground. The physician who knowingly tends patients who have a fatal, highly infectious and communicable disease for which there is no vaccine or antibiotic is not obviously looking out for his own welfare. Against this kind of obvious objection, proponents of the teleological defense could argue that the physician would feel that he could not bring about his own welfare if he did not attend such patients. But this raises the question of the very *status* of the claims made by the teleological defense about human motivation. In particular, we may legitimately ask how it is that the proponent of the teleological defense is able to conclude what the physician's motivations are. To put the point another way, can we even imagine any evidence which *in principle* could count against the teleological view? If we cannot, which proponents of the view have tended to assert, then the teleological defense is not a *testable* view of human motivation. That alone is a very severe defect in the defense.

Without further explanation, then, the teleological defense of egoism is wanting. Proponents of this defense have not provided such an explanation.

A second type of psychological argument is sometimes offered in defense of egoism. This argument begins in the same fashion as the teleological defense; that is, it asserts that person is motivated to produce a certain situation if and only if he believes it will be, or will produce, a desirable state of himself; and he will be motivated to produce one situation in preference to another if and only if he believes it will be, or will produce, on the whole, a more desirable state of himself. It follows from this, the argument continues, that a person can act only in such a way as he thinks will promote his own welfare. Nonegoist theories imply that sometimes a person may have to sacrifice his own welfare. Thus if nonegoist theories are right, it is sometimes a person's duty to do what he _cannot_ do. But clearly it is never our duty to do what it is impossible to do. Thus nonegoist theories are just mistaken, the argument concludes.

This argument, which we shall call the "causal impossibility" defense, is unsound for several reasons. First, it not so clear that it is never a person's duty to do something if it is impossible only _because he does not desire_ to do it. What is clear is that we do not have a duty to do something if we cannot do it _no matter how hard we try_. Second, if the causal impossibility argument were sound, it would show that it is not a person's duty to do _anything_ he does not do. But this is a patently ridiculous consequence of the theory: we clearly hold there are many cases of people simply failing to do their duty. Thus, the way in which people actually use mandatorial terms is inconsistent with the causal impossibility defense.

_Arguments against egoism as such_. On the whole the existing arguments for egoism are wanting. This does not mean, however, that egoism is just false, and it may be that the arguments against egoism (as opposed to arguments against particular defenses of egoism) are as problematic as those we have considered in its defense. If that were so, we could find ourselves in a situation in which we could not show that egoism is false, although we could show that certain specific defenses of it are unsound. We must therefore look at objections which have been offered to the egoist thesis per se, and determine whether they fare any better than the semantic, teleological, or causal impossibility defenses.

_A paradox_. It is been claimed that egoism is paradoxical. On the one hand, this argument begins, the egoist claims to believe the thesis of egoism. If he does, then it would seem that he should act in such a way as to try to get others to believe the thesis, i.e., the egoist must proselytize. To the extent that he can convince others of the egoist thesis, however, the egoist acts against his own principle: if others believe the egoist thesis, they will tend to look after themselves, and not after him. Thus, to hold the egoist thesis is to act against it.

At the heart of this argument, which we will call the "paradox" objection, lies a familiar emotivist claim: to believe a view is, among other things, to be strongly inclined to persuade others to adopt that view. The egoist can dodge the paradox objection simply by rejecting this emotivist way of looking at things. Thus the paradox objection is not as forceful as it might seem.

*The "inconsistent-directives" objection.* To overcome this difficulty, some critics have modified the paradox objection as follows. Suppose two rival surgeons desire to perform a promising new operation on a patient suffering from an otherwise fatal cancer. The egoist thesis, the argument goes, directs each surgeon to destroy the credibility of the other, and thus obtain the exclusive opportunity to perform the operation. With the credibility of each destroyed, there is no plausibility that either will be allowed to do the operation, however. Thus, the egoist thesis gives incompatible directives.

This argument, which we shall call the "inconsistent-directives" objection, is not sound, primarily because it misrepresents what the egoist thesis says. The egoist thesis says that each surgeon should do what will in the circumstances maximize his own welfare. If the surgeons engage in mutual defamation, at least one, and probably both of them will lose their credibility. Moreover, if one of the surgeons happens to have the stronger hand in such a conflict, it can hardly be in the interest of the weaker to engage in an activity which will result in his own loss of credibility. Probably some sort of agreement will serve the best interests of each. In any case, there is not inconsistency of directive, contrary to what the objection says. Thus the inconsistent-directive objection is not forceful.

*Can egoism even be consistently applied?* There is a third line of objection which has been leveled against the egoist thesis. According to the Standard Method, we examine whether our value-related principles cohere with our qualified feelings of obligation and our qualified demands that other people act in a certain way. Let us ask what our motivation could be like in order for our qualified attitudes to cohere with the principles of egoism. If the two were to cohere, we would have to be partial to the people doing certain acts, but only because of the acts that those people do. We would have to demand, that is to say, that a person do that which maximizes his own welfare while at the same time, we must have no interest in that person in relation to the acts of others. Egoism thus has the consequence that the acts of Person A must be ignored when Person B is deliberating. That is, so long as Person A is acting, his interests are the only ones that can count; but when we consider the acts of others, the interests, qualities, character, and virtue of Person A must not matter at all. This leads to the paradoxical, if not contradictory, view that we must both consider and not consider the interests of a given person when trying to decide what obligations we have. To cast this point somewhat more concretely, suppose there were a physician who was completely indifferent to the welfare of his terminally ill patients. On the other hand, egoism directs us to look at the physician's obligation only with regard to whether the physician's behavior promotes his welfare. As long as we consider the physician's behavior by itself (without regard to his effects on others), it is possible we could find that behavior promotes his own welfare and thus find, if the egoist thesis were true, that the physician is meeting his obligations. On the other hand, if egoism is true, when we view the matter from the perspective of his terminally ill patient, whether the physician is doing his duty is of no consequence. This result is patently incompatible with the way we in fact deliberate. We are, in fact, that is to say, concerned about the effects of the actions of others on us, and we do demand that they refrain from harming us. Thus egoism is just incompatible with the way people in fact deliberate about the value of health.

To this objection, no egoist has ever given a convincing reply. Until one is offered, we must regard the egoist theory, on the whole, as unsubstantiated.

## Act-utilitarianism

If egoism is just mistaken, as the arguments which we have considered suggest, we are faced with the result that it may sometimes be our obligation to do that which does not contribute to our own health or welfare.

What general principles should guide us, then, in determining what our obligations concerning matters of health, at least, are?

One very widely held answer to this question runs roughly as follows. "It is our duty to perform a specific act on a specific occasion if and only if so doing will produce a state of conscious beings that is of maximum intrinsic worth, as compared with what would have been produced by other acts the agent could have performed instead." This idea is known as *act-utilitarianism*. According to act-utilitarianism, *if we regard basic* health of intrinsic worth, then it is our duty to act in each case in which we are faced with a deliberation concerning health, to do that which will maximize the health of all of us; such acts may, as far as act-utilitarianism is concerned, involve even the death of some for the sake of others.

We must first be clear what is suggested by such a formula. In this attempt, we immediately run into a small difficulty. Should we take act-utilitarianism to be saying that (1) a person's duty is to perform that act whose performance will *actually* produce a state of conscious beings of maximum intrinsic worth, as compared with what would in face be produced by other actions he could perform instead? Or should we take it to mean that (2) a person's duty is to perform that act whose performance will, *on the available evidence*, produce a state of conscious beings of maximum intrinsic worth? The distinction is important. An example will help show this. Suppose a surgeon is trying to decide whether to perform an operation on a patient's brain. If the operation works, it is well known, the patient will be completely cured, whereas if it does not, he will permanently be confined to bed. The operation, moreover, is a dangerous one: it has a mortality rate of 70 percent. Is the act-utilitarian saying that the surgeon's duty is to perform the operation if and only if *in fact* so doing will produce the best results (i.e., he should perform the operation if and only if the patient will survive)? Or is the act-utilitarian saying that it is the surgeon's duty to perform the operation if and only if so doing is the best "bet," that is, has *maximal net expected utility* (= $(P_s)(V_s) - (P_f)(V_f)$, where P is probability, V is value, s stands for "success", and f stands for "failure")?

There is something to be gained if we choose (1) as the meaning of the act-utilitarian thesis. This interpretation is conceptually clean and clear. Such an interpretation has two major drawbacks, however. We frequently do not have all the information required to make a decision in accordance with (1). Every operation, for example, has unknown risks associated with it. Every decision which we may make in the next decade or so about ceilings for radiotoxic agents in drinking water will be attended by the risks of uncertainty about the consequences of these ceilings for public health. If we were to choose (1) as the meaning of the act-utilitarian thesis, therefore, we would be adopting a principle that is at best awkwardly applicable to the deliberations about the value of health which we often face. Secondly, as a matter of fact most people have taken the act-utilitarian thesis in the sense of (2), above. By defending (1) as the "proper" interpretation of the act-utilitarian thesis, we would thereby be condemn its adequacy for describing the way people actually deliberate--a very serious strike against the proposal at the outset. For these two reasons, we will take the act-utilitarian to mean something like (2). Henceforth, then, act-utilitarianism will be taken to be the following thesis: "If doing A has, among all the things X can do, maximal net expected utility, then it is X's duty to do A."

The job of clarifying the act-utilitarian thesis is by no means ended with this decision. We must still make some sense out of the phrase "maximal net expected utility." To to this, it will suffice to understand what someone might mean by the phrase "greater expected net utility," for with this notion in hand, we can define "maximal net expected utility" in a natural way: a set of acts has maximal net expected utility if and only if the net utility of those acts is equal to or greater than the utility of any other set of acts. (Note that there could be more than one act which has maximal net expected utility.) We will make the following proposal for the meaning of maximal utility: Action A has greater net expected net utility than action B if and only if the consequences of doing A are preferred, in the sense of the Standard Method, to the consequences of doing B.

Act-utilitarianism has two great virtues. First, it can be stated simply; the alternatives to act-utilitarianism appear to involves complicated sets of rules which may be difficult to identify. And, second, it appears to accord with most of the intuitions which we might have about value-related matters.

<u>Some problems faced by act-utilitarianism</u>. These are non-trivial considerations in favor of act-utilitarianism. Nevertheless, the thesis has some problems which we must examine.

Let us imagine a small hospital with extremely limited financial resources which is completely occupied by poor, terminally ill, deeply suffering, bedridden, elderly patients, all under the care of a single physician. Let us suppose that the physician wishes to place another of his patients, a young pregnant woman, in the hospital, because he believes that if he does not, she may die during delivery. If she is in the hospital, the physician believes, she has a good chance of living. The physician deliberates whether it is his duty to hasten the death of one of the other patients to make room for the expectant mother. "After all," he reasons, "all the patients currently in the hospital have lives of little worth even to themselves, whereas the young woman and her child have a much better life to look forward to if I can save her. I need only one bed. If I discreetly relieve one of the current patients of his suffering, he will no doubt be better off for it. What greater good could I do?"

The physician's decision clearly runs counter to our intuition about what to do. The dilemma is not an easy one on any theory, of course, but the distinctive problem that it poses for the act-utilitarian is that nowhere in the example does there seem to be any room for the angst which attends our deliberation about the matter. Even if we should decide that the physician's action is right, the act-utilitarian thesis simply fails to capture this essential element of our deliberation. Although it may not lead to incorrect actions if adopted, it cannot provide a complete account of how we reason about our obligations in relating to the value of health.

Consider a second example. Suppose that a widower acquaintance has a five-year-old child which he has supported on a rather meager income. Realizing that he cannot purchase life insurance or save against the future, he asks you whether you will look after his child if anything should happen to him. You promise to do so. One afternoon, you receive a call informing you that the acquaintance has just died in an auto accident.

Did the making of such a promise create an obligation to look after the child? Most people would agree that the promise does create the obligation. But it is just possible that on the act-utilitarian view, there is no such obligation. The act-utilitarian holds, that is to say, that the promise is binding only if the fact of having made it affects the relative expected utilities of various courses of action. If it turns out that the fact of having made the promise does not affect the relative expected utilities, say, because no one knows about the promise, and hence will not be disappointed by failure to keep it, then the promise is not, according to act-utilitarianism, binding. Thus, the act-utilitarianism view collides with how we actually think about obligations. The act-utilitarian is faced with a number of such difficulties, most of which involve our beliefs about our obligations to our children, our parents, and our spouses, because, and only because of the special relation we have to these people. Moreover, we think that we have an obligation to seek the welfare of others if we have unjustly ignored them; it is not enough to merely read about their suffering in the newspaper.

What can the act-utilitarian say about such problems? He might say that there is no discrepancy between our actual obligation and the implications of act-utilitarianism when <u>all</u> the consequences of an action are taken into account. The act-utilitarian is likely to say "Don't forget the effect on your own character of looking after the child, and the increased tendency to take promises less seriously if you should decide not to follow through." But by this, the act-utilitarian surely does not mean that we should keep all our promises in an undiscriminating way: obviously, if we have promised to tell the truth and then find ourselves in a situation in which telling the truth would result in the deaths of innocents, we are not, even on the act-utilitarian view, obligated to keep our promise. What we must learn is a habit of keeping those promises that on the act-utilitarian principle we ought to keep.

The problems faced by act-utilitarianism do not end here. We do believe, in general, that if there were no economic or social impediments, we would want to distribute basic health care equally to everyone, and not arbitrarily certain favor certain individuals over others, even if playing favorites did not diminish the overall amount of good or benefit in the world. But on this issue, the act-utilitarian thesis says that any distribution of health care is as good as any other, provided merely that the total amount of benefit is the same for each of these distributions. Thus the act-utilitarian thesis does not describe the way we actually think about matters of health-care distribution.

Consider yet another example which poses a serious difficulty for act-utilitarianism. Suppose that a community is faced with a severe shortage of antibiotics during an epidemic of an otherwise fatal disease which can marginally be controlled by that drug. The citizens of the community are asked to use only the dose which health authorities believe will, on the whole, save most of the population. A certain wealthy citizen, however, is not convinced that this allocated dose for sure will save him. He approaches his weak-willed physician-friend with a bribe, and asks for twice the allocated dose, "for safety's sake." The request is reasonable, he thinks, because one dose more or less will not significantly affect the survival of the entire community, and he will certainly be more likely to live. Thus, he reasons, the maximum amount of good will be done if he appropriates the additional dose.

On the act-utilitarian view of matters, the wealthy man's reasoning is sound. But intuitively, we reject it. Why? Among other reasons, we believe that if a health risk has to be taken, it should be shared equally. Act-utilitarianism does not capture this feature of our beliefs and so is deficient in that respect.

And this example raises a further difficulty. Everyone in the community could argue that he deserves a double dose, and that his allocation will not have an effect on overall survival of the community. But if everybody followed this reasoning successfully, the community would indeed perish. Thus, the act-utilitarian thesis leads to a course of action that is to no one's benefit.

## Formalism

Act-utilitarianism clearly has limitations. At the very least, it does not correctly predict the way people think about obligations in certain situations, and this alone is a very severe defect of the theory. Because of this, some authors have urged that a more nearly correct theory of obligation will have to take into account certain exceptions to the act-utilitarian view. Whether an act is right or wrong, they say, is not determined solely by its relative net expected utility, although this may play a role in what is relevant to the rightness of an action. There are general principles, this view maintains, different from the act-utilitarian principle, which prescribe which acts are right or wrong. These rules have an unqualified claim on us, no matter what else may be true about our situation. One of the the most convincing formulations of this type of theory, called _formalism_, was propounded by W. D. Ross.[11]

Ross's theory attempts to address all obligations we might have, not just those which involve matters of health. Nevertheless, such matters, or their consequences, play a very direct role in his theory. The heart of the theory consists of a set of primary rules, which Ross claims describe _prima facie_ obligations--obligations which have claim on us no matter what. Among those obligations, Ross asserted, are the following. (1) There is an obligation to make reparation to people for any injury we may wrongfully have caused them. (2) There is an obligation to render services to others in return for services rendered us by them. (3) There is an obligation to assist in a distribution of happiness in accordance with merit. (4) There is an obligation to do whatever good we can for others, that is, to improve the virtue, intelligence, and pleasure of others. And, (5) there is an obligation not to injure other people.

On Ross's theory, our overall duty in a particular case is to act in accordance with these prima facie obligations, provide that they do not conflict. If they conflict, Ross maintained, one's obligation is to act in accordance with the "more stringent" prima facie obligation. This, in essence, is Ross's theory.

_Problems in formalism_. There are a number of difficulties which formalism faces.

First, we often find ourselves in circumstances in which Ross's rules conflict. Although the theory urges that we apply the "more stringent" of our prima facie obligations in the case of such conflicts, it does not define precisely what this stringency order is, nor it is not intuitively obvious what that order should be. Ross himself sometimes suggested that there were "secondary," higher-order rules which provided a resolution of conflicts, but he never provided a clear and complete statement of what these rules were, nor did he clearly justify the particular examples of such rules he gave.

Ross's theory has a second very disturbing feature: even when we happen to know, independently of the theory, what course of action would maximize the welfare of sentient beings, the theory sometimes does not prescribe what course of action we should take. Yet we in fact believe that what will maximize the welfare of sentient beings should have an important influence on what we decide our obligations should be. The theory thus fails to predict what we will believe in certain circumstances.

Third, the theory may just overprescribe our obligations. There is substantial anthropological evidence, for example, that in certain cultures the elderly are obligated not to live beyond the time when they can be useful to others. In some instances, they are left to die. Ross's theory, however, categorically implies that we must not injure others. Because of this, some authors have questioned whether Ross's theory, or any other theory which contains a categorical prescription against not injuring others if they have not injured us, can describe the way all cultures actually think about obligations.

Ross's system is thus incomplete in the sense that it is not possible to infer, from the principles he explicitly states, what our duty is in a particular situation, except when our prima facie obligations do not conflict, even when full factual information is available and when we know what action would maximize the welfare of sentient beings. Furthermore, it may not accord with the values held by all cultures, and hence cannot be taken as an accurate description of how people actually think about obligations.

## Rule-utilitarianism

The problems faced by act-utilitarianism and formalism have driven many writers to a compromise theory, which we shall call rule-utilitarianism. The central thesis of rule-utilitarianism can be stated as follows: It is obligatory overall for an agent to perform an act A if and only if the prescription that it be performed (i.e., "Do A!") follows logically from a complete description of the agent's situation conjoined with a set of ideal prescriptions for his community; these ideal prescriptions must be set of universal imperatives (of the form "Do A in circumstances C!") which (1) contain no proper names, (2) supply direction for any decision one may have to make (that is, the set of prescriptions is complete), (3) is as economical in distinct imperatives as completeness will permit, (4) is such that a conscientious effort to obey it, by everyone in the agent's community, would have greater net expected utility than similar effort to obey any other set of imperatives. For example, suppose we wish to determine whether a ceiling of 10 pCi/l for uranium in drinking water should be set. To make this determination, the rule-utilitarian thesis requires us to consider all complete, maximally parsimonious, universal sets of imperatives, universal conscientious effort to conform with which will have maximum net expected utility in our community. Given the description of our situation, and given these universal rules, we must determine whether the imperative "Do not allow the uranium alpha concentration in drinking water to exceed 10 pCi/l!" follows as a logical consequence. If it does, then our overall obligation is to set the ceiling at the indicated level; if not, then the ceiling is not an obligation.

There are several advantages which rule-utilitarianism has over formalism and act-utilitarianism.

First, it ties value-issues to the issue of general welfare, thus correctly predicting this part of what people believe about obligation.

Second, the implications of rule-utilitarianism are closer to reflective conclusions about our obligations than the implications of formalism and act-utilitarianism are.

Third, rule-utilitarianism leaves open the possibility that the system of prescriptions that defines whether an agent's act is right or wrong will differ from one agent to another, depending on the community to which he belongs. It also leaves open the possibility that one set of prescriptions will suffice for all cultures.

On the whole, then, the rule-utilitarian theory is simpler, more complete, and more nearly consistent with the way people actually think about obligations than are formalism and act-utilitarianism.

## JUSTICE IN THE DISTRIBUTION OF HEALTH RISKS

To this point, we have ignored a very important feature of the way we think about our obligations in matters of health: somehow, we believe "justice" should result from any principles of obligation which can demand our respect. And justice, it would seem, is intimately bound up with the notion of a "fair" distribution of benefits and risks. Most people, for example, believe that if health risks, such as those arising from the nuclear generation of electricity (or from coal-fired generating plants) must be incurred, the benefits and risks arising from such operations should somehow be justly distributed.

## What does "just" mean?

We thus need to get some idea of what the word "just" means. One might suppose that is an adjectival term because it appears in phrases like "It is wrong to distribute the risks of nuclear power in a unjust way." Such an example would suggest that "is wrong" implies "is unjust." But it is clear that we can act wrongly without acting unjustly: incest may be wrong, but it is not, strictly speaking, unjust.

Does "is unjust" imply "is wrong?" At first, it might not seem so. Consider, for example, the following. Suppose that a new drug for curing cancer comes on the market. The drug is very effective: a microgram ensures a cure in 99% of all cases. The initial cost of the drug is quite high because the manufacturer has spent hundreds of person-years developing and testing it. The first thousand patients to use the drug pay $10,000 per microgram. The manufacturer recovers its development costs far faster than it had imagined, and it decides to lower the price of the drug to a penny a microgram; all patients after the first thousand receive the drug at the vastly reduced price. We might say that there is a kind of injustice in this situation. We would not be particularly inclined, however, to say that in this case the manufacturer acted wrongly. Thus, the example seems to suggest that "is unjust" does not imply "is wrong."

The example is somewhat misleading, however. What we can say about it is that a "certain injustice" accompanied the manufacturer's actions, but he was not strictly speaking, unjust in engaging in that action. Given this, the example does not stand in the way of inferring that "is unjust" involves something like "is wrong."

## Distribution of health risks and benefits and the social order

Questions about the justness of a policy or action are often questions about the fair allocation of risks and benefits. Suppose, for example, we ask whether a nuclear generating facility or a uranium mine operator justified in increasing the probability that we and our children will suffer deleterious heritable genetic disorders, leukemia, and bone cancer? Ordinarily, after understanding the issue, we would probably say that the generating facility or mine operator is justified if everyone were at equal risk, and if some net benefit derived in spite of the enhanced exposure to ionizing radiation.

This is how we ordinarily answer such questions. It is not so clear, however, that this constitutes a complete answer. What the ordinary answer does not address is the question of whether the effects of ionizing radiation are just too treacherous, regardless of whether everyone is equally at risk. The latter question calls into account whether artificially increased exposure to ionizing radiation is just at all. Such questions cannot be answered by looking at the issue of comparative risk alone, but call into question the entire way in which we allocate risks and benefits. Properly understood, such a question could involve questions about the whether good health is intrinsically desirable, and whether existing institutional arrangements, such as the economic system and government, are overall just. Our aim here is to see what it would mean to try to answer such a question.

## The utilitarian's view

The utilitarian approach to answering questions of this sort has played a very large role in Western thought. A good case could be made, in fact, that much of modern health protection law derives directly from the utilitarian way of seeing things.

Can utilitarianism adequately answer the question of what a just distribution of health risks and benefits is? Since we argued above that rule-utilitarianism is most defensible of all the theories of obligation we have considered, a forceful way to attack this problem would be to see what rule-utilitarianism has to say about it. If rule-utilitarianism cannot handle the question, then presumably every other theory of obligation we have considered will be an even less defensible account.

What bearing, then, can rule-utilitarianism have on the distribution of health risks and benefits? There are a number of connections. First, the distribution of health risks and benefits is partly a matter of law, involving taxation, what is in the public interest, what the relation between individuals and/or states and a central government should be, who should own what property, and so on. Laws are enacted by humans. Thus, the utilitarian formula bears on them, because it is a formula stating the duty of any person in a position to enact laws: as a principle about right acts, it is a principal about right enactments. Second, rule-utilitarianism bears on those persons in a position to influence the distribution of health benefits and risks, even though these risks and benefits may not be a matter of law but are a matter of convention or custom only, for it is a principle about right acts, regardless of whether they are formally part of the law. Third, and not least, rule-utilitarianism bears on private citizens, because they must repeatedly make allocations of health benefits and risks in everyday life.

What, then, does the rule-utilitarian formula say about the proper principles for the distribution of health benefits and risks? In general, of course, it holds that we are to distribute according to those rules, a universal conscientious effort to follow which will maximize net expected utility (in the case in point, _health_ _benefits_). And maximizing net expected utility, rule-utilitarians claim, _is_ what we mean by a "just" distribution.

This formula, we should note, is in one way a mercilessly egalitarian one. It counts the utility or health benefit of everyone as equal. No exception is made. In another respect, however, the rule-utilitarian formula is not egalitarian. It does not recognize _per capita_ _equality_ of health benefits and/or risks as important as such. It allows, in principle, a grossly inequitable distribution of risk, provided merely that the _overall_ utility is maximized by a distribution. If, for reasons which may not now be fathomable to us, we should discover that the maximal expected net utility were to require that a small group of people live in excellent health at the price of the health of all others, rule-utilitarianism would hold that justice lies in bringing this state of affairs about.

Rule-utilitarianism does not prescribe specific rules for obtaining a just distribution of health benefits and risks; rather, it provides only a criterion by which we can judge whether a rule will produce a just distribution. A number of specific candidates for rules which will produce a just distribution in the sense that the rule-utilitarian proposes have been articulated. Perhaps the most important of these argues that an _equitable_ distribution of health benefits and risks will in fact produce maximal net expected net utility. If this is true, of course, one of the most serious objections to rule-utilitarianism can be met.

We need to look at this important claim in some detail. As it is typically articulated, the argument for it runs as follows. Let us suppose that we can distribute health benefits by allocating money, to be spent solely for health benefits, to individuals. Whether individuals actually are given this money is immaterial to the argument. By "health" benefits, we shall mean "net expected health benefits," on strict analogy with "net expected utility," as defined above. First, the argument maintains, after an individual spends some certain amount on health, the health benefit per monetary unit (dollar) declines. Second, the argument holds, we cannot compare the height of the allocation-health-benefit curves of different individuals, for a given amount allocated to health.

Suppose, then, that we had allocation-vs.-health benefit curves for persons A and B, for allocations ranging from $0 to $100. Let us suppose that both these graphs are plotted on the same set of vertical (health benefit) axes, but have the following difference in their horizontal (allocation) axes: from left to right, the values on the horizontal axis for A's graph increase; going right to left, the values on the horizontal axis for B increase. Let us superpose these two graphs so that the vertical axes of the graphs coincide, and so that the $0 point on A's horizontal axis coincides with the $100 point on B's horizontal axis, and conversely. Thus, according to the first premise of this argument, A's curve decreases, going left to right, while B's curve increases, going left to right.

Since we cannot compare the benefits derived for any specific allocation, we may choose the shape of these curves to be the same, that is, mirror images of one another as they appear on the graph. The area under each curve represents the total health benefit of the amount spent up to the cost designated by that coordinate. For example, the total health benefit of the first $60 of A's expenditures is represented by the area under A's curve up to the point on that curve directly above the $60-mark on the horizontal axis of A's graph.

Now suppose, this argument continues, that the curves for A and B accurately represent the allocation-vs.-health-benefit relations for the two people. The curves for the two people, presumably, will cross at some point, which we may suppose is off-center. This point represents the optimal distribution of costs allocated to A and B _jointly_, because if costs are distributed in this way, the area under the curves jointly will be maximized. If we divided equally, we would lose all the area between the two curves between, say, the $50-mark and the point of intersection. Now unfortunately, we don't know where these two curves intersect. And since we don't, if we try to distribute unequally, we shall make mistakes about half the time--that is, half the time we will correctly give extra to the person with the higher utility curve and hence gain, and half the time we will give extra to the person with the lower health-benefits curve, and thus lose. As a result, on the average we will lose health benefits if we depart from equal distributions.

An example will make this point clearer. Suppose we give an extra $10 to A, and $10 less to B. This may be a gain in total health benefit; suppose that it is a gain. But suppose--which is equally probable, given our ignorance about the point of intersection, we reverse the roles of A and B. Then we lose in total utility. We now make an observation. The amount lost is greater in the latter case than in the former. (To see this, just try a few graphs out on paper.) Thus, if we make one arrangement one week, and the other arrangement the next, the total health benefit is lower than that which we would have obtained by an equitable distribution in each of the two weeks. Because we do not, in general, know the relative positions of the curves of the two individuals, on the average any deviation from an equal distribution will work out at a loss, although we cannot say how large the loss will be. From this we may conclude, proponents of this argument hold, an equitable distribution of health benefits and risks will produce maximal net expected health benefits.

This argument can be generalized for any number of individuals or allocations; an equal distribution in health costs thus has maximal net expected utility. Therefore, rule-utilitarianism strongly supports the view that equality should be the rule guiding the distribution of health benefits.

Problems with the marginal-utility argument. What are we to make of this complex argument? First, the foregoing argument does not show precisely how much utility is probably lost by any given amount of inequality. The conclusion to which it commits us, therefore, is only that equality of health benefits and risks is one thing with good consequences, other things being equal.

Second, we must be careful not to underestimate the amount of health benefit that is lost by inequalities of income. It hardly makes sense, for example, to deny that there is a great difference between what an extra $100 will gain for a man who already has a private medical staff, and what it will gain for a poor man who has nothing to spare for medical services.

Third, it is sometimes possible to show, without highly subjective reasoning, that health benefit is increased by certain inequalities of distribution. For example, a person in pain will be more benefit from the purchase of drugs than can the average person from the same amount of money. So, in principle, we may admit that unequal distribution of health services may be justified on the basis of partially known utility curves.

Fourth, there is at least in some instances a further justification for an inequitable distribution of health risks. There may be efficiencies to be gained only by an inequitable distribution of risk. Quite clearly, for example, people living near a uranium mine are more likely to be exposed to radioactive materials than are people who live at great distances from the mine. Reducing the exposure to zero, it may be argued, results in gross diseconomy. If the risk is not large, one can argue that overall maximal utility demands an inequality of risk.

There is a much more serious objection to the proposal that an equitable allocation of health benefits produces maximal net expected health benefits, however. The "marginal utility" argument in fact rests on the assumption that we can adjust the shapes of the allocation-vs.-health-benefit curves arbitrarily, and at the same time find a definite (numerical) order in the areas under the curves; what the exact magnitude of the difference between the areas is, is of course of no consequence in the argument. Now it is possible to draw utility curves for which this assumption is just false. In certain circumstances, therefore, the equitable distribution argument rests on a contradiction, i.e., we must both be able to know something about the relative magnitude of the benefits curves, and at the same time, not know anything about those magnitudes.

Proponents of rule-utilitarianism have to date have had nothing to say about this problem. Nor have they provided any other defense of the the claim that an equitable allocation of money for health benefits yields a maximal net expected benefits. Until they do, we must regard the claim that an equitable allocation of money for health leads to maximal net expected health benefits as unsubstantiated.

In all likelihood, then, rule-utilitarianism needs to be augmented by some special justice-handling rules if it is to accord with the way in which people actually think about obligation. These exceptions probably will include consideration of equality of distribution, and of what we know about utility curves in special cases.

## SUMMARY

In this chapter, we have surveyed the complex value-related issues involved in radiation protection policy; on the whole, it appears that a rule-utilitarian approach augmented by a set of justice-determining considerations handles those issues better than any others we have examined.

In the next chapter, we examine how these considerations are reflected in U.S. water law.

NOTES

1. Throughout this chapter, my debt to Richard Brandt's *Ethical Theory* (R. B. Brandt, *Ethical Theory*, Prentice-Hall, 1959) is deep. Professor Brandt's work concerns the entire topic of value theory, of course, whereas this chapter is concerned primarily with health protection. My contribution here, if anything, has been to adapt Professor Brandt's erudite review to the theory of health policy. I do take issue with him in a few places, most notably concerning the strength of the "marginal utility" argument for an equitable distribution of goods and/or services. All of the examples in this chapter are mine. If some appear too macabre, however, I would be happy to credit them, too, to Professor Brandt.

2. E. Westermarck, *Ethical Relativity*, Harcourt, Brace and Company, 1932.

3. R. B. Perry, *Realms of Value*, Harvard University Press, 1954.

4. C. L. Stevenson, *Ethics and Language*, Yale University Press, 1944.

5. G. E. Moore, *Principia Ethica*, Cambridge University Press, 1903.

6. R. M. Hare, *The Language of Morals*, Clarendon Press, 1952.

7. Brandt, *op. cit.*

8. C. L. Hull, *Principles of Behavior*, Yale University Press, 1952.

9. S. Freud, *New Introductory Lectures on Psychoanalysis*, W. W. Norton, 1933.

10. W. Köhler, *The Place of Value in a World of Facts*, Liveright Publishing Company, 1938.

11. W. D. Ross, *The Right and the Good*, Clarendon Press, 1930.

# 6
# The Legal Milieu

## INTRODUCTION

The preceding chapters survey the complex scientific and value-related issues which have affected and will affect our understanding of natural radioactivity in water supplies. That understanding is embodied in equally complex federal law. This chapter is an introduction[1] to that law.

## EARLY FEDERAL WATER POLLUTION LAW

The development of American federal water law is rooted in American history and culture. To the colonists, America appeared a land of unlimited natural resources. Water and land were paradigms of this abundance: clear streams traversed seemingly boundless forests everywhere. Anyone who tried to suggest that the waters and land would not always be so would have been thought deprived of the testimony of his senses. To have urged that strong federal control of water use was necessary for the common good, moreover, would have invited immediate suspicion from the first Euro-Americans for other reasons: many of them had left the Old World to escape what they perceived to be capricious, unjustifiably powerful oligarchies. These two perceptions--that natural resources were inexhaustible and that strong central governments were suspect--persisted throughout the nineteenth and early twentieth century. From the outset, therefore, there was little to support, and much to discourage, the development of federal water-use control.

In the long run these attitudes produced such drastic consequences that the public could not afford to tolerate them. Agricultural runoff contaminated drinking water supplies throughout rural America.[2] Even the largest lakes rapidly eutrophied from residential and manufacturing waste.[3] And careless handling of human sewage spread death in the cities.[4]

Although awareness of the causes and consequences of water pollution grew throughout the first half of this century, federal water pollution law did not. Even as late as 1972, the legal mechanisms available for handling water pollution came largely from traditional property law, augmented only by the doctrine of nuisance, the public trust doctrine, and the doctrine of strict liability. Given their origins, it is not surprising that these mechanisms were ill-suited for water pollution management. In this section, we survey the distinctive limitations of that law.

## Ownership in traditional property law

Central to traditional property law is the concept of ownership of interest. Under and because of the common law there are potentially several interests in a parcel of land. These interests include license, easement, and leasehold. Ownership of interests can be divided among several people. The most inclusive interest in land is the fee simple absolute, an interest of nominally perpetual duration which covers any potential use. The only limitations imposed on a fee simple absolute use under the common law are prohibitions against creating a nuisance (see "Nuisance," below), prohibitions against extremely hazardous activities which create a high risk of harm to other persons or property (see "Strict liability," below), or those prohibitions imposed by duly authorized and enacted governmental regulation (e.g., zoning).

Under traditional property law, ownership of interests determines what use of the land is authorized and by whom. Only the lessee may possess the land, for example; only the easement holder may use the right-of-way. All others may be stopped from using the land. Furthermore, ownership of interest determines who can litigate interference with the use and enjoyment of the land.

Only if a person has an interest recognized as such under traditional property law does he have the standing to sue under that law. Private individuals usually do not have such an interest in publicly held lands or resources. This has made it especially difficult for private individuals to use traditional property law to litigate the public interest in pollution cases. Even when individuals can meet the standing-to-sue requirement, they may not be able to use traditional property law to stop a polluting activity for several reasons.

First, in traditional property law, standing to sue is determined by the specific interest of the interest owner, and not every interest confers standing for every other interest; nor does a given interest necessarily confer standing to sue in a fashion that provides for effective remedies. The owner of an easement, for example, has standing to sue under traditional property law only insofar as his right-of-way suffers interference. If he holds an easement on a segment of a river which is being polluted with cadmium, he cannot sue merely because he believes that the cadmium levels are harmful to human health, because this is irrelevant to his ability to exercise his right-of-way.

Second, the focus of common property law is the use and ownership of land. Land, at least as conceived by that law, is localized. That is, land has certain definite and closed boundaries, in terms of which common property law defines its scope. Water, unfortunately, is very unlike land in this respect: it knows no boundaries other than the planet itself.

For each of these reasons, traditional property law has proven a particularly awkward idiom in which to litigate the public interest in water pollution.

## Nuisance

A _private nuisance_ is the unreasonable and intentional interference with the use and enjoyment of real property. Nuisance requires that "substantial harm" result from the interference.

Nuisance law recognizes that some socially beneficial activities may cause some injury; more specifically, it holds that the harm to the individual must be "unreasonable" compared to the social benefit if the activity in question is to be actionable. The most clearly actionable activities under this doctrine are accordingly those which produce "unreasonable" harm to one person only for the private benefit of another. Unreasonableness is determined by the judge's or jury's findings of facts that weigh the seriousness of the harm to the individual against the social utility of the defendant's activities. Social utility must take into account the _purpose_ of the defendant's activities, including, for example, whether they are done for spite or for commercial reasons.

There are obviously formidable problems in using private nuisance to litigate the public interest in water pollution.

First, the person alleging harm must convince the judge and/or the jury that the harm done to him is unreasonable compared to the social benefit of the polluter's activity. Some cf the conceptual problems inherent in such a comparison are discussed in detail (see the sections on hedonism and utilitarianism) in

Chapter 5; whether there is even any general, consistent meaning which can be assigned to the notion of "comparative utility" is a thorny question.

Second, it is frequently difficult to impossible to show what the purpose of the defendant's activities is; trying to determine whether something is done out of spite, for example, involves determining matters of intention, and substantial practical doubt will always attend any such determination.

Third, successful use of nuisance law requires a plaintiff to demonstrate a tight connection between his case and others which have already been decided under the doctrine. There is obviously a substantial risk that a plaintiff will lose in any such undertaking, even if his case is only slightly different from others which have been decided favorably.

But perhaps the most serious problem which any attempt to litigate the public interest in water pollution through the doctrine of private nuisance will face stems from the very role which the doctrine was intended to play. It is and always has been intended as a mechanism for mitigating harm to an individual or small group of _individuals_ who claim to be harmed, who are willing to, _and_ who can (financially) pursue the litigation. As such, the doctrine tends to consider the _public_ interest only accidentally, that is, in so far as it happens to bear on the harm done to those who can afford to, and who will, pursue their claim in court. Indeed, the public's interest as such may be considered in nuisance litigation only to the extent that it affects the determination of the social utility of defendant's action.

In general, then, the doctrine of private nuisance is poorly suited to the task of litigating the public interest in water pollution.

_Public_ nuisance law proves even more poorly suited for litigating the public interest in water pollution cases than its private counterpart. A private party typically does not have the standing to sue to abate a public nuisance; this only a designated public legal official can do. That official is typically elected, and whether he is willing to risk bad publicity and the loss of support which might result from an unpopular pursuit of a public nuisance case may have little to do with the public interest in a given case.

Public nuisance law, therefore, has proven an unreliable mechanism for litigating the public interest in water pollution cases.

## _The public trust doctrine_

Under the public trust doctrine, public waterways and immediately adjacent lands cannot be used in any way which would be adverse to the public interest. The doctrine has been applied in situations involving commerce, navigation, or fishing, although there is no in-principle restriction of the application of the doctrine to these activities. Because it is a judicial doctrine, however, its use requires a case-by-case application and extension. Thus, the doctrine has been extremely slow to extend to water pollution applications other than those affecting commerce, navigation, or fishing.

Compared to other legal strategies, including the obviously cumbersome approach of developing new statutory remedies, use of the public trust doctrine has proven a weak device for litigating water pollution cases.

## Trespass

Trespass is the direct, intentional, and physical entry onto, or use of, another's land without the landowner's permission. The invasion must interfere with the owner's interest in the exclusive possession of the land.

Because it depends very narrowly on specific matters of ownership of interest, and on matters of physical invasion by persons, using the doctrine of trespass to litigate the public interest in water pollution proves difficult on several grounds. First, typical water pollution involves the entry of a chemical, microbial, or radiotoxic agent into water, not the entry of a person as such. Attempts to extend the trespass doctrine from entry of persons to entry of arbitrary objects or agents has not in general been successful. Second, the introduction of a pollutant into water typically involves an intervening force or agency such as weather, the flow and drainage of surface water not directly controlled by the polluter, and so on; these features tend to compromise the directness and intentionality of the invasion of the pollutant, and hence make it more difficult to build a trespass case against the polluter. And third, trespass requires the plaintiff to show a tight causal connection between the polluter's activity and the plaintiff's possessory interests. This means that if there are several potential sources of the same kind of pollution which could contribute to the problems experienced by the plaintiff, the plaintiff must establish the polluter's precise contribution to the whole. If a number of sources may be discharging the same pollutant into a waterway, this may prove to be a practically unsatisfiable requirement: literally dozens of industries along a river, for example, may discharge the same cleaning solvent into the water used by the plaintiff. Identifying which industry contributed what may be beyond the state of technology.

Even in those trespass cases in which relief is granted, the relief may allow only monetary damages to the injured individual, and hence will not necessarily stop the polluting activity.

On the whole, then, trespass is a particularly poor mechanism for litigating the public interest in water pollution problems.

## Strict liability

The doctrine of strict liability holds that no one may engage in activities which could result in a high risk of harm to other persons or their property.

In many water pollution cases, the pollutant (such as mining or agricultural runoff) is not considered to be extremely hazardous, nor is it perceived as involving an abnormal use of natural resources. To this extent, it has been difficult to handle many water pollution control cases through the doctrine of strict liability.

There are some notable exceptions to this, however. Common law strict liability has been changed significantly by statutes in some states. For example, in 1979 the Texas Court of Criminal Appeals upheld a criminal conviction for water pollution based on the state water code; the issue in the case was whether the polluter had to have knowledge of the pollution to be culpable.[5] The court held that ignorance of the pollution was no excuse.

## Constitutional issues

Early (pre-1972) federal water pollution law also generated at least two serious constitutional issues.

First, any regulatory action by the state or federal government which affects an individual landowner has the potential to be construed as the unjust taking of property or the taking of property without just compensation. For some purposes, such as national defense, a reasonably strong case can be made that the federal government should have the power to take property to protect the common interest. A regulation of the use of land (or possibly other resources) such as zoning is also considered a constitutionally valid limitation of personal interests and rights; as such, zoning losses are not a compensable loss under the Constitution, although the argument seems less forceful in this case. For yet other purposes, such as control of navigable waterways, that rationale is hardly plausible. In general, the courts have been justifiably reluctant to extend federal regulatory authority without clear legislative direction.

Second, any assumption of authority or jurisdiction by the federal government runs the risk of compromising the constitutional allocation of state and federal powers. This issue has generated substantial litigation, particularly in the area of surface mining regulation (see the "Federal Surface Mining Control and Reclamation Act," below).

The contention surrounding constitutional issues is not likely to resolved soon; to the extent that water pollution law involves this dispute, it will accordingly be weakened.

## Administrative mechanisms

Administrative agencies whose charter involved water use regulation proliferated throughout the 1930s; their spread is hardly diminished today. With the rise of these agencies, new complexities in litigating water pollution arose. Prior to the 1970s, hearings held by these and (other agencies) were often restricted to the parties immediately affected and were often held on little, if any, notice. Frequently, before an agency decision could be reviewed in the courts all administrative review and other appeals had to be exhausted. Exhaustion of such remedies was often time-consuming and expensive. And almost invariably, the polluter's action had to threaten imminent harm to the party seeking review, because federal courts would rarely render opinions on a matter of future or possible harm. Thus, review of an agency's action was frequently allowed only after actual harm was done.

Furthermore, prior to 1972 a party seeking review of an agency's action typically had to be someone with a sufficient stake in the outcome of the case so that the issues would be fully controverted. There had to be a clear, actual controversy between the parties for the courts to settle before review was allowed. In practice, this often meant that harm already had to be done to the claimant before the action could be reviewed.

These features of early (pre-1972) administrative water pollution law made using administrative mechanisms to handle water pollution problems particularly difficult.

In general, then, pre-1972 water pollution law reflected the interests and concerns of a much earlier period in American history. That law was, as a result, ill suited to handle the water quality regulation needs of a rapidly growing, highly industrialized country.

## THE CLEAN WATER ACT

The federal government was initially slow to respond to the need for a strong, consistent legal mechanism for controlling water pollution. But when it finally began to address the problem in 1972, Congress radically altered federal involvement in water use control. Technically, that legislation appeared as amendments to the Federal Water Pollution Control Act (FWPCA) of 1948.[6] In reality, the 1972 amendments wholly supplanted the 1948 act and all intervening amendments to it. Indeed, the 1972 amendments now are the Federal Water Pollution Control Act.[7]

In 1977 Congress again substantially amended the FWPCA.[8] In what follows, we will either identify the 1972 and 1977 amendments by year, or will refer to them collectively as the "Clean Water Act" (CWA).

### Scope of the CWA

The scope of the CWA is vast. Although the language of the act suggests that its application is limited to "navigable" waters, the definition of that term in the act is "waters of the United States."[9] In *United States v Holland*[10] the Supreme Court ruled that Congress intended this phrase to apply to all waters which could be considered under the commerce power, not merely those which are in fact navigable.

Now federal authority under the commerce power is great. Under this power, activities can be regulated because of their effect on interstate commerce. In particular, sources of pollution can be regulated if they are engaged in interstate business or if they put products in interstate commerce. Any activity which occurs wholly within one state but which can affect interstate waters is subject to the commerce power. Given water's propensity to respect no boundaries, it is difficult to imagine a use of any body of water, save possibly of farm ponds, playa lakes, or wetlands confined wholly to one state, which could not be reached by the federal commerce power.[11]

### Funding of publicly owned treatment works

The CWA establishes an extensive system of funding for the construction and operation, and hence practical control, of publicly owned treatment works (POTWs).[12] (POTWs by definition are wastewater treatment plants; typically they discharge into some body of water which would fall under the commerce power or which would satisfy the "navigability" test.) One of the principal objectives of this feature of the CWA is to provide a way of upgrading existing wastewater treatment systems to at least the secondary treatment[13] level. Under the act, funding can be obtained for up to 75 percent of the cost of construction and operation of a new plant. Existing plants can receive funding of up to 50 percent if they were built after 1966 according to CWA standards; plants built prior to 1966 can obtain a reimbursement of up to 30 percent of their costs, provided that they meet CWA standards. These features alone provide a strong incentive for a POTW to meet CWA standards; the CWA funding schedule may make it more cost-effective for operators of older (pre-1966) plants to construct new, advanced technology facilities rather than to try to upgrade older ones.[14] Moreover, the funding features of the CWA gives POTWs a substantial financial advantage over privately owned treatment works, and thus help consolidate federal control of waste water treatment.

The CWA attempts, by means other than funding support, to encourage POTWs to adopt and develop new technology. The administrator of the act is charged with promoting management practices which integrate waste water treatment as such with recycling or other technology. Planning of the waste treatment facilities, for example, must take into consideration the development of open space and recreational needs. In fact, if the technology proposed in a POTW's application is deemed particularly comprehensive in this sense by the act's administrator, the POTW may be eligible for as much as 85 percent federal support.[15]

The CWA tightly couples funding of POTWs to state-wide water pollution control planning processes, which are also prescribed by the act. For a POTW to qualify for funding under the act, in particular, the state in which the POTW is located must have adopted a water quality control planning process which includes meeting the standards of the CWA; the POTW must be able to achieve the best practicable waste treatment under this plan (the plan and attainment standards are discussed below).[16]

The act's administrator must make certain determinations before approving any grant to a POTW under the CWA. First, he must determine that the POTW conforms to any applicable state water quality plan under the act. Second, he must determine that the POTW is regulated by an area-wide water quality management plan (discussed below) defined by the act. Third, the administrator must determine that the POTW has priority construction status under the state plan. Fourth, the POTW applicant must show the administrator before construction begins, that the plant can be operated efficiently with trained and competent personnel. Fifth, the administrator must determine that the capacity of the POTW can meet the needs of the area it will serve, and that it will have sufficient reserve capacity. And sixth, the administrator must ascertain that the specification for bids for the POTW do not contain any proprietary, exclusionary, or discriminatory standards other than ones based on performance.[17]

The CWA requires that the non-federally-funded portion of the cost of building and operating the POTW is to be paid for by its users. Under the act, the POTW must adopt a system of charges from the recipients of its services to pay for this non-federally-funded portion. If a POTW fails to institute such a system of charges, its funding may be denied or revoked.[18]

## Area-wide planning process

A major feature of the CWA is the area-wide water quality control planning process prescribed by the act.[19] This planning process, commonly known as the "Section 208" or "area-wide" planning process, is aimed at effectively constraining both point and non-point sources of pollution. In principle, the act can be astoundingly thorough in meeting this aim. Under Section 208, the administrator is required to identify those areas in a state which have substantial water quality control problems. Section 208 requires the governor of each state to designate a single, representative organization capable of developing an effective area-wide waste treatment plan for these problem area(s); the organization so designated should include elected officials. If the governor does not designate such an organization, local government units (at the county, district, or municipal level) may do so. Under Section 208, the state is also to act as the planning agency for any part of the state not designated as a problem area. All such designations and plans must be approved by the EPA. The EPA can reject any area-wide plan and the designation of any area-wide management agency if the designated agency lacks the authority, or merely fails, to implement the plan developed under Section 208. If such a failure occurs, the act empowers the EPA to directly create and administer the entire water quality control program for the state.[20] Exercise of this power is by no means a mere abstract possibility: the EPA has in fact for brief periods forcefully intervened when states have failed to meet Section 208 requirements.

The plan developed by the area-wide planning agency, furthermore, must identify what wastewater treatment works will be needed to handle the waste that will be generated in the area covered by the plan during a twenty-year period, and the plan must be updated annually. The plan has to project future needs arising from such diverse concerns as land acquisition and storm runoff systems, and must address financial contingencies. It must establish construction priorities and schedules for all treatment works in the area. It must also establish a program for regulating the location, modification, and construction of any POTW in the area that may result in a discharge into any waters in the area.[21]

A Section 208 plan must include a process to identify and regulate pollution from agriculture, silviculture, construction activity, mining, and salt water intrusion from any source which might arise from reduction of fresh water flows. Additionally, the plan must include processes meant to dispose of any waste generated in the area which might affect ground and surface water quality. Since it is difficult to imagine a human activity which would not potentially affect ground and surface water quality, the scope of activities which would be covered by an area-wide plan and which could be regulated by the designated area-wide management agency or the EPA is in principle nearly universal.

## The point/nonpoint source distinction

At the heart of the effluent limitations imposed by the CWA lies a distinction between point and nonpoint sources of pollution: effluent limitations are imposed on point sources that discharge pollutants into "navigable" waters, but these limitations do not apply to nonpoint sources. Point sources are defined by the CWA to be "any discernible, confined and discrete conveyance...," and are regulated by Section 1311 of the act.[22] Point sources include POTWs and most industrial sources. (Non-point sources include certain, but not all, agriculture- and silviculture- related sources.)

Because it plays such a fundamental role in the CWA, the point/nonpoint distinction has been the subject of substantial legal contention. Typically at issue is whether such activities as construction, mining, agriculture, and silviculture are nonpoint sources in certain circumstances. The courts have tended to construe the notion of a point source very broadly, and hence have extended the scope of the CWA through their interpretation of this distinction. In Consolidation Coal Co v Costle,[23] for example, the Fourth Circuit ruled that discharges from pumped, siphoned, or drained coal or refuse storage areas are point discharges. This interpretation was broadened by the Tenth Circuit, which ruled in United States v Earth Sciences, Inc[24] that overflow from a confined, recirculating wastewater system caused by a rapid snowmelt (which could not be handled by the recirculating capacity of the system) was a point source discharge; Earth Sciences was accordingly held accountable. The concept of a point source was further extended in United States v Oxford Royal Mushroom Products, Inc,[25] in which the court ruled that the accidental runoff of wastewater sprayed onto fields and intended to be wholly absorbed by those fields was a point source discharge.

## Level of effluent control required

The CWA requires that the EPA define effluent limitations for industrial point sources. The act suggests that this be done for classes of, not individual, point sources. Each member of such a class must then use the best practicable control technology currently available (BPT) for all "non-toxic" and "conventional" pollutants (these pollutants are defined by EPA convention) by 1 July 1977 for that class.[26]

The EPA determines what is meant by "BPT;" there is no challenge which can be brought to the agency's interpretation of that term on the grounds that it does not accord with some intuitive or pre-established meaning. Nevertheless, the EPA must attempt to be reasonable in setting BPT attainment standards, specifically taking into account a number of factors such as what consumers are willing to pay for the goods and services whose prices would be affected by the standards, what financing methods are typically open to members of a point-source discharge class, what kind of government involvement would be required in the administration and enforcement of these standards, and what the tradeoffs between these costs and public health should be. All these decisions are typically made under considerable uncertainty about the effects of any standards on any of these matters. Whether the CWA requires the EPA to make such a determination is beyond doubt. Whether these tradeoffs can be decided wisely or even consistently by anyone is a very difficult question (see Chapters 3 through 5 for a discussion of the problems inherent in such a determination).

If an industrial source discharges into a POTW, the 1972 amendments also require that the source satisfy applicable toxic and pretreatment requirements.

The CWA effluent standards for POTWs are nominally less rigorous than those for industrial point sources. At the least, the act requires that POTWs meet secondary treatment standards by 1 July 1977, or after that date, any stricter limitation required, such as water quality standards, treatment standards, or schedules of compliance under state or federal law. By 1 July 1988, POTWs must meet the requirements of the Section 208 plan under which they operate, unless the POTW is operating under a compliance schedule in effect on 29 December 1981; in that case, the Section 208 plan requirements must be met by 1 July 1983, unless delays of financial assistance under the CWA or other conditions beyond the control of the owner or operator make compliance by that date impossible. POTWs must apply the best practicable waste treatment technology over the life of the facility.[27]

The 1972 amendments also require that point source pollution control technology improve beyond BPT. Toward this end, the act requires the EPA to again classify point-sources into categories. Discharge standards are then to be defined by the EPA for these sources on a category basis (rather than on a source-by-source basis). Each member of a category of point sources is then required to apply the best available technology economically achievable (BAT) for all pollutants designated.[28] BAT, like BPT, standards are to be determined by the EPA. As defined, BAT standards can require the elimination of discharges if the administrator finds a standard technologically and economically achievable for the category or class of which a given point source is a member. Thus, BAT standards are nominally stricter than BPT.

The cost-effectiveness of moving from BPT to BAT attainment standards, as required by the 1972 amendments, was of some concern to Congress. In 1977, therefore, Congress amended the BAT attainment requirements for all nontoxic pollutants, to the later of (1) three years after the standard is set by the administrator for (BAT) or (2) 1 July 1984; in any case, the attainment date was set by the 1977 amendments at no later than 1 July 1987. In addition, the 1977 amendments created a new classification of pollutants and a new effluent limitation standard. This new class of pollutant was called conventional.[29] Roughly speaking, conventional pollutants are those most like to be encountered by POTWs and for which reasonably economical control technology is available. The 1977 amendments designated, for example, biological oxygen demand, suspended solids, fecal coliform, and pH as conventional pollutants. The 1977 amendments also directed the administrator to identify other conventional pollutants in terms of their chemical and biological characteristics. The CWA administrator is to prepare and regularly revise this list of conventional pollutants. For classes of point sources which emit conventional pollutants, the administrator is to establish a best conventional pollutant control technology (BCT) requirement by 1 July 1984.[30]

Like BPT and BAT, BCT standards are set by the EPA. And again, the BCT standards are set for classes of, not individual, point sources. In setting BCT requirements for industrial point sources, the CWA requires the administrator to perform a cost/benefit analysis which compares the costs to industry of meeting a given BCT standard to the benefits that standard achieves. For POTWs, the administrator is required to perform a cost/benefit analysis which compares the cost/benefit tradeoff for conventional pollutants discharged to that attained by industrial sources; in making this comparison, the administrator is directed to consider the age of the POTW equipment involved, the treatment process employed, the engineering aspects of the application of various types of control techniques, process changes, non-water quality environmental impact factors such as energy demands, and other factors the administrator deems appropriate.

With the 1977 amendments, BCT thus replaced BAT as the standard for conventional pollutants; BAT remains the standard for toxic and nonconventional (including radioactive) pollutants.

## Development of standards

Section 1314[31] of the CWA requires the EPA to conduct studies on the effects of pollutants on water quality. This data must be gathered in consultation with state authorities and other interested persons. These studies and data must be used to establish measures needed to protect water quality. They should also specify the the conditions needed for the protection and propagation of fish, shellfish, and wildlife.

Under Section 1314, furthermore, the EPA must issue effluent limitation guidelines. These guidelines are not the technical standards themselves, but are decision-making considerations to be used to define how effluent standards will be set and modified. The guidelines should identify the technology available to control pollutants for point sources classes other than POTWs. The guidelines should specify the factors which states can consider when determining control measures. Those factors have to include the cost of the technology for achieving effluent reduction, the benefits of the reduction, the age of the equipment and the source, the process used, and energy requirements for the process. Furthermore, the guidelines must identify the control measures that can eliminate discharges of pollutants and effluent reduction attainable by application of the BCT.[32]

The CWA also authorizes the EPA to set new performance standards for categories of point sources. These new standards must identify the maximum effluent reduction achievable through use of the "best available demonstrated control technology, processes, or operating methods, or other alternatives, including, where practicable, a standard permitting no discharge of pollutants."[33] The administrator is required to publish a list of categories of sources for which the new source standard will be developed and promulgated.

The CWA further prescribes toxic effluent standards. A list of prohibited toxic pollutants is to be revised and updated on a regular basis. Toxic pollutants are subject to the BAT standards.

The administrator must similarly issue pretreatment standards for the introduction of pollutants into the POTWs. These standards have to apply to pollutants that are not susceptible to treatment by the treatment works, or to pollutants which would interfere with POTW operation. An industrial point source using a POTW must comply with these pretreatment requirements.[34]

Effluent limitation lists must be revised at least every five years. The revised limitations must include a general prohibition against discharging any radiological agent.[35]

## Variances

Section 1311(c) of the act allows the administrator to change the compliance schedule for industrial point sources in certain well defined conditions.[36] Such changes require the applicant for the variance to show that the altered standard requires it to use the best technology within its economic capability. The source must also show that the variance will still allow reasonable progress toward the elimination of the discharge of pollutants.

Section 1311(g)[37] defines a second mechanism for allowing an industrial point source to obtain a waiver from published standards. The administrator, with the concurrence of the state, can modify a standard for individual pollutants other than toxic, conventional, or thermal components of a discharge. The modified standard must require the use of the more restrictive of BPT or another applicable water quality standard. The alteration of a standard under this Section may not impose a stricter standard on another source. Furthermore, the modified standard cannot endanger public drinking water supplies, create an unreasonable risk to public health, nor endanger recreational uses of the water supplies such as fishing.[38] Compared to 1311(c) variances, this mechanism provides little relief for most point sources. It does, however, alleviate some administrative overhead if a source can show that it already complies with a standard which is stricter than those imposed by the CWA.

Section 1311(h)[39] empowers the administrator to allow similar variances for POTWs. To obtain a 1311(h) variance, a POTW applicant must show that there is a well defined water quality standard which it will meet for the specific pollutant. No modification of a standard under this section can interfere with the attainment or maintenance of water quality for public water supplies and recreational purposes. If granted such a variance, the POTW must agree to continuously monitor water quality to determine the effect of its discharge on the aquatic biota. The applicant must also show that the modified standard will not impose additional requirements on any other point or nonpoint source and that all applicable pretreatment requirements will be enforced for all sources. In addition, the 1311(h) variance applicant must adopt, as far as is practicable, a schedule of activities that will eliminate discharges of toxic pollutants from nonindustrial sources into the treatment works. This modified standard must not result in new or substantially increased discharges beyond that allowed in the permit from the point source for the pollutant being modified. If granted a 1311(h) variance, the POTW must agree to use available Section 1281[40] funds to comply with effluent reductions required under that Section or under Section 1311.[41]

The question of exactly who can issue what variances to what standards under the CWA was, not surprisingly, a matter of contention almost from the passage of the act. At the outset, the EPA took the position that the CWA did not allow the cost to individual point sources of achieving the standards it set, nor did it allow the mechanisms of Section 1311(c) as decision-making considerations in determining a variance from BPT attainment requirements. In Appalachian Power Co v Train,[42] however, the Fourth Circuit rejected this stand, ruling that the agency must consider total cost to the individual point source under Section 1314(b)(1)(B)[43] as well take into account Section 1311(c) variance considerations.[44]

The Fourth Circuit's opinion in Appalachian Power appeared to set rather severe limits on the EPA's authority to establish BPT as the minimum attainment standard to which industry must conform. In 1973 before the Fourth Circuit, the National Crushed Stone Association further challenged the very way in which EPA established BPT standards;[45] the court reviewed this contention under the "arbitrary and capricious" doctrine. The court attempted to determine whether the specific BPT standards which the EPA had set for the cement and crushed stone industry were adequately justified; to help answer this question, the court examined whether an earlier case, Portland Cement Association v Ruckelshaus,[46] showed that the EPA had provided for its standards. The court concluded that the EPA had failed to adequately explain its reasons for the specific BPT requirements it had established, and remanded the total suspended solids limitations, the recycling provisions, and the no-discharge provisions of the BPT standards for the crushed stone industry to the EPA for further consideration. The court then invalidated the variance provisions in the (Section 1314) guidelines EPA had established.

The Fourth Circuit's view of the limitations of EPA's power to set BPT standards stood for nearly seven years. In December 1980 the United States Supreme Court reviewed the Fourth Circuit's National Crushed Stone Association[47] decision. The Court upheld EPA's variance guidelines, which took into account variances only by classes of sources, specifically disregarding the economic capability of an individual source to meet BPT standards. The Supreme Court further concluded that the Fourth Circuit was wrong to require Section 1311(c) to be applied to BPT variances, because Section 1311(c) allows, the Court concluded, only variances from the 1987 BAT requirements.

The Supreme Court's decision in this case merits careful inspection, because it demonstrates just how pervasive the Court perceived federal involvement in water pollution control should be under the CWA.

First, the Court noted that the general provisions and language of BAT and 1311(c) are very similar. 1311(c) requires use of technology within the owner's economic capability and requires reasonable further progress toward reduction in discharges, whereas BAT requires that a point source use the best available technology economically achievable which results in reasonable further progress toward reduction of pollutants. Thus, by analogy, the Court reasoned, 1311(c) variances refer to BAT, not BPT standards.

Second, the Court argued, the phrase "reasonable further progress" in 1311(c) implies the existence of a prior standard. For BAT, that prior standard is BPT. There is no federal standard prior to BPT. Thus, if the language of 1311(c) is even sensible, the Court argued, it must refer to BAT.

Third, the Court noted, BPT does not talk about maximum financial commitment to pollution control, even if it is affordable; 1311(c) does contain such language. Thus 1311(c) must refer to something other than BPT. The only available federally defined candidate is BAT.

Fourth, the Court concluded that legislative history strongly suggests that 1311(c) was intended to refer to variances from BAT, not BPT.

Fifth, and perhaps most importantly, the Court held that to allow individual economic capability to play a role in BPT variances would compromise the technology-forcing aspects of BPT. Specifically, the Court noted that BPT is intended to upgrade the poorest performing elements in an industry by requiring them to achieve the average of the best existing performance within the industry or cease operation. The effect of this over the long run, if there is any improvement in the best pollution control technology economically achievable in a class of point sources, is to force each member of the point-source class regulated to improve its control technology.

The Court accordingly concluded that the EPA had been correct in assuming that 1311(c) did not apply to variances from BPT standards: the Fourth Circuit ruling was reversed. This review thus firmly established BPT as the baseline from which, and BAT as the goal toward which, industrial point sources must move.

## The role of the states in CWA implementation

The authors of the CWA were aware that the act would be effective only if the states were allowed, and strongly encouraged, to play a major role in its implementation. Toward this end, the CWA contains a number of features to enhance state participation.

First, the CWA gives the states the option of retaining water quality standards which were in effect just prior to the 1972 amendments, provided those standards were as stringent, or more stringent than the standards of the 1972 amendments.[48] Although this feature of the amendments appears to offer the states little advantage, it effectively waives a rather serious administrative burden for any state which already has strong enough standards in place. While the bill which became the 1972 amendments was in Congress, furthermore, it served notice to the states that they had a definite schedule for establishing local water quality standards and administrative mechanisms.

If the intrastate standards maintained by a state just prior to the passage of the amendments were not as stringent as those established by the amendments, the state was given six months to bring its standards in line with the ones established by the amendments. If the state's interstate standards were not as stringent as those established by the amendments, the state had 90 days to adopt national standards.

In any case, the EPA approval was required for any state standard which was in place just prior to the 1972 amendments. If a state failed to adopt changes if recommended, the administrator was required to impose these changes on the state.[49]

Under the CWA the governor or the state water quality board is required to review the state's water quality standards at least every three years. Any change to the standards which the state proposes must be submitted to the act's administrator for approval. The standards must indicate the designated uses of the waters involved and give the criteria for those uses and for the application of the standards. The standards must protect the public health and welfare and enhance the quality of the water.[50]

The states are also encouraged to play a principal role in the administration and the enforcement of the formal discharge permit mechanism prescribed by the CWA. The mechanism for discharge permit issuance is technically known as the National Pollutant Discharge Elimination System, or "NPDES" for short.[51] A permit issued under this mechanism is accordingly known as an "NPDES permit." NPDES permits may, and typically do, contain conditions. These conditions may include a requirement to satisfy all applicable standards under the act or any other condition that the administrator deems necessary to fulfill statutory requirements. Among these conditions, the administrator may include requirements for monitoring, reporting, data collecting, or discharge control technology.[52]

The NPDES permit system forms the foundation of point-source control under the act. The CWA requires any point source which discharges into "navigable" waters to obtain a certificate from the state in which the discharge will occur; failure to obtain the certification can result in stiff federal civil and/or criminal penalties. The applicant must provide the state certificate to the federal licensing or permitting authority before a license or permit can be issued. The certificate from the state or interstate water pollution control agency must assure that the applicant's discharge will comply with all applicable effluent limitations, water quality, and other standards of the CWA. Any state-issued NPDES permit is subject to EPA veto.[53] Section 1341[54] of the CWA forces a resolution on any NPDES permit application: if a state does not act on a request for certification within one year, the state certification requirement is deemed waived, and the federal license or permit may (but need not) be issued.

The licensing or permitting agency must notify the EPA when it receives an application and/or certification. If the discharge could affect waters in another state, the EPA must notify that state. The affected state must then inform EPA as to whether its water quality standards would be violated. If they would, the EPA must hold a public hearing to determine appropriate handling of the application.

## Enforcement

Although the CWA is a federal statute, and hence nominally to be enforced by the judicial branch, the EPA itself is authorized to enforce the CWA in certain circumstances. If a violation of the conditions of an NPDES permit occurs, or if there is a violation of pretreatment standards, the EPA can do one of two things. First, it can notify the violator and the state of the violation. If within 30 days the violation is not remedied, the EPA can issue an order requiring compliance with the applicable limitation or condition. Second, the EPA may bring a civil action for appropriate relief to remedy the violation. The suit may be brought in the federal district court in which the violator is located, resides, or is doing business. Civil penalties may include fines of up to $10,000 per day of violation.[55]

Similar options are available to the EPA if violations are widespread throughout a state. If widespread violations occur in a state, the federal government may assume enforcement of the act in that state.[56]

Criminal penalties are also authorized under the act. Willful or negligent violation of effluent limitations, standards, or other requirements such as permit conditions can be punished with fines of $2,500 to $25,000 per day of violation, imprisonment, or both.[57]

## Citizen suit and judicial review provisions

Pre-CWA federal water law typically did not give private citizens standing to sue. Even when the standing-to-sue requirement could be met, the costs of litigation were often prohibitive. Moreover, the grounds on which the standing- to-sue requirement could be met often precluded the possibility of effectively controlling the polluter's activity: activities were often actionable for reasons having little to do with the damage, potential or real, being done. The CWA directly addresses these problems; Section 1365[58] allows any citizen to bring a civil action against any person or governmental unit that is violating an effluent limitation, standard, or an order implementing the limitation or standard under the act. The Section also authorizes suit against the administrator for failing to perform a mandatory duty under the act. Section 1365 suits must be brought in the federal district court where the violation occurs. The court is authorized to award litigation costs, including attorney and expert witness fees, to any party as a part of its decision.

Section 1369[59] further authorizes any citizen to petition in the appropriate court of appeals for a review of certain EPA activities under the act. Effluent limitations, treatment and toxic standards, and other regulations can be challenged under this Section. In addition, EPA approval of a state NPDES program or denial of a state NPDES permit is subject to judicial review under Section 1369. Costs including expert witness and attorney fees may be awarded to any party by the court.

## Relation of the CWA to other federal water law

A fundamental question arises when any federal act is passed: If the law in existence before the act was passed and the act itself both address the same causes of action and their jurisdiction, what is the status of the causes and jurisdiction when both old and new law coexist? Ohio v Wyandotte Chemical Corporation[60] is an important case in the evolution of this relationship in federal water law. In Wyandotte, the state of Ohio sought to invoke the original jurisdiction of the United States Supreme Court to abate a nuisance: an Ohio and Canadian corporation was discharging mercury in Lake Erie. The Court invoked the abstention doctrine and elected not to assume jurisdiction over the cause of action, although it in fact had subject matter and personal jurisdiction over the parties. In making this ruling, the Court noted that Ohio law would probably govern the dispute whether it were to be decided by a state or by a federal court. It further argued that Ohio would have no difficulty obtaining jurisdiction over the Canadian defendants. The Court concluded that since a mechanism for handling this case existed without the Court's involvement, the interests of original jurisdiction were best served by holding that an Ohio state court was the appropriate forum for the dispute.

By itself, this conclusion showed that the Court intended to take a constructionist approach to original jurisdiction--not a particularly significant move. But in a surprising footnote[61] to Wyandotte, the Court remarked that concurrent state and federal jurisdiction in the case did not exist, because on the facts of the case, neither diversity nor federal question jurisdiction was available. Since this case was about as clear a candidate for the application of federal common law to water pollution as one could hope to encounter, the footnote appeared to sharply circumscribe the role of federal common law in water pollution cases.

Wyandotte was decided before the 1972 amendments were passed. In the Supreme Court's eyes, those amendments drastically changed the status of at least the federal jurisdiction question: in 1972 in Illinois v Milwaukee,[62] the Court put aside the Wyandotte footnote. In Milwaukee, Illinois sought to bring an original action in the Supreme Court against the City of Milwaukee for creating a public nuisance: the state alleged that Milwaukee was discharging raw or inadequately treated sewage or other waste material into Lake Michigan. Milwaukee contended that a suit against a municipality in a state was a suit against the state (of Wisconsin) itself, and hence a state, not federal, court had exclusive jurisdiction. The Supreme Court rejected that contention, holding that state courts had concurrent jurisdiction with federal district courts. The Court further held that a federal jurisdiction question under Section 1331[63] of the CWA would be appropriate in a federal district court. (The case ended up in the Seventh Circuit; see below for further twists in the plot.) The Court concluded that pollution of interstate or navigable waters is a cause of action arising both under the laws (apart from the CWA) of the United States and under Section 1331 of the CWA per se. On this interpretation, then, 1331 establishes causes of action based on either statutory or common law. The concept underlying this ruling became known as the "Illinois doctrine."

At first, the Illinois doctrine appeared to have very broad implications: it seemed that the doctrine could be used to establish federal jurisdiction and cause of action for very nearly any water pollution case. This appearance was short-lived: Illinois strategies, the courts soon decided, are not available without qualification. In Massachusetts v United States Veterans Administration,[64] for example, Massachusetts sought to obtain federal jurisdiction under Section 1331 of the CWA after the courts denied the state's complaint federal jurisdiction under Section 505[65] of that act: the state contended that the facts alleged in the complaint were sufficient to support a nuisance cause of action under the Illinois doctrine. The First Circuit questioned the state's claim that federal jurisdiction existed for its claim under 1331, however, on several grounds. First, the court noted that there was conflict among the circuits concerning whether nuisances based on alleged violations of the CWA are exclusively reviewable under that act or could be reviewed under federal question jurisdiction. Second, the court questioned whether Illinois was sufficiently like the case brought by Massachusetts: whereas Illinois involved a dispute between two states (at least), the complaint brought by Massachusetts involved acts occurring entirely within one state. Furthermore, the court suggested, the CWA might have preempted part of the federal common law of nuisance.

Although it noted these problems in applying the Illinois doctrine, the First Circuit did not fundamentally base its decision to bar the suit on these grounds. Instead, it invoked the doctrine of sovereign immunity: because the suit was against an agency of the federal government and sought monetary penalties against that agency, the action could not be maintained without the consent of Congress. Massachusetts argued that this consent was implicit in the "citizen suit" clauses of the CWA (Section 1365), but the First Circuit rejected this claim, arguing that the clause should not be as expansively construed as Massachusetts urged.

The suggestion in U.S. Veterans Administration that a federal common law action based on the Illinois doctrine had to involve interstate pollution was furthered in Board of Supervisors v United States.[66] The plaintiffs claimed that waste and coal dust runoff from a prison complex resulted in a nuisance from water pollution. The district court held that for the pollution action to stand independent of violations of any federal statutes, it must allege that the pollution is interstate. The plaintiffs were accordingly allowed to amend their suit to include the interstate nature of the pollution.

At first the courts tended to hold that the Illinois doctrine could be used only for public, rather than private causes of action. In Committee for Jones Falls Sewage System v Train,[67] for example, the Fourth Circuit rejected an effort by private individuals to abate an alleged nuisance. The court explicitly ruled that "...the doctrine of Illinois v Milwaukee has not been extended beyond the abatement of public nuisances in interstate controversies where the complainant is a state and the offenders are creating extra-territorial harm."[68]

The courts subsequently allowed private parties to bring Illinois doctrine actions even though a Section 1365 action was time-barred. For example, in National Sea Clammers Association v City of New York,[69] the Third Circuit ruled that Section 1365(e)[70] saves for private parties causes of action independent of the ones created by Section 1365. The court held that these saved causes of action must arise under general federal question jurisdiction. In this case, the court also ruled that a cause of action existed under Section 1331 for the plaintiff. In addition, the court determined that federal common law action in nuisance existed for the plaintiff because the alleged harm involved pollution of interstate waters; the fact that the pollution was in interstate waters, the court reasoned, creates an overriding federal interest in a uniform rule or decision. The court also ruled that private individuals could bring an action to abate the public nuisance under the federal common law; private litigants, however, must still show special individual harm different from the harm experienced by the public in general.

The courts have also held that a state may sue its own residents under federal common law nuisance action established or sustained by the CWA. For example, in Illinois v Outboard Marine Corporation,[71] the Seventh Circuit allowed a state to sue its own residents under the federal common law nuisance action: the company was discharging highly toxic polychlorinated biphenyls (PCBs) into a tributary of Lake Michigan. The court held the state had the statutory right to sue under Section 1365, provided that the state's special interest could not be adequately represented by the federal government.

The case which began the definition of the relation between the CWA and federal common water law continued to evolve. In 1980, eight years after the original suit was filed, the Seventh Circuit issued its opinion in Illinois v Milwaukee.[72] The court concluded that Illinois did establish that Milwaukee was discharging untreated sewage into Lake Michigan, which constituted a cause of action in nuisance under the federal common law. The court further concluded that the CWA did not preempt federal common law actions for nuisance. Furthermore, the court concluded that compliance with effluent limitations or permit provisions of the CWA is not a defense to the federal common law action for nuisance. The common law action, it held, is not based on the statute. Thus, if a nuisance is created, a court may take whatever remedial steps it believes are necessary to alleviate or abate the situation. In addition, the court concluded that the original trial court (which first heard Illinois) was not limited to the law of Illinois in deciding the dispute, and, more specifically, that in interstate water pollution cases the trial court is not limited to the law of the complaining state in deciding relief to be granted.

In the subsequent appeal of Illinois, the appellate court concluded that the effluent limitations prescribed by the state and sought by Illinois could not be sustained without more evidence showing that if those standards (more stringent than those set by EPA) were not met, the results would be injurious to the health of the residents of Illinois.[73]

The Seventh Circuit's view of Illinois did not stand long. In 1981 in City of Milwaukee v Illinois,[74] the Supreme Court held that the CWA indeed preempted the federal common law action in nuisance; further, the Court concluded that the Seventh Circuit had erred in holding that standards more stringent than those of the CWA could be imposed. In part, the Court reached this decision by arguing that Congress had intended the CWA to supplant existing federal common law with a uniform set of statutory regulations.

The extent to which the CWA preempts state and federal common law has been addressed in subsequent cases. In Illinois v Outboard Marine Corporation,[75] on remand from the Supreme Court, the Seventh Circuit concluded that federal common law action of nuisance was preempted by City of Milwaukee. However, in Scott v City of Hammond,[76] the court ruled that the CWA does not preempt a cause of action for nuisance based on state law against out-of-state dischargers.

On the whole, then, the CWA is comprehensive and controversial. Fundamental interpretation issues, including matters of jurisdiction, permissible causes of action, and reasonableness of standards remain before the courts. Because of this, the CWA will continue to dramatically affect the evolution of federal water quality law.

## THE SAFE DRINKING WATER ACT

In 1974 Congress enacted a program designed to set standards for acceptable levels of contaminants in drinking water, to allow enforcement of those standards by the states, and to protect (pretreatment) drinking water supplies from underground injections. This act is known as the Safe Drinking Water Act (SDWA).[77] The act applies to public drinking water systems and sources of piped water for human consumption in each state, provided that the systems have at least 15 connections or serve at least 25 individuals. The act does not apply to any system engaged only in the distribution and storage, but not in the collection and treatment of water; the regulations also do not apply to a system if it does not sell water or if it is an interstate carrier but does not provide passenger service.

### Standards of the act

The SDWA prescribes the development of three sets of standards for the quality of drinking water. The first of these are interim standards to be promulgated in the nine months following 16 December 1974; these interim standards[78] were to be put into effect within eighteen months after promulgation. The act further prescribes that a set of revised national primary drinking water standards,[79] designed to protect human health from the adverse effect of water contaminants, are to gradually supplant the interim standards. Yet a third set, so-called secondary drinking water regulations,[80] are prescribed to protect the "public welfare" from the adverse effects of water contamination.

In setting the interim standards, the SDWA required the EPA to consider technology existing on the effective date of the act as well as the cost of meeting those standards.

In setting the national primary standards, the SDWA requires the EPA to identify: (1)the maximum acceptable level of contaminants consistent with public health; (2)technology controls that can reduce levels of contaminants; and (3)criteria and means to monitor systems. The EPA's revised national standards are to be based on a National Academy of Sciences report, which must identify the methodology, data, and calculations used in establishing the recommended standards.[81] These national primary standards must themselves be subject to review and revision. The revised standards must take into account the best technology or treatment techniques and consider, but not necessarily defer to tradeoffs between, the cost of the best, and other, control measures.[82]

The revised national primary standards are thus not necessarily tempered by the cost to drinking water suppliers of meeting those standards. In this sense, the SDWA is much stricter than the CWA, which requires the EPA to directly consider cost/benefit tradeoffs.[83] This feature of the SDWA is the public's boon and potentially a drinking water supplier's bane. A drinking water treatment works is typically built to meet the primary standards in effect at the time. If these standards change, under the SDWA the supplier may then be required to upgrade his system to meet the revised standards at his own expense.

The SDWA also attempts to protect public water supplies from contamination through underground injection processes by requiring the EPA to promulgate underground injection control rules (UICs).[84] Any person who injects any material underground which could affect (pretreatment) drinking water supplies must apply for, and obtain, a permit to inject; exempted from this requirement are certain oil or natural gas recovery techniques, provided that these activities do not affect underground sources of drinking water. The applicant must show that the underground injection will not endanger drinking water sources. In practice, this may be a particularly difficult requirement to meet, because the hydrology of a region may not known in enough detail to credibly establish the relation of given pretreatment drinking water sources to the distribution of the injected materials (see Chapter 8 for an example of this problem). UIC rules are to be administered by the states, unless the EPA, in exceptional circumstances, intervenes. The state programs for administering the UICs must be promulgated according to EPA regulations.

## Variances

The SDWA authorizes the states to allow variances and exemptions from it under limited circumstances.[85] Variances may be granted from the national primary drinking water regulations if (1) because of the nature of its water source the public water system cannot meet the requirements for a contaminant despite use of the best technology, treatment, and other measures economically available, or (2) because of the nature of the water source, the treatment is not needed. In any case, the variance may be granted only if it does not present a threat to public health. Any variance granted must have a compliance schedule that requires incremental progress toward achieving the requirements. All variance applications are to be followed by notice and the opportunity for public hearings.[86]

The state must notify the EPA of all variance applications and all those granted. The notification must give the statutory reason for variance, and must include supporting documentation for the variance. The EPA retains the right to review any variance granted and reject it.[87]

## Enforcement

The state has the primary responsibility for enforcement of the SDWA, provided that the administrator has determined that the state has adopted a program to administer and enforce drinking water regulations that are at least as strict as the applicable national interim or primary standards. Such a program must have effective procedures for enforcing the regulations, including monitoring and inspection procedures and maintenance of necessary records and reports. A state program has to be approved by the EPA before it can become effective.[88]

The act allows the EPA to assume enforcement if the state fails to do so.[89] In those cases the EPA is authorized to bring civil action against the owners and operators of the violating drinking water supply systems. These actions may be brought in district court to require compliance with the standards.

There are at least two cases which show that the EPA is ready to exercise this power.

In United States v Neskowin Enterprises, Inc,[90] the court concluded that the state of Oregon had failed to enforce regulations against the owner of a public water system; the court found that the defendant had willfully allowed serious threat to the public health from microbial contaminants in the water it supplied. Neskowin had also failed, the court found, to notify the public of the threat. EPA assumed enforcement in the case; a fine of $100/day was imposed on the owner for each day of violation, totalling $26,400. In United States v Alder Creek Water Company,[91] the court held that the defendant, a public water supplier, had failed to comply with the EPA regulations; it ordered the defendant to add and maintain new equipment, maintain accurate and legible records of maintenance of its chlorination practices, and to sample and test the water it produced on a regular basis. All such records were required to be submitted to the EPA.

FEDERAL SURFACE MINING CONTROL AND RECLAMATION ACT

In 1977, Congress passed the Federal Surface Mining Control and Reclamation Act (FSMCRA),[92] with the intent of regulating surface mining activities and postmining reclamation operations. The statute addresses, among other things, environmental quality problems and water pollution. The FSMCRA is expressly made supplemental to the CWA and other environmental quality statutes. Under the act, state environmental quality laws remain in effect, and the states are given the authority to set more stringent performance standards for surface mining than are established under federal law. The act also preserves remedies and private rights that exist under state or common law.

## Implementation

The FSMCRA is implemented in much the same way as the CWA. The states are given a strong role in its administration and enforcement within well defined bounds. Section 503[93] of the FSMCRA, in particular, circumscribes the contents of a state administrative and enforcement program required for federal approval. First, the state program must contain law authorizing regulation of surface mining operations according to the requirements of the FSMCRA. Second, the program must include a state law authorizing sanctions for violations of state laws, regulations, or conditions on permits issued under the act. These sanctions must satisfy the minimal requirements of the FSMCRA for civil and criminal actions, suspensions, revocations and withholding of permits, and cease and desist orders by a state regulatory body. Third, there must be a provision for a state regulatory authority for the act, with sufficient administrative and technical personnel and adequate funding to carry out the statutory requirements. Fourth, there must be a state law establishing, maintaining, and enforcing a permit system which can effectively regulate surface mining and reclamation operations on state lands. Fifth, the state program must contain rules and regulations consistent with federal regulations issued under the act.

The EPA must concur in writing with those features of the state FSMCRA program that concern the CWA.[94] Approval for any proposed state plan must be preceded by at least one public hearing within the state.

The act contains a safety net: if a state (a) fails to submit an approved program, or (b) fails to have a previously rejected program accepted, or (c) fails to implement, enforce, or maintain an approved program, the federal government may assume direct control of the state program.[95]

FSMCRA regulations are implemented largely through a complex permit system.[96] All surface mining activities must operate under a permit. The duration of permit cannot be greater than five years unless an applicant establishes that because of a time limit, necessary financing cannot be obtained. A permit must terminate within three years of issuance if mining has not commenced by that time, although "reasonable" extensions may be granted if the permittee has been unable to commence operations for reasons beyond his control. FSMCRA permits must be reviewed on a periodic basis. Renewals must be preceded by public notice and opportunity for public hearings.[97]

There are extensive requirements on the contents of a FSMCRA permit application. The application must contain the names and addresses of permit applicants, all legal owners of the surface and mineral property, record owners of the surface and subsurface areas adjacent to the permit area, a description of the mining processes to be used, the starting and termination dates, and additional information, such as the impact of the mining activity on water quality.[98]

Performance standards for the permittee are set forth in Section 515[99] of the FSCRMA. These include standards for water impoundments, hydrologic balance, mine waste piles, and spoil deposits. Elaborate monitoring and record-keeping requirements are also imposed.

In general, the FSMCRA imposes very strict requirements on all surface mining operations, particularly if they might affect water quality.

## SUMMARY

In this chapter, we have surveyed the most important features of federal law concerning natural radiocontamination of water supplies. The heart of that law lies in the CWA, the SDWA, and to a lesser extent, the FSMCRA. Since these acts are in principle vast in scope and are very new, a number of fundamental legal issues, including reasonableness of standards, jurisdiction, permissible causes of action, and constitutional matters, are still in or will be brought to the courts. As a result, current federal water quality law is complex, controversial, and continuously evolving. Substantial uncertainty therefore attends any attempt to predict the long-term obligations or liabilities which a POTW or a drinking water supplier may face under this law. In the next chapter, we examine how one city, in the face of these complexities, dealt with the prospect of radiocontamination of its drinking water supplies by a uranium surface mine.

NOTES

1. I make no representation that anything stated in this chapter constitutes legal advice. Readers who need advice on water law should consult an attorney who specializes in this area.

The states, which I mention in this chapter only insofar as they play into federal law, may and typically do impose their own distinctive restrictions on water use. For example, water use rights in the Southwest may derive in part from treaties with native American groups (e.g., with the Utes in Colorado) or from Spanish colonial law (e.g., in New Mexico).

For a more detailed discussion of some of the issues raised in this chapter, see F. Skillern, Environmental Protection: The Legal Framework, McGraw-Hill, 1981, with supplementation through 1984 (hereafter called Skillern, Environmental ...). I roughly follow Skillern's treatment in this chapter; most of the case and statute citations are the same as his. I tend to emphasize certain issues more than Skillern does, however, including:

a) the strong dependence of BPT, BAT, and BCT standards on the decisions of the EPA, and the corresponding independence of these standards on any other decision-making mechanism;

b) the deep conventionality of BPT, BAT, and BCT standards;

c) the administrative relief which the CWA provides for those states which already had water quality standards as strong as those prescribed by the CWA;

d) the importance of the Supreme Court's decision in National Crushed Stone Association v EPA in the development of CWA litigation;

e) the relation of federal common law to the CWA;

f) the independence of the SDWA national primary drinking water standards and the costs of meeting those standards.

Any errors of interpretation are solely my responsibility.

2. This problem is hardly solved yet. Six years ago, Fountain Valley School, a residential high school located about 10 miles from Colorado Springs, Colorado, found its spring-fed drinking water supplies contaminated by fecal coliform presumably percolating from cattle-separation pens. The school immediately advised its 300 residents to boil all drinking water for at least thirty minutes until further notice. Within two weeks, the school was permanently connected to a nearby chlorinated public water supply.

3. See, for example, A. M. Beeton, "Eutrophication of the St. Lawrence Great Lakes," Limnology and Oceanography 10, 1965, pp. 240-254.

4. B. Commoner, _The Closing Circle_, Alfred A. Knopf, 1971.
5. _American Plant Food v Texas_, 587 SW2d 679, 14 ERC 1244 (Tex Crim App 1979).
6. Pub L No 80-845, 62 Stat 1155 (1948).
7. 33 USC Section 1251 _et seq_ (1976).
8. Pub L No 95-217, 91 Stat 1566 (1976) (codified at 33 USC Section 1251 _et seq_ (Supp II 1978)).
9. 33 USC Section 1362 (7) (1976).
10. 373 F Supp 665, 6 ERC 1388 (MD Fla 1974).
11. 33 USC Section 1362 (7) (1976).
12. 33 USC Section 1281 _et seq_ (Supp II 1978).
13. _Ibid._
14. The city of Pueblo, Colorado, made just such a determination; construction of a new facility began in May 1985.
15. 33 USC Section 1281 _et seq_ (Supp II 1978).
16. 33 USC Section 1288 (1976) and (Supp II 1978).
17. 33 USC Section 1281 _et seq_ (Supp II 1978).
18. 33 USC Section 1284(b)(1) (Supp II 1978). See also _Middlesex County Utilities Authority v Borough of Sayreville_, 18 ERC 1001 (3d Cir 1982), and _City of New Brunswick v Borough of Milltown_, 17 ERC 1937 (3d Cir 1982). In both cases the Third Circuit upheld EPA's authority to deny funding to a POTW which had not adopted a user charge system.
19. 33 USC Section 1281 _et seq_ (1976) and (Supp II 1978).
20. 33 USC Section 1288 (1976) and (Supp II 1978).
21. See notes 19 and 20.
22. The definition is in 33 USC Section 1362 (14) (Supp II 1978); "Section 1311" is formally "33 USC Section 1311 (Supp II 1978)."
23. 604 F2d 239, 13 ERC 1289 (4th Cir 1979).
24. 599 F2d 368, 13 ERC 1417 (10th Cir 1979).
25. 487 F Supp 852, 145 ERC 1321 (ED Pa 1980).
26. 33 USC Section 1311 (Supp II 1978).
27. 33 USC Section 1281 (1976) and (Supp II 1978).
28. 33 USC Section 1311 (1978) and (Supp II 1978).
29. 33 USC Section 1314(b)(4)(A) (Supp II 1978).
30. 33 USC Section 1314(b)(4)(B).
31. 33 USC Section 1314 (Supp II 1978).
32. _Ibid._
33. 33 USC Section 1316(a)(1).
34. 33 USC Section 1317(Supp II 1978).
35. 33 USC Section 1314 (Supp II 1978).
36. 33 USC Section 1311(c) (1976).
37. 33 USC Section 1311(g) (Supp II 1978).
38. _Ibid._
39. 33 USC Section 1311(h) (Supp II 1978).
40. 33 USC Section 1281 (1976).
41. 33 USC Section 1311 (Supp II 1978).
42. 545 F2d 1351, 9 ERC 1274 (4th Cir 1976).
43. 33 USC Section 1314(b)(1)(B).

44. 33 USC Section 1311(c) (1976).
45. 601 F2d 111, 13 ERC 1277 (4th Cir 1979).
46. 486 F2d 375, 5 ERC 1593 (DC Cir 1973).
47. 101 S Ct 295, 15 ERC 1209 (1980).
48. 33 USC Sections 1312-13 (Supp II 1978).
49. Ibid.
50. Ibid.
51. 33 USC Section 1342 (Supp II 1978).
52. Ibid.
53. 33 USC Section 1369(b)(1)(F).
54. 33 USC Section 1341(a) (Supp II 1978).
55. Skillern, Environmental ..., p. 163.
56. Ibid.
57. Ibid.
58. 33 USC Section 1365 (Supp II 1978).
59. 33 USC Section 1369 (1976).
60. 401 US 493, 2 ERC 1331 (1971).
61. Ibid.
62. 406 US 91, 4 ERC 1001 (1972).
63. 28 USC Section 1331 (1976).
64. 541 F2d 119, 9 ERC 1507 (1st Cir 1976).
65. 33 USC Section 1365 (1976).
66. 408 F Supp 556 , 10 ERC 1881 (ED Va 1976).
67. 539 F2d 1006, 9 ERC 1212 (4th Cir 1976).
68. 539 F2d 1009, 9 ERC at 1214.
69. 616 F2d 1222, ERC (3rd Cir 1980).
70. 33 USC Section 1365(e) (1976).
71. 619 F2d 623, 14 ERC 1281 (7th Cir 1980).
72. 599 F2d 151, 13 ERC 1049 (7th Cir 1979).
73. See note 72.
74. 101 S Ct 1784, 15 ERC 1908 (1981).
75. ____ F2d ____, 18 ERC 1091 (7th Cir 1982), on remand from the Supreme Court.
76. 18 ERC 1041 (ND Ill 1981).
77. Safe Drinking Water Act, 42 USC Section 300f-300j-9 (1976).
78. EPA Final Rule: Interim Primary Drinking Water Regulations Amendments, 45 Fed Reg 57332 (27 Aug 1980).
79. 40 CFR pt 142 (1981).
80. 40 CFR pt 143 (1981).
81. 42 USC Section 300g-1 (1976).
82. Ibid.
83. Ibid.
84. 42 USC Section 300h-3 (1976).
85. 42 USC Section 300g-4 (1976).
86. Ibid.
87. Ibid.
88. 42 USC Section 300g-2 (1976).
89. Ibid.
90. F Supp , 14 ERC 1636 (D Ore 1980).
91. F Supp , 14 ERC 1413 (D Ore 1979).
92. 30 USC section 1201 et seq (Supp II 1978).
93. 30 USC 1253 (Supp II 1978).
94. Ibid.

95. 30 USC Section 1254 (Supp II 1978).
96. 30 USC Section 1256 (Supp II 1978).
97. *Ibid.*
98. *Ibid.*
99. 30 USC Section 1265 (Supp II 1978).

# 7
# One City's Experience

## INTRODUCTION[1]

In early 1979, Cyprus Mines Corporation (a wholly owned subsidiary of Amoco), and Wyoming Minerals Corporation (a wholly owned subsidiary of Westinghouse), applied to the Water Quality Control Division of the Colorado Department of Health for a National Pollutant Discharge Elimination System (NPDES) Permit[2] to develop a suite of surface ("open pit") uranium mines and a uranium mill in the drainage of Tallahassee and Cottonwood Creeks. These creeks are tributaries of Currant Creek, which is a tributary of the Arkansas River. The proposed operations, which came to be called the "Hansen Project," were to be located about 20 air miles northwest of Canon City, Colorado (see Figure 7-1).

The dimensions of the largest of the mines (called the "Hansen Mine" or "Hansen Pit"), even by modern standards, were impressive. Cyprus geologists estimated that about 300 feet of overburden would have to be removed just to reach the ore body. The ore body itself extended an additional 500 feet below that,[3] and was perhaps a half-mile wide and a mile long (see Figure 7-2).[4]

Because the mines would interrupt water-bearing strata, Cyprus engineers proposed to drill a set of "mine-dewatering" wells on the perimeter of the regions to be excavated, to dilute water pumped from these wells (at roughly 500-1,000 gallons per minute) with water of unspecified origin, alpha activity, and quantity, and to discharge the resulting mixture into Middle Tallahassee Creek.[5,6] This mode of operation, the engineers estimated, would continue over the first two to three years of the Hansen Project's operation. Thereafter, the Cyprus permit application maintained, mine-dewatering water would be diverted to the mill, which would "consume" the "entire" dewatering output of the mines.[7]

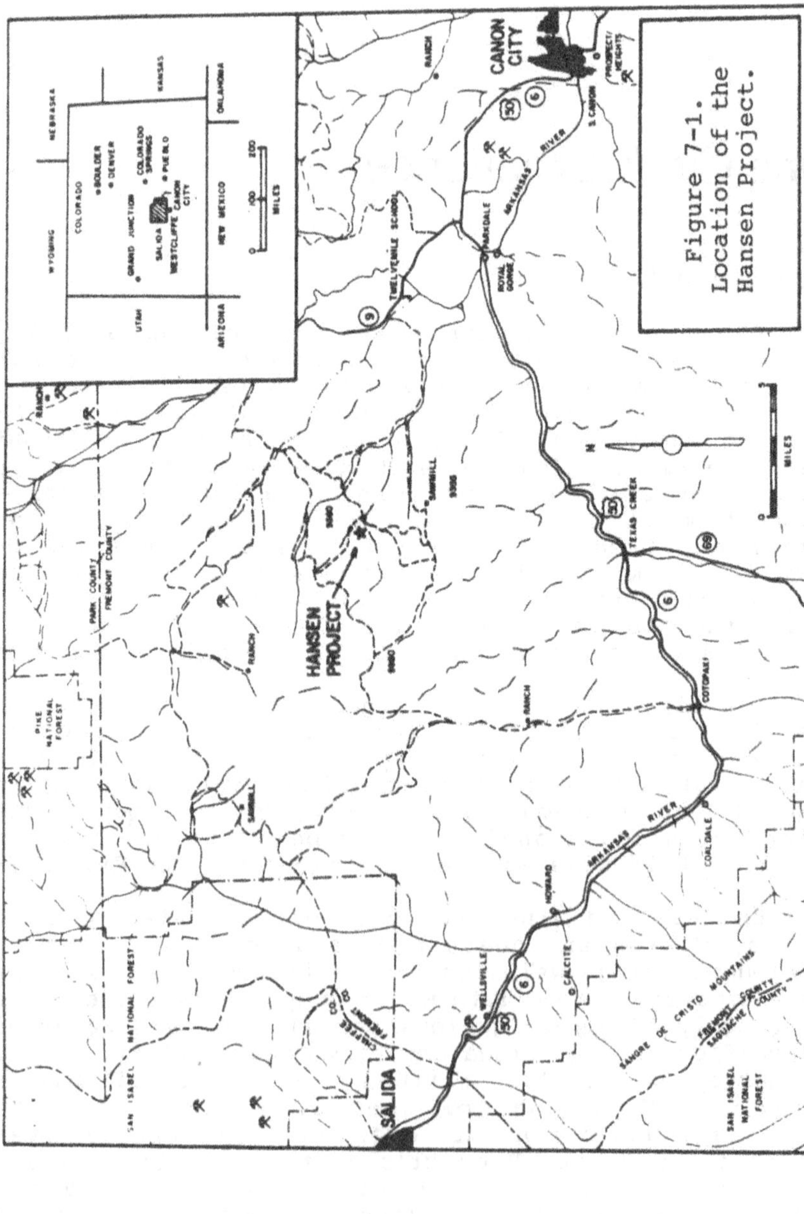


Figure 7-1. Location of the Hansen Project.

The actual excavation activities of the Project were expected to contribute an unspecified quantity of radioactive material to Middle Tallahassee Creek through airborne particulates, and to surface and subsurface waters of unknown hydrology and connection to Middle Tallahassee Creek and the Arkansas River.[8] Mine tailings were also expected to contribute to the radioactive burden of the Arkansas through similar, and poorly understood, mechanisms.

Ore from the Hansen mine, and possibly from other mines, was to be converted to uranium "yellowcake" at the mill, which was to be built by Cyprus near North Tallahassee Creek. The mill was expected to contribute to the radioactive load of the Arkansas, also: Cyprus engineers estimated that 99.9% of the radium in the water discharged by the mill would be removed, while the remaining 0.1%, it was assumed, would be discharged into the Arkansas River through North Tallahassee Creek or into holding ponds. (No indication of the fate of uranium in the mill effluents was given in the permit application, however.[9]) Likewise, tailings from the mill were expected to be a source of radioactively contaminated run-off or leachate. Under certain infrequent conditions, an unspecified quantity of partially, or even untreated, water of unknown activity was to bypass the mill altogether.[10]

## CONCERN ABOUT LIABILITY UNDER FEDERAL WATER LAW

From the outset of its involvement in the Cyprus/Hansen NPDES permit application, the city of Colorado Springs announced that it did not seek "to obstruct the development of the mine" or make "value judgments about the wisdom of nuclear power."[11] But city engineers and the city attorney believed that mine and/or mill operations could increase the concentration of radioactive species in the Arkansas River, and beginning in 1983, the Arkansas would supply about 15% of the city's water through the Fountain Valley Conduit (see Figure 7-3).[12]

Because the city of Colorado Springs owns and runs both the drinking water and wastewater treatment works which service most of the Colorado Springs area, the council and the city attorney were deeply concerned that the city could become subject to a variety of stringent and costly federal water quality sanctions, were the radioactive burden of the Arkansas to increase.[13]

## DIFFICULTIES IN ANTICIPATING URANIUM STANDARDS

No matter which federal regulations might ultimately be invoked or revised to control uranium in

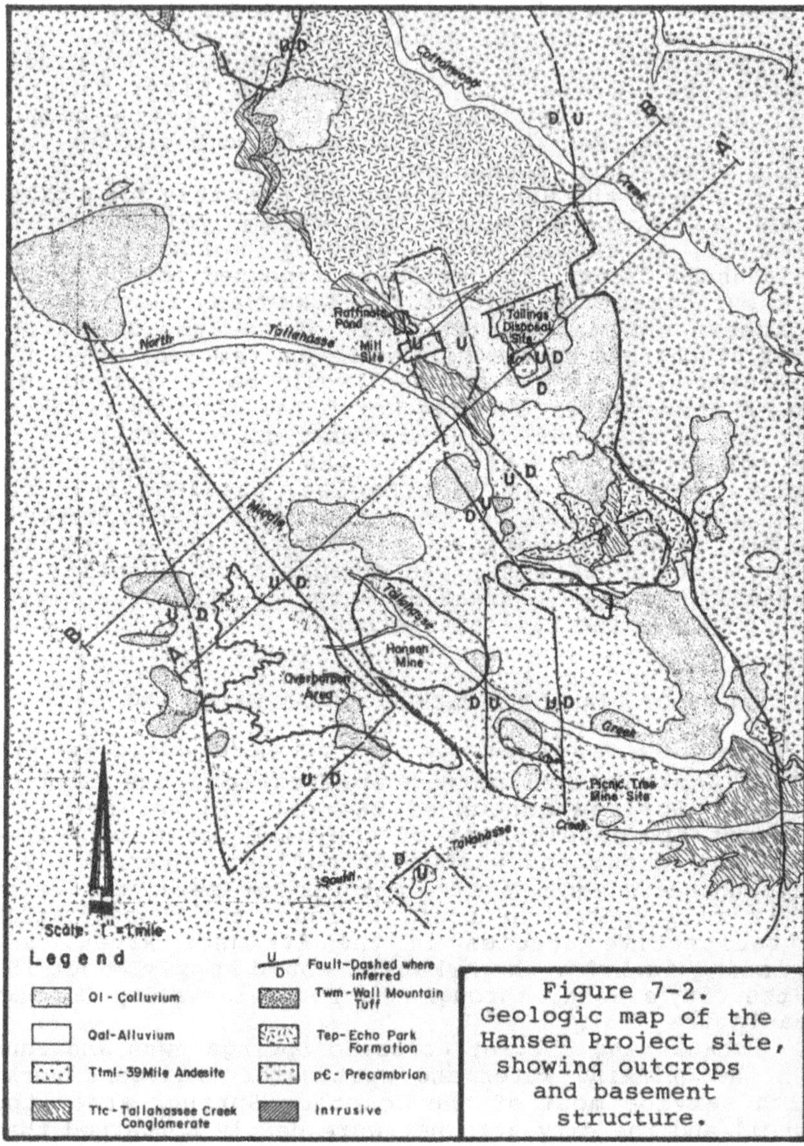


Figure 7-2. Geologic map of the Hansen Project site, showing outcrops and basement structure.

water, there was, the council and city attorney believed, profound uncertainty about what standard might be set for that element. Correspondence between the EPA and the Colorado Department of Health (CDH) had suggested a ceiling (for drinking water) of not more than 10 pCi/l[14] for uranium-derived alpha activity. The Colorado Mining Association wanted a ceiling not lower than 40 pCi/l.[15] A consultant for the CDH had recommended 5 pCi/l.[16] The CDH itself had published a memorandum supporting a 40 pCi/l standard.[17] And, two consultants hired by the city, the Pueblo (Colorado) Regional Planning Commission, and the Pueblo Board of Water Works had suggested a standard of 30 pCi/l,[18] and 10-40 pCi/l,[19] respectively. The range of these values is at least eight-fold, complicating the task of guessing what radiotoxicity standard for uranium in water might be set.

Were the standards of the CWA for wastewater to be extended to include regulation of uranium, they would probably depend, in part, on a characterization of a "natural" concentration of uranium in stream water. But the job of determining this concentration is not an easy one. First, it is likely that the concentrations of uranium in stream water, even in the absence of human activity, vary over a wide range. Secondly, it is often difficult to identify or characterize the dynamics of these variations, although they probably depend on a complex of hydrogeochemical parameters.[20] Thirdly, even if the variables which affect the concentration of uranium are known, it may be difficult to collect a sufficient number of samples to render nominal observed values statistically respectable. Thus, a policymaker with even the most neutral interests is not likely to find the task of setting fair, rational wastewater or drinking water quality standards based on natural concentrations of uranium an easy one.

The task of determining "the" uranium-dependent background alpha activity for the Arkansas exhibits all these aggravations and more. In the first place, it is unclear where the sample or samples to be used in determining the "background" levels of uranium in the Arkansas should be taken. And the choice of sampling sites and schedule is crucial. Along the river at a given time, one can easily observe a 10-fold range in uranium-derived alpha activity.[21] Even at a given point on the river, a 10-fold variation over time in uranium alpha levels can be observed.[22] Furthermore, there is no guarantee that even these ranges are "real:" across the state of Colorado stream gross alpha activities vary from less than 0.01 pCi/l[23] to over 400 pCi/l,[24] a 40,000-fold range (see Figure 7-4).[25]

Secondly, it is not obvious how the background is to be computed. An average value derived from several samples will vary, depending on the geographic and

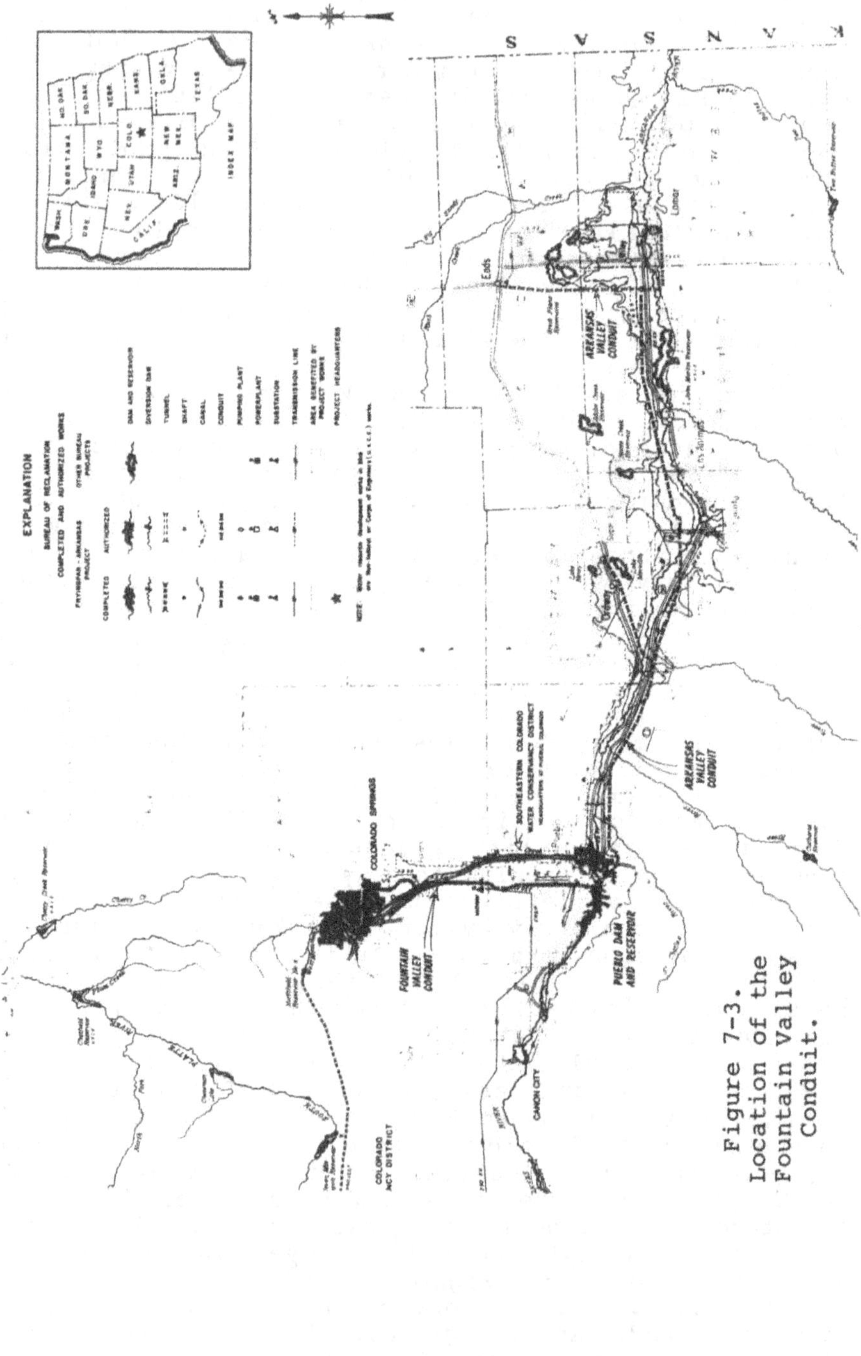


Figure 7-3. Location of the Fountain Valley Conduit.

temporal distribution of those samples. If the entire United States were used as a basis for the determination, a value of 5 pCi/l for uranium-derived alpha activity, for example, would seem reasonable (see Figure 7-5).[26] In the USGS Pueblo Quadrangle (which includes the proposed mine and mill site), that average is also plausible. But USGS data taken at Pueblo near the site of the Fountain Valley Conduit diversion point (the Pueblo Reservoir dam) show alpha levels ranging from 6-32 pCi/l, with an average of about 23 pCi/l.[27]

The problem is in fact more perverse than this. The confidence which one can place in the belief that an observed average value of a physical quantity (such as uranium-derived alpha activity) is close to the "real" average depends on at least two things:

1. The number N of samples used to compute the average (mean); and,

2. How far removed, on the average, an observed value of the alpha activity is from the average value of that activity in the samples (the standard deviation).

Neither of these desiderata were thoroughly addressed in the Cyprus permit application (nor did the law require them to be). For several places along the Arkansas, fewer than a dozen samples had been taken (this was true in particular for uranium-derived alpha activity at the point at which the Fountain Valley Conduit originates). Furthermore, the standard deviation of the USGS gross alpha values for the Pueblo Reservoir sampling site is 6.9 pCi/l, a value so large (30%) relative to the mean (about 23 pCi/l) that little confidence can be placed in the belief that the mean is also a "typical" concentration.[28]

Specification of a background says nothing, unfortunately, about how verification of compliance with a water quality regulation is to be accomplished. Commonly, a sampling and analysis for this purpose is contained in the NPDES permit. With the schedule specified, the question of how the statistic(s) derived from this sample are to be used to ascertain compliance must still be specified. Because uranium-derived alpha is not presently included in the list of substances regulated by the CWA or SDWA, such scheduling and computational issues were not addressed in the Cyprus permit application. All these issues rendered the permit application more difficult to evaluate.

Figure 7-4. Summary of Uranium Hydrogeochemical and Stream Sediment Data for All Waters, NTMS Quadrangles, Colorado

| Quadrangle | Number of Samples | Range Min. - Max. | Mean | Standard Deviation | Anomaly Threshold | Anomaly % |
|---|---|---|---|---|---|---|
| Sterling | 1653 | 0.01 - 364.0 | 14.52 | 31.4 | NA | NA |
| Limon | NA | | | | | |
| Lamar | 1043 | 0.01 - 257.4 | 17.63 | 38.53 | 50 | 9 |
| La Junta | 1351 | 0.01 - 748.4 | 7.22 | 26.49 | 20 | 4.9 |
| Greeley | 1210 | 0.10 - 340.0 | 15.73 | 26.82 | 70 | 3.8 |
| Denver | 1264 | 0.01 - 147.4 | 5.3 | 11.7 | 30 | 2.7 |
| Pueblo | 861 | 0.01 - 253.5 | 5.88 | 16.72 | 36 | 2.1 |
| Trinidad | 1060 | 0.01 - 88.3 | 4.05 | 8.79 | 20 | 5.4 |
| Craig | 1234 | 0.01 - 856.4 | 2.66 | 26.1 | 10 | 3 |
| Leadville | NA | | | | | |
| Montrose | 1365 | 0.01 - 209.5 | 1.84 | 9.49 | ** | ** |

| | | | | | | |
|---|---|---|---|---|---|---|
| Durango | 1518 | 0.01 - 25.7 | 0.84 | 1.81 | 5 | 2.2 |
| Vernal | NA | | | | | |
| Grand Junction | NA | | | | | |
| Moab | 442 | 0.01 -1874.2 | 8.46 | 8.46 | 10 | 3.8 |
| Cortez | 598 | 0.01 - 241.5 | 3.80 | 14.69 | 20 | 2.3 |
| Total | 13599 | 0.01 -1874.2 | 7.37 | | | |

$$\text{Mean for Colorado} = \frac{\text{sum (\# of samples per quad) x (mean per quad)}}{\text{total \# of samples for Colorado}}$$

$$= 7.37 \text{ ppb uranium}$$

------------

All uranium concentrations in ppb (parts per billion)

NA = Not Available

** = Anomaly thresholds classified relative to geological and physiological subregions within quadrangle and not according to whole quadrangle.

CONCERNS ABOUT THE "ZERO-DISCHARGE" PROMISE

Whatever concentrations might be established as the uranium-derived alpha background of the Arkansas, and whatever compliance and verification schedule might be developed for the Hansen Project, Cyprus was willing to try to design the mine/mill to be a "zero-discharge" operation. But desire to achieve such a goal is one thing, and achieving that goal is another. If Cyprus were to make good its claim to be able to achieve zero-discharge for various radioactive species in its effluents, the city reasoned, Cyprus would have to answer four difficult questions:

1. What would be the nature, magnitude, and mechanism of the introduction of radioactive species into the effluents of the Hansen Project?

2. What would be the hydrological relation of Hansen effluents to local surface and subsurface water and to the hydrology of the Arkansas basin?

3. How much would Hansen mine dewatering processes contribute to the radioactive load of the Arkansas?

4. How much would Hansen mill operations contribute to the radioactive load of the Arkansas?

City engineers and consultants could not ascertain that Cyprus could answer any of these with sufficient confidence. Their apprehension came from several sources.

Throughout 1978, Cyprus had drilled test wells in the vicinity of the proposed dewatering wells. The results from Test Well 21-122 from the Echo Park formation,[29] a well "close to" the ore body were particularly difficult to understand. In particular, the mean gross alpha activity as of the well's completion (31 May 1978) was determined by Cyprus to be 1830 pCi/l.[30] Mean radium-226 alpha activity for that same set of samples is noted to be 2300 pCi/l. By definition, since radium-226 activity is just alpha emission, radium activity for that set of samples should be equal to or less than gross alpha for those samples. But in fact the mean of the radium-226 activity reported for the set is 25% greater than the mean of the gross alpha reported for that set. An even worse discrepancy could be found in data for the 24-hour test for Well 21-122 (21 June 1978). This set of samples shows a mean value for gross alpha of 2400 pCi/l, whereas the radium-226 alpha mean value for the same set of samples is 3400 pCi/l; in short, the mean of the radium-226 determination is 42% greater than the gross alpha. But perhaps most disturbing is a comparison of the means for

Figure 7-5. Concentration of Uranium in Surface, Ground, and Domestic Waters in the United States

| Water | Surface Water (in pCi/liter) | Ground Water (in pCi/liter) | Domestic* (in pCi/l) |
|---|---|---|---|
| Range | 0.07 - 653 | 0.01- 582 | 0.01 - 653 |
| Average | 1 | 3 | 2 |
| Median Concentration | 0.1 - 0.2 | 0.2 - 0.5 | 0.1 - 0.2 |
| Modal Concentration Range | 0.2 - 0.5 | 2 - 5 | 0.1 - 0.2 |
| Total Number of Samples | 34,561 | 55,433 | 28,239 |

* Those surface and ground water supplies identified as domestic water sources. These are not necessarily tap water samples.

all Echo Park gross alpha to radium-226 activity. In a sufficiently large number of samples, "wild" data (far from the mean or average) tends to have relatively little effect on the mean itself. Yet the mean for all Echo Park gross alpha determinations reported by Cyprus is 2115 pCi/l, while radium-226 activity was reported at 2733 pCi/l for the same set.[31] The latter value is simply 30% too high to be reconciled with gross alpha activity of the same test well samples.

There was a further puzzle in the Cyprus characterization of the Echo Park alpha activity. For any sample, one can calculate the distribution of decay daughters and emissions, given a decay series (or set of the same), the time at which decomposition began, the quantity of one nuclide is each of those series at a given time, and a given decay mechanism. The distribution of daughters and activities is then a function of time. Computations done by a city consultant[32] showed, however, that no credible natural mixture of radionuclide parents could have produced the proportions of gross alpha and beta activity reported by Cyprus.

There were further concerns about the activities which Cyprus reported in its test well data. Examination of Cyprus field data by a consultant[33] revealed that only 11 samples of water from the Echo Park wells had been tested for uranium alpha. From the standpoint of good statistical methodology, 11 samples is about a tenth of the minimum number which should have been tested to provide a basis for inference compatible with the level of precision which proposed federal standards would surely demand.[34] Thus, on statistical considerations alone, the Cyprus NPDES permit data raised too many questions to provide a sound basis for city acceptance.

Based on these analysis, the city concluded that at least one of the following was true:[35]

1. The statistical methodology or practices used by Cyprus to determine gross alpha and beta activities in water near the ore body were insufficient to provide a reliable characterization of those activities; or,

2. The analytical practices used by Cyprus to determine gross alpha, beta, uranium, or radium-226 concentrations in the test well water were inadequate to reliably characterize those values; or,

3. The identification of alpha and beta sources had not taken into account a large contributor to alpha and beta activities.

CONCERNS ABOUT THE MINE-DEWATERING PROPOSALS

The proposed mine-dewatering process generated further concerns. Cyprus test well data strongly suggested that some of the dewatering well water would have to be diluted prior to discharge. The credibility of this claim depended on whether water of appropriately low alpha activity ($\leq$ 10 pCi/l) would be available in appropriately large volume. What that volume must be can easily be approximated. If FRI is the flow of the river, CRI is the background activity of a radioactive species in the river, FE is the mine-dewatering effluent flow discharged into the river, and CE is the initial concentration of the given species in the effluent, then the resulting concentration of CRF of the species in the river is given by Equation 7-1:

$$CRF = \frac{(FRI)(CRI) + (FE)(CE)}{FRI + FE} \qquad (7\text{-}1)$$

FRI, FE, CE and CRI are determined by field observations. If one assumes, as Cyprus engineers did, that nominally ten dewatering wells would operate at approximately equal flow (about 100 gal/min each), the combined flow of the nine wells containing low alpha activity (about 10 pCi/l) with one well containing 2500 pCi/l (akin to Test Well 21-122) alpha activity would yield water at 250 pCi/l at 1000 gal/min. If the flow of the Arkansas were assumed to be 129.6 cubic feet per second and the background uranium-derived alpha in the Arkansas were 15 pCi/l, Equation 7-1 would give a concentration of uranium-derived alpha[36] in the Arkansas of 15.2 pCi/l, about a 1% increase in that activity. Cyprus's claim here, given its assumptions, was thus credible: a 1% increase in the concentration of uranium alpha in the Arkansas probably would not be measurable.[37]

But, city engineers asked, were the assumptions underlying this calculation reasonable? In particular, was it known that whether water would be available with sufficiently low activity to support the dilution? This concern spanned two issues:

1. What would be the activity of the water in each of the dewatering wells?

2. How much water would be available from "low" (less that 10 pCi/l) alpha activity wells?

For neither of these questions did Cyprus have an answer. The hydrology of the mine site was unknown[38] and an adequate characterization of the alpha concentration in even the test wells had yet to be achieved.

There was no guarantee, in fact, that a dewatering rate of about 1000 gal/min was sufficient to clear the mine of water as excavation proceeded. Perhaps a value ten times this would be required. And if that were so, even on the assumption that the combined dewatering well output had an alpha activity of 250 pCi/l, the resulting alpha activity in the Arkansas would be, by Equation 7-1, 16.7 pCi/l, a nominal increase of 10% over ambient activity (assumed to be 15 pCi/l) and potentially 1-2 pCi/l over several candidates proposed as a federal ceiling.[39] A better characterization of the mine dewatering hydrology was thus, in the city's eyes, essential. The Cyprus data did not provide that characterization.

## CONCERNS ABOUT MILL IMPACT

Cyprus's characterization of the impact of the mill raised additional questions:

First, it was not clear from Cyprus's application how much water would be needed by the mill.[40] Without this data, it was difficult to assess Cyprus's claim that the mill would consume all mine-dewatering water produced after the third year of mine operation.

Second, the Cyprus application contained insufficient data on how much ore the mill would actually process.[41] City consultants and Cyprus engineers felt it unlikely that the mill would be cost-effective to run on the output of the Hansen and Picnic Tree mines alone. Ore mined elsewhere, in all likelihood, would have to be brought in for processing,[42] but how much was unknown. Without a sense of the scale of the mill operation, no clear idea of the impact of the mill could be made.

Third, it was unclear from the Cyprus application how frequently and at what rates "bypass" or "upset" (untreated, or partially treated) water would be discharged, or where it would be discharged if it were.[43] Although city engineers considered it an unlikely scenario, a permit which allowed Cyprus to bypass water around the mill could, in principle, give Cyprus carte blanche to discharge what and where it wished. Accordingly, the city engineers recommended that the permit require bypass or upset water to be diverted to holding ponds, then treated at a "convenient" time.[44] But it was difficult to tell from the permit application how large the holding ponds should be. Cyprus had cited a nominal volume of 1.9 million gallons but had provided inadequate supporting calculations or assumptions for this value. Without a fairly precise estimate of the volume of the holding ponds, it was not clear that Cyprus could support a zero-discharge operation of the mill.

Fourth, Cyprus had claimed that the mill would remove 99.9% of the radium in the water which it discharged from the mill.[45] It was unclear, however, what water this was. Was it from North Tallahassee Creek, from the Echo Park formation dewatering wells, from some process within the mill itself, or from just what? And, even if 99.9% of the radium were removed from all these sources, it was unclear what relation this had to uranium-derived alpha activity.

Most surprisingly, it was not clear from the application that Cyprus even intended to build a mill.[46] If the mill were not built, all the claims which Cyprus had made about mine water control procedures which depended on the existence of the mill were simply gratuitous.[47]

## CONCERNS ABOUT DRINKING WATER DECONTAMINATION

All the administrative, legal, and technological concerns surrounding the proposed Cyprus permit mining and milling operations would have been moot if uranium could be removed inexpensively by city water treatment facilities. But the only economically feasible technology for reducing the very "low" concentrations of uranium that would be produced by the Hansen Project would involve at least charcoal filtration of all water treated by the city.[48] On the scale of operation envisioned by city engineers, annual replacement of these filters would be required, costing about $200,000 per year.[49] Over the life of the Project (about 20 years), the total cost of replacing the filters would be at least $4,000,000.

The disposal of these filters would generate problems in its own right. Assuming that the average consumer used 100 gallons of water per day, and that 2 pCi/l would have to be removed from that water, treatment works servicing a city of 250,000-300,000 people (the nominal size of Colorado Springs) would have to dispose of filters contaminated with about 1 Ci of uranium per year. A source of that activity would require very special, costly handling. Because of the long half-life of U-238,[50] moreover, the contaminated filters would require storage in an area well removed from human contact for literally thousands of years.

## POSTSCRIPT

The Cyprus permit was granted in 1981, subject to the zero-discharge condition. Within months, however, the price of uranium yellowcake dropped so low that the proposal became unprofitable. The mill and mine were never built.

## NOTES

1. My thanks to Tad Foster of the city attorney's office, Colorado Springs, Bob Schukle of the Colorado Department of Health, and Charles Weir of the Fountain Valley Authority for assistance in obtaining documentation for this chapter.
2. See Chapter 6.
3. Cyprus Mines Corporation and Wyoming Minerals Corporation, Hansen Project Environmental Report, 1979 (hereafter referred to as Cyprus Mines, Hansen Project), pp. 2.1-2, 2.4-23.
4. Adapted from Wright Water Engineers, Inc. (2420 Alcott Street, Denver, Colorado 80211), Hansen Project Augmentation Plan Report, October 1979 (hereafter referred to as Wright Water Engineers, Hansen Project), p. 5.
5. Ibid., p. 2.
6. Cyprus Mines, Hansen Project, pp. 3.6-28, 3.6-29. See also City of Colorado Springs, Statement of the City of Colorado Springs Concerning the NPDES Permit Application of the Cyprus Mines Corporation-Hansen Project, Permit No.: CO-0036510, Fremont County, 15 May 1980 (hereafter referred to as City of Colorado Springs, Statement), p. 4.
7. Ibid.
8. In fact the only detailed hydrology which had been done in the area was on the Four Mile Creek drainage, which is more than 10 miles from the Hansen site. Cyprus assumed that the Four Mile Creek study was of a region sufficiently analogous to the Hansen site to allow extrapolation. Ironically, this same report argues that there is extreme local variation in hydrology; extrapolations of more than a mile, the report advises, are not reliable. See Wright Water Engineers, Hansen Project. See also Cyprus Mines, Hansen Project, pp. 2.6-10, 2.6-20.
9. NPDES Permit Application, Summary of Rationale, Cyprus Mines Corporation-Hansen Project, Permit No.: CO-0036510, Fremont County, Colorado, 1981.
10. City of Colorado Springs, Statement, pp. 4, 7.
11. Ibid., p. 4.
12. Adapted from U.S. Bureau of Reclamation, Map of Fryingpan-Arkansas Project, Colorado Lower Missouri Region, Map No. 382-706-2648, revised April 1977. In fact, Colorado Springs did not receive water from the conduit until mid-1985.
13. City of Colorado Springs, Statement, p. 1.

14. Letter, W.L. Lappenbusch, U.S. Environmental Protection Agency, to F.A. Traylor, Colorado Department of Health, 6 July 1979. See also C.R. Couthern, W.L. Lappenbusch, and J.A. Cotruvo, Health effects guidance for uranium in drinking water, Health Physics, 1979.

15. Colorado Mining Association, Minutes of the meeting of the Colorado Mining Association, Radiation Protection Subcommittee, 25 June 1980, Denver, Colorado, p. 3.

16. Letter, J.W. Healy, Los Alamos Scientific Laboratory, to A.J. Hazle, Colorado Department of Health, 28 February 1979.

17. Colorado Department of Health, Uranium in Water Guidance, 8 July 1979.

18. J.O. Ledbetter, Recommendation for a Uranium Standard for the Arkansas River Basin, in Engineering-Science (2785 North Speer Boulevard, Denver, Colorado 80211), ed., Recommendation for a Uranium Standard, (hereafter referred to as Engineering-Science, Recommendation for a Uranium Standard), p. 2-6.

19. M.E. Wrenn, Written statement of Dr. McDonald E. Wrenn submitted after oral testimony, Engineering-Science, Recommendation for a Uranium Standard, p. 3-9.

20. J.K. Felmlee and T.H. Weaver, Radium and uranium concentrations and associated hydrogeochemistry in ground water in southwestern Pueblo County, Colorado, Open-File Report 79-974, USGS, Department of the Interior, 1979.

21. See Figure 7-4.

22. Ibid.

23. Ibid.

24. Ibid.

25. K.J. Schiager, Data, assumptions, calculations, and summary statements, Exhibits in support of oral testimony on behalf of Cotter Corporation for the Public Rulemaking Hearing before the Colorado Water Quality Control Commission, 14 July 1980, p. III-2.

26. J.S. Drury, S. Reynolds, et al., Uranium in the U.S. surface, ground, and domestic waters, EPA 570/9-81/001.

27. S.S. Shannon, Hydrogeochemical and stream sediment reconnaisance of the Pueblo Quadrangle Colorado, GJBX-135(78)(LA-7341-MS), 1978.

28. Engineering-Science, Recommendation for a Uranium Standard, data for sampling station AD-8, Reference 38 16 00.0 104 36 00.0 4 (3 pp.).

29. Adapted from Cyprus Mines, Hansen Project, p. 2.6-7.

30. Adapted from Cyprus Mines, Hansen Project, p. 3.6-36.

31. Adapted from Cyprus Mines, Hansen Project, pp. 3.6-72, 5.2-7.

32. The author.

33. The author.

34. See note 14.

35. City of Colorado Springs, _Second Statement by the City of Colorado Springs Concerning Proposed NPDES permit of Cyprus Mines Corporation-Hansen Project_, 23 May 1980 (hereafter referred to as City of Colorado Springs, _Second Statement_), pp. 1-3. The analysis which led to these conclusions was performed by the author.

36. Cyprus Mines Corporation, _Estimated radiological impact of the Hansen Project water discharge on the Arkansas River_, 4 September 1980, p. 1.

37. _Ibid._, p. 4.

38. See note 8.

39. See note 14.

40. Cyprus Mines, _Hansen Project_, Section 3.0.

41. Wright Water Engineers, _Hansen Project_, p. 1, assumes 3500-4000 tons of ore per day.

42. _Ibid._

43. The question was not resolved by the time the permit was issued.

44. See note 35.

45. Cyprus Mines, _Hansen Project_, Section 3.0.

46. City of Colorado Springs, _Second Statement_.

47. _Ibid._

48. Engineering-Science, _Recommendation for a Uranium Standard_, pp. 2-5, 2-6.

49. City of Colorado Springs, _Second Statement_, p. 6.

50. See Chapter 4.

# 8
# Summary: The Determinants of Evolving Regulation

Current views about the effects of natural radioactivity in water supplies on human well-being concern two distinct questions:

1. How do we describe the physical and biological effects of the exposure of humans and their environment to natural radioactivity in water supplies?

2. How do we assign significance to the biological and physical effects of natural radioactivity in water supplies on human well-being?

The biological effects of ionizing radiation depend on environmental dynamics, on whether the dose is internal or external, on the nature of the species, on the nature of the nuclide and its intrinsic decay regime, on metabolic factors, on individual variation, on the spatial and temporal distribution of the dose within the body, and on the linear energy transfer of the energy involved. Currently, there are no well controlled experiments on the effects of chronic low-dose-rate exposure regimes for man; all the reliable experimental data we have comes from relatively high-dose, high-dose-rate, short-term exposures of small experimental mammals, primarily mice. Extrapolation from one irradiation regime to another must therefore explicitly identify the extrapolating assumptions (e.g., linearity of the dose-effect relation, similarity of effects across species, and so on) involved.

Explicitly identifying these caveats, the consensus scientific opinion holds that the most serious long-term effect of external exposure to ionizing radiation will be damage to the genetic material of fathers; that estimate says 10 to 150 offspring per million population per $10^{-2}$ Gy of chronic paternal exposure to low-dose, low-dose-rate, low-LET (gamma-, x-, and beta-) external radiation will suffer from seriously debilitating or lethal genetic disorders. For internal exposures arising from the ingestion of natural radioactivity in drinking water, the primary health detriment will come from radiation-induced leukemia and bone cancer. Best estimates suggest that 10 and 150 leukemia- and 10 - 150 bone-cancer deaths will occur per generation per million population per 10 picoCuries uranium or radium alpha per liter drinking water.

None of this is to say, however, that any of these estimates are correct, or even conservative, and the fact that no well controlled experimental evidence is available for exposure to chronic low-dose-rate ionizing radiation must temper any policy decision.

There is thus substantial uncertainty in our current knowledge of the physical and biological effects of natural radioactivity in water supplies. Yet even if there were no uncertainty about these matters, it is not obvious what radiation protection policy should be set, because such decisions intrinsically involve questions about the importance of human health relative to other things in which we have an interest, such as the cost of electrical energy. One of the most deleterious features of natural radiocontamination of water supplies which will arise from the activities involved in the nuclear generation of electrical power, low-dose-rate uranium-produced alpha radiation, is not currently regulated. The question of whether it should be, and if so, in what way, clearly involves issues which go beyond the mere scientific account of these effects: to the extent that radiation protection policy is not capricious, it tries to capture what is in the interest of the public's health, and this effort raises difficult questions about values. On the whole, it appears that a rule-utilitarian approach augmented by a set of justice-determining considerations handles those issues better than any other.

The uncertainties in our knowledge of the physical and biological effects of natural radioactivity in water supplies, and the contention which surrounds our understanding of the value-related issues intrinsic to radiation protection standards, are embodied in complex federal water pollution law. The heart of that law lies in the CWA, the SDWA, and to a lesser extent, the FSMCRA. Since these acts are in principle vast in scope and are very new, a number of fundamental legal issues, including reasonableness of standards, jurisdiction, permissible causes of action, and constitutional matters, are still in or will be brought to the courts. As a result, current federal water quality law is complex, controversial, and continuously evolving. Substantial uncertainty therefore attends any attempt to predict the long-term obligations or liabilities which a POTW or a drinking water supplier may face under this law. In light of recent cases, these liabilities can arise even after the fact of nominal compliance, potentially at great expense to the POTW or purveyor.

If this assessment is even roughly correct, it paints a sober picture. Our biological welfare, like that of the animals from which we developed, depends on a genetic equilibrium reached with our environment over millions of years. The mechanisms of genetic adaptation available to us are ill-suited to handle rapid change. Yet in the last forty years, dozens of radioactive sources from electrical-power-generation, industrial, military, and medical activities have been been added to the ionizing radiation burden which we already carry. The dose rate increment to humans produced by human redistribution of natural radioactivity now roughly equals the dose rate to humans produced by natural processes; we have, in effect, doubled the dose rate background in which we evolved. The effects of these new, human-induced exposures on our equilibrium with our environment and our hope for rationally controlling them are almost completely unknown.

# Appendixes

A. Radioactive Decay Properties of the $^{238}U$ Series

B. Radioactive Decay Properties of the $^{40}K$ and $^{232}Th$ Series

C. Weighting Factors Recommended by the ICRP for Calculation of Effective Dose Equivalent

D. Intake in Normal Areas of $^{238}U$, $^{232}Th$, and Their Decay Products

E. Average Activity Concentrations in Tissues and Annual Absorbed Doses Resulting from Internal Irradiation by Radionuclides from the Uranium-238 Subseries

F. Average Activity Concentrations in Tissue Resulting from Internal Irradiation by Thorium-230

G. Average Activity Concentrations in Tissues Resulting from Internal Irradiation by Thorium-232

H. Average Activity Concentrations in Tissues and Annual Absorbed Doses Resulting from Internal Irradiation by Radium 226 and Its Short-lived Decay Products

I. Approximate Frequency of Lethal or Profoundly Debilitating Genetic Damage in a Population Exposed to Low Dose, Low-Dose-Rate, Low-LET Irradiation at a Rate of 1 Gy per Generation

## APPENDIX A. Radioactive Decay Properties of the $^{238}U$ Series

| Nuclide | Half-life | Major radiation energies (MeV) and intensities: alpha | beta | gamma |
|---|---|---|---|---|
| $^{238}U$ | $4.5 \times 10^9$ y | 4.15(25%)<br>4.20(75%) | - | - |
| $^{234}Th$ | 24.1 d | - | 0.10(21%)<br>0.19(79%) | 0.06(3.5%)<br>0.09(4%) |
| $^{m234}Pa$ | 1.17 min | - | 2.29(98%) | 0.77(0.3%) |
| 99.9% 0.1% | | | | 1.00(0.6%) |
| $^{234}Pa$ | 6.75 h | - | 0.53(66%)<br>1.13(13%) | 0.10(50%)<br>0.70(74%)<br>0.90(70%) |
| $^{234}U$ | $2.5 \times 10^5$ y | 4.72(28%)<br>4.77(72%) | - | 0.05(0.2%) |
| $^{230}Th$ | $8.0 \times 10^4$ y | 4.62(24%)<br>4.68(76%) | - | 0.07(0.6%)<br>0.10(0.1%) |
| $^{226}Ra$ | 1602 y | 4.60(6%)<br>4.78(95%) | - | 0.19(4%) |
| $^{222}Rn$ | 3.82 d | 5.49(100%) | - | 0.5(0.07%) |

| | | | | |
|---|---|---|---|---|
| ↓ | | | | |
| $^{216}Po$ | 3.05 min | 6.00(100%) | 0.33(0.02%) | - |
| 99.98% 0.02% | | | | |
| $^{214}Pb$ | 26.8 min | - | 0.65(50%)<br>0.71(40%)<br>0.98(6%) | 0.30(19%)<br>0.35(36%) |
| $^{218}At$ | 2 sec | 6.65(6%)<br>6.70(94%) | ? (0.1%) | - |
| $^{214}Bi$ | 19.7 min | 5.45(0.01%)<br>5.51(<0.1%) | 1.0(23%)<br>1.5(40%)<br>3.26(19%) | 0.61(47%)<br>1.12(17%)<br>1.8(17%) |
| $^{214}Po$ | 164 microsec | 7.69(100%) | - | 0.8(0.01%) |
| $^{210}Tl$ | 1.3 min | - | 1.3(25%)<br>1.9(56%)<br>2.3(19%) | 0.3(80%)<br>0.8(100%)<br>1.3(21%) |
| $^{210}Pb$ | 21 y | 3.72(<0.0001%) | 0.016(85%)<br>0.061(15%) | 0.5(4%) |
| $^{210}Bi$ | 5.01 d | 4.65(<0.01%) | 1.61(100%) | - |
| 100% <0.01% | | 4.69(<0.01%) | | |
| $^{210}Po$ | 138.4 d | 5.31(100%) | - | 0.8(0.01%) |
| $^{206}Tl$ | 4.19 min | - | 5.71(100%) | - |
| $^{206}Pb$ | Stable | - | - | - |

## APPENDIX B. Radioactive Decay Properties of the $^{40}K$ and $^{232}Th$ Series

| Nuclide | Half-life | Major radiation energies (MeV) and intensities: alpha | beta | gamma |
|---|---|---|---|---|
| $^{40}K$ → 10.7% → $^{40}Ar$; 89.3% → $^{40}Ca$ | 1.26 x $10^9$ y | - | 1.32(89%), | 1.46(11%) |
| $^{40}Ar$, $^{40}Ca$ | Stable | | | |
| $^{232}Th$ ↓ | 1.41 x $10^{10}$ y | 3.95(24%)<br>4.01(76%) | - | - |
| $^{228}Ra$ ↓ | 5.8 y | - | 0.055(100%) | - |
| $^{228}Ac$ ↓ | 6.13 h | - | 1.18(35%)<br>1.75(12%)<br>2.09(12%) | 0.34(15%)<br>0.91(25%)<br>0.96(20%) |
| $^{228}Th$ ↓ | 1.91 y | 5.34(28%)<br>5.43(71%) | - | 0.08(1.6%)<br>0.21(0.3%) |
| $^{224}Ra$ ↓ | 3.64 d | 5.45(6%)<br>5.68(94%) | - | 0.24(3.7%) |
| $^{220}Rn$ ↓ | 55 sec | 6.29(100%) | - | 0.6(0.07%) |
| $^{216}Po$ ↓ | 0.15 sec | 6.78(100%) | - | - |

| | | | | |
|---|---|---|---|---|
| ↓ | | | | |
| $^{212}Pb$ | 10.6 h | - | 0.346(81%)<br>0.586(14%) | 0.24(47%)<br>0.30(3.2%) |
| ↓ | | | | |
| $^{212}Bi$ | 60.6 min | 6.05(25%) | 1.55(5%) | 0.04(2%) |
| ↙ 64% ↘ 36% | | 6.09(10%) | 2.26(55%) | 0.73(7%)<br>1.62(1.8%) |
| $^{212}Po$ (from 64%) | 304 nanosec | 8.78(100%) | - | - |
| $^{208}Tl$ (from 36%) | 3.10 min | - | 1.28(25%)<br>1.52(21%)<br>1.80(50%) | 0.51(23%)<br>0.58(86%)<br>0.86(12%)<br>2.6(100%) |
| $^{208}Pb$ (from $^{212}Po$ and $^{208}Tl$) | Stable | - | - | - |

# APPENDIX C.
## Weighting Factors Recommended by the ICRP for Calculation of Effective Dose Equivalent

| Tissue | Reference risk coefficient ($10^{-2}$ $Sv^{-1}$) | Weighting factor $w_T$ |
|---|---|---|
| Gonads | 0.40 | 0.25 |
| Breast | 0.25 | 0.15 |
| Red Bone Marrow | 0.20 | 0.12 |
| Lungs | 0.20 | 0.12 |
| Thyroid | 0.05 | 0.03 |
| Bone Surfaces | 0.05 | 0.03 |
| Remainder | 0.50 | 0.30 |

Notes. 1. The reference risk coefficient is the average probability per unit dose equivalent over both sexes and all ages of induction of a fatal tumor or a hereditary effect in the first two generations. 2. For the tissues lumped under "Remainder," a weighting factor of 0.06 applies to each of the five remaining organs or tissues receiving the highest dose equivalents; exposure of all other organs can be neglected. When the gastrointestinal tract is irradiated, the stomach, small intestine, upper large intestine and lower large intestine are treated as four separate organs.

# APPENDIX D.
## Intake in Normal Areas of $^{238}U$, $^{232}Th$, and Their Decay Products.

| Source | Annual intake (Bq) | |
|---|---|---|
| | Inhalation | Ingestion |
| $^{238}U$ series | | |
| $^{238}U$ | 0.01 | 5 |
| $^{234}Th$ | 0.01 | 5 |
| $^{234}Pa$ | 0.01 | 5 |
| $^{234}U$ | 0.01 | 5 |
| $^{230}Th$ | 0.01 | - |
| $^{226}Ra$ | 0.01 | 15 |
| $^{210}Pb$ | 4 | 40 |
| $^{210}Po$ | 0.8 | 50 |
| $^{232}Th$ series | | |
| $^{232}Th$ | 0.01 | - |
| $^{228}Ra$ | 0.01 | 15 |
| $^{228}Ac$ | 0.01 | 15 |
| $^{228}Th$ | 0.01 | 15 |

# APPENDIX E. Average Activity Concentrations in Tissues and Annual Absorbed Doses Resulting from Internal Irradiation by Radionuclides from the Uranium-238 Subseries

| Organ or tissue | Activity concentration of $^{238}U$ ($mBq\ kg^{-1}$) | Annual absorbed doses ($10^{-7}$ Gy) | | | |
|---|---|---|---|---|---|
| | | $^{238}U$ alpha | $^{234}Th$ beta,gamma | $^{234m}Pa$ beta,gamma | $^{234}U$ alpha |
| Gonads | 5 | 1.1 | 0.02 | 0.2 | 1.2 |
| Breast | 5 | 1.1 | 0.02 | 0.2 | 1.2 |
| Lungs | 5 | 1.1 | 0.02 | 0.2 | 1.2 |
| Cortical bone | 150 | | | | |
| Trabecular bone | 150 | | | | |
| Red bone marrow | 5 | 2.1 | 0.11 | 1.4 | 2.4 |
| Bone lining cells | | 17 | 0.34 | 4.3 | 20 |
| Thyroid | 5 | 1.1 | 0.02 | 0.2 | 1.2 |
| Kidneys | 50 | 11 | 0.15 | 2.1 | 12 |
| Other tissues | 5 | 1.1 | 0.02 | 0.2 | 1.2 |

Note. $^{238}Th$, $^{234m}Pa$, and $^{234}U$ are assumed to be in radioactive equilibrium with $^{238}U$ in all organs and tissues.

APPENDIX F.
Average Activity Concentrations in Tissues Resulting from Internal Irradiation by Thorium-230.

| Tissue | Activity concentration (mBq $kg^{-1}$) | Annual absorbed doses ($10^{-7}$ Gy) alpha |
|---|---|---|
| Gonads | 0.3 | 0.07 |
| Breast | 0.3 | 0.07 |
| Lungs | 20 | 4.7 |
| Cortical bone | 20 | |
| Trabecular bone | 70 | |
| Red bone marrow | 0.3 | 5.6 |
| Bone lining cells | | 74 |
| Thyroid | 0.3 | 0.07 |
| Kidneys | 10 | 0.07 |
| Other tissues | 10 | 2.4 |

## APPENDIX G.
Average Activity Concentrations in Tissues Resulting from Internal Irradiation by Thorium-232.

| Tissue | Activity concentration (mBq $kg^{-1}$) | Annual absorbed doses ($10^{-7}$ Gy) alpha |
|---|---|---|
| Gonads | 0.15 | 0.03 |
| Breast | 0.15 | 0.03 |
| Lungs | 20 | |
| Cortical bone | 6 | |
| Trabecular bone | 24 | |
| Red bone marrow | 0.15 | 1.7 |
| Bone lining cells | | 20 |
| Thyroid | 0.15 | 0.03 |
| Kidneys | 3 | 0.6 |
| Liver | 2 | 0.4 |
| Other tissues | 0.15 | 0.03 |

**APPENDIX H. Average Activity Concentrations in Tissues and Annual Absorbed Doses Resulting from Internal Irradiation by Radium-226 and Its Short-lived Decay Products.**

| Organ or tissue | Activity concentration (mBq $kg^{-1}$) | | Annual absorbed doses ($10^{-7}$ Gy) | | | | | |
|---|---|---|---|---|---|---|---|---|
| | $^{226}Ra$ | $^{222}Rn$ | $^{226}Ra$ alpha | $^{222}Rn$ beta, gamma | $^{218}Po$ beta, gamma | $^{214}Pb$ beta, gamma | $^{214}Bi$ beta, gamma | $^{214}Po$ alpha |
| Gonads | 2.7 | 0.9 | 0.7 | 0.3 | 0.3 | 0.01 | 0.03 | 0.04 |
| Breast | 2.7 | 0.9 | 0.7 | 0.3 | 0.3 | 0.01 | 0.03 | 0.04 |
| Lungs | 2.7 | 0.9 | 0.7 | 0.3 | 0.3 | 0.01 | 0.03 | 0.04 |
| Cortical bone | 170 | 60 | | | | | | |
| Trabecular bone | 170 | 60 | | | | | | |
| Red bone marrow | 2.7 | 0.9 | 2.0 | 0.8 | 0.9 | 0.2 | 0.6 | 1.1 |
| Bone lining cells | | | 22 | 9.1 | 9.9 | 0.7 | 1.7 | 13 |
| Thyroid | 2.7 | 0.9 | 0.7 | 0.3 | 0.3 | 0.01 | 0.03 | 0.4 |
| Other tissues | 2.7 | 0.9 | 0.7 | 0.3 | 0.3 | 0.01 | 0.03 | 0.4 |

Notes. 1. The values reported in this table include doses resulting from the formation of radon-222 and its short-lived decay products in the body by decay of radium-226 but does not take into account the doses arising from inhalation of radon-222 and its daughters. 2. $^{218}Po$, $^{214}Pb$, and $^{214}Po$ are assumed to be in radioactive equilibrium with radon-222 in all organs of the body.

APPENDIX I. Approximate Frequency of Lethal or Profoundly Debilitating Genetic Damage in a Population Exposed to Low Dose, Low-Dose-Rate, Low-LET Irradiation at a Rate of 1 Gy per Generation.

| | Induced number of cases per $10^6$ births | |
|---|---|---|
| Condition | First generation | Equilibrium |
| Autosomal dominant | 1400 | 9000 |
| X-linked (both sexes) | 125 | 500 |
| Chromosomal | 240 | 400 |
| Irregularly inherited | 450 | 4500 |
| TOTAL | 2215 | 14400 |

# Index